Perspectives on Environmental Impact Assessment

Perspectives on Environmental Impact Assessment

Proceedings of the Annual Training Courses on Environmental Impact Assessment, sponsored by The World Health Organization, Regional Office for Europe, Copenhagen, Denmark at the Centre for Environmental Management and Planning, University of Aberdeen, Scotland, 1980-1983

Managing editors

BRIAN D. CLARK and ALEXANDER GILAD

CENTRE FOR ENVIRONMENTAL MANAGEMENT AND PLANNING

WORLD HEALTH ORGANIZATION REGIONAL OFFICE FOR EUROPE

Editors

RONALD BISSET and PAUL TOMLINSON

Centre for Environmental Management and Planning, University of Aberdeen

D. REIDEL PUBLISHING COMPANY

DORDRECHT / BOSTON / LANCASTER

Library of Congress Cataloging in Publication Data

Main entry under title:

Perspectives on environmental impact assessment.

Includes index.
1. Environmental impact analysis–Congresses. 2. Environmental health–Congresses. I. Bisset, Ronald. II. Tomlinson, Paul. III. World Health Organization. IV. University of Aberdeen. Centre for Environmental Management and Planning.
TD194.6.P47 1984 333.7'15 84-4709
ISBN 90-277-1753-2

Published by D. Reidel Publishing Company,
P.O. Box 17, 3300 AA Dordrecht, Holland.

Sold and distributed in the U.S.A. and Canada
by Kluwer Academic Publishers,
190 Old Derby Street, Hingham, MA 02043, U.S.A.

In all other countries, sold and distributed
by Kluwer Academic Publishers Group,
P.O. Box 322, 3300 AH Dordrecht, Holland.

Printed in The Netherlands

TABLE OF CONTENTS

Brian D. Clark

PREFACE AND ACKNOWLEDGEMENTS

Environmental Impact Assessment (EIA) is a young and dynamic subject. The environment is now firmly on the political agenda in both developed and developing countries and there is a growing realization that procedures, methods and techniques to assess the likely environmental, health and social impacts of projects, plans and policies must now be given the same emphasis that economic and technical assessments were accorded in the past.

This volume is a collection of papers on a wide range of EIA topics produced by an international group of experts drawn from developed and developing countries. Perspectives are presented from the viewpoint of government, industry, national and international agencies and academia. The 31 papers are arranged in four sections; (A) Objectives and Procedures, looking at the development of EIA in selected countries; (B) The Role of Environmental Health Assessment, covering medical effects of pollution and risk and hazard; (C) Assessment methods & techniques including developments in monitoring; and finally (D) Case Studies ranging from EIA for oil, gas, coal, water and tourism developments.

This book therefore should be of interest to all those concerned with EIA whether their interests relate to the role that EIA can play in rational decision making or whether they require an understanding of the procedures, methods and techniques that are available to introduce, implement and evaluate EIA systems. As such the book should prove of value to politicians and decision makers, technical experts concerned with environmental management in government, industry and consultancy, and environmental educationists now increasingly concerned not only with introducing EIA training at both undergraduate and postgraduate levels but also in mounting basic and applied research programmes on the subject. Everyone concerned with the new art or science of EIA is on a learning curve and it is confidently hoped that sections of this book will benefit those on different parts of the curve!

The volume is a direct outcome of the now annual Aberdeen based WHO International Training Course on EIA

B. D. Clark et al. (eds.), Perspectives on Environmental Impact Assessment, ix–xi.

which is organized by the Centre for Environmental Management and Planning (CEMP), formerly the PADC unit of Aberdeen University. A caravan visit by Dr Alex Gilad, then the peripatetic WHO (Euro) Regional Officer for Environmental Systems Management, to the far north of Scotland introduced him to the work of PADC and he suggested that a training course on EIA, with emphasis on health assessment, be instituted. Since this date in the late 1970's four courses have been mounted attended by over 500 faculty and participants from 86 countries. This volume comprises a selection of papers presented at courses between 1980 and 1983. It is hoped that a further volume of papers will be published: many papers not included are worthy of publication but we have pushed to the limit the tolerance of our publishers concerning the length of the volume!

The production of this volume would not have been possible without the support and dedicated hard work of many people. To Sandra Ralston a great debt of gratitude is owed for the calm and efficient manner in which she manages to get faculty and participants to Aberdeen. We are grateful to the faculty of the early courses for their forbearance in the delay to publication and to authors selected from the July 1983 course who had to produce papers at break neck speed! Ron Bisset and Paul Tomlinson have borne the brunt of the work of coordinating and editing the volume, and, given their other commitments, they have undertaken this task with a cheerfulness and commitment that belies the many problems involved. Latterly Belinda Foster has had the dubious privilege of putting the volume "to bed" and to her I am indebted for her ruthless determination to meet the publishers final deadline. That time could be found to produce the volume at all is due to the work of Julie Martin and help with 'things' afforded by the 'night shift' Marion Chalmers! The typing has been undertaken by Pearl Allan and Kathleen Brebner and their ability to cope with illegible handwriting and impossible deadlines is gratefully acknowledged.

Without the support of WHO (Euro) and in particular for their provision of fellowships for participants, this volume would never have seen the light of day. To Dr Alex Gilad and Eric Giroult, WHO (Euro), we are more than grateful for their encouragement and constructive criticisms and I hope that the WHO Course will continue to prosper as an internationally recognized annual event. Our heartfelt thanks must also be accorded to Ian Priestnall and Michiel ten Raa of our publishers, Reidel, for their patience and tolerance.

Only they fully appreciate the multifarious Scottish/Dutch interpretations that can be placed on the term "deadline"!

Finally it is interesting to note that one further outcome of the annual training courses, mainly at the request of participants, has been the bi-monthly EIA Worldletter jointly produced by CEMP, Aberdeen University, and the Environmental and Ground Water Institute of Oklahoma University as a means of establishing a network of contact on relevant EIA matters.

A quotation in a book on Iran published some years ago states that "anyone who knows anything or all about Iran must be grossly misinformed". It is hoped that this volume will make some contribution to a better understanding of EIA at a time when there appears to be misinformation and lack of knowledge about the aims, scope and objectives of EIA and its potential as a tool for sound environmental management.

Brian D. Clark
Executive Director
Centre for Environmental
Management and Planning (CEMP)
Department of Geography
University of Aberdeen
Aberdeen
Scotland.

Managing Editor

J. I. Waddington

INTRODUCTION TO THE PUBLICATION

Health is fundamental for the happiness and wellbeing of individuals and nations. It is the essential, and probably the most important ingredient of quality of life.

It has often been said that "an ounce of prevention is worth a pound of cure", but this old slogan, although often repeated, has seldom been practiced. Until very recently and to a large extent at the present time, the bulk of the technological, human and economic resources devoted to promotion of health have been invested in services and facilities for the cure of diseases, rather than for their prevention. Only recently has it been realized that the vast expenditure on curative health care in most highly developed countries contributes relatively little to the standard of health of the population as a whole. As a consequence, health professionals are learning to think less in terms of therapy and more in terms of preservation of that "state of physical, mental and social wellbeing" which we in WHO define as health.

In parallel with this trend, the accumulated scientific, clinical and epidemiological evidence increasingly points to the importance of environmental factors as determinants of health. In the field of environmental protection, as in the field of health protection, timely identification and institution of preventive measures are often much easier and much less costly than post-factum corrective and curative action. A few thousand dollars worth of planning is often more effective than tens or hundreds of thousands of dollars worth of clean-up equipment, and it is therefore always more economical and more expedient to introduce environmental concerns as early as possible into the planning process.

In many countries throughout the world, there is a growing awareness of the possible harmful side-effects that may be caused by the development of major projects. At the same time there is a realization that many development programmes and policies may lead to short-term conflicts between economic objectives and the need to preserve and enhance environmental quality.

B. D. Clark et al. (eds.), Perspectives on Environmental Impact Assessment, xiii–xv.

The experience of highly industrialized countries demonstrates that single-minded pursuit of economic development is self-defeating because, by disregarding the other components of what is commonly called "the quality of life", it creates conditions which are not acceptable to large sectors of the population.

In the recent past a number of projects, for example, major dams, have had unexpectedly deleterious social, environmental and health consequences. As a result, many government department and agencies are investigating the impacts of specific projects and are examining the role impact analysis could play in project planning.

The process of environmental impact analysis has been developed, tested and institutionalized in several countries. The objective of the process is a prior identification and definition of likely environmental impacts of projects such as public works, industrial developments and tourist developments, as well as the impact of policies and legislative proposals. The environmental impact analysis process also includes the definition of alternative courses of action which would achieve comparable economic objectives while eliminating some or all of the detrimental environmental consequences. Identification of preventive or precautionary measures, which would minimize the unavoidable impacts, form an integral part of the process.

The aim should be for a balanced appraisal in which economic, technical, social, environmental and health aspects are fully evaluated. Thus viewed, environmental impact analysis emerges as one of the most powerful planning tools for the prevention of environmental pollution and degradation.

One of the most important components of any environmental impact assessment system is health, and the human health component of EIA is the primary focus of interest for the World Health Organization. It is clear, however, that assessment and evaluation of health effects must be based on, and integrated with, the process of environmental impact assessment. Without an adequate environmental impact assessment system, predictions of health effects are likely to be incomplete and inaccurate. For this reason the WHO Regional Office for Europe has been involved in a number of activities aimed at the promotion of EIA methodology, at both the country and regional levels, including sponsorships of the Seminars on Environmental Impact Assessment organized yearly by the University of Aberdeen since 1980.

It is hoped that the seminars contribute to the

participants' perception of the concept and the methodology of environmental impact assessment systems in general, and to the health component thereof in particular, and that the current publication will serve as a useful reference material for the students and the practitioners in this field.

J. I. Waddington

Director, Environmental Health
World Health Organization
Regional Office for Europe
Copenhagen
Denmark

PART A:

OBJECTIVES AND PROCEDURES

Brian D. Clark

ENVIRONMENTAL IMPACT ASSESSMENT (EIA): SCOPE AND OBJECTIVES

1. INTRODUCTION

At a time of economic recession it is an encouraging sign that many countries in both the developed and developing world now recognize that major development projects may have harmful environmental impacts. Increasingly the environment is being thought of as an economic resource and not as a dispensable luxury. The simulation models of the Club of Rome which propounded an almost apocalyptic view of impending global disaster have probably been the most important single influence in fostering public concern for the effects of continued economic growth upon the physical environment. It is in this broader context of concern that Environmental Impact Assessment (EIA) has been evolved.

EIA as an approach to the evaluation of development actions evolved in the early 1970s, in response to a number of stimuli. First was the growing scale, pace and associated repercussions of resource development schemes such as large dams, motorways, and nuclear power plants, which were built in post-war years. Unforeseen harmful impacts occurred which reduced predicted benefits. Second was an upsurge in environmental activism as the public became increasingly aware of the environmental consequences of development actions. Finally, there was considerable evidence of inadequacy in existing appraisal techniques. Projects were assessed mainly on grounds of technical and economic feasibility, and potential environmental, social and health impacts were rarely examined explicitly or rigorously. Even when considered, the assessment usually took the form of cost-benefit analysis, which attempts to express all impacts in terms of resource costs valued in monetary terms and many environmental, social and health impacts do not readily lend themselves to economic analysis. They may be difficult to quantify as in the case of disturbance to the special cultural and social patterns of native peoples. They may also be long term and indirect. For example, the Aswan High Dam had deleterious

B. D. Clark et al. (eds.), Perspectives on Environmental Impact Assessment, 3–13.

secondary effects such as a decrease in agricultural productivity. In addition, there was a failure to consider the policy context in which proposals were put forward. A mechanism for addressing more fundamental questions, regarding for instance the need for a development, possible alternatives, and appropriate levels of safety and environmental protection, was needed.

Stages in the development of EIA and its relationship to other forms of evaluation have been well summarized by Garner and O'Riordan (1982).

1) No formal accounting; decisions made on basis of interest group lobbying and engineering feasibility; relatively little political review of development agency budgets; primary emphasis on economic development.
2) Conventional cost benefit analysis; emphasis on the efficiency criterion and engineering feasibility; major concern still economic development.
3) Innovative cost benefit analysis; use of multiple objectives and discount rates, imaginative proxy-pricing mechanisms; economic development one of a number of objectives sought.
4) Environmental impact assessment mainly concerned with describing the repercussions of proposals on bio-physical processes; economic development still a primary objective.
5) Environmental impact assessment with more attention given to describing and evaluating the repercussions of proposals on both social processes and cultural norms as well as bio-physical systems; economic development a principal but not sole objective.
6) Environmental impact assessment made at a two-tier level, first to look at generic issues associated with big projects as a class, second to investigate how an 'approved' scheme can best be designed with minimal net social and bio-physical disruption. EIA visualised as a creative participatory environmental management activity.

2. DEFINITIONS OF EIA AND KEY ISSUES

It is important to stress that there is no general and universally accepted definition of EIA. The following examples selected at random from a number of authorities illustrate the great diversity of definitions:-

(i) "... an activity designed to identify and predict the impact on man's health and well being, of legislative proposals, policies, programmes and operational procedures, and to interpret and communicate information about the impacts" (Munn 1975).

(ii) "... to identify, predict and to describe in appropriate terms the pros and cons (penalties and benefits) of a proposed development. To be useful, the assessment needs to be communicated in terms understandable by the community and decision-makers and the pros and cons should be identified on the basis of criteria relevant to the countries affected" (UNEP 1978).

(iii) "... an assessment of all relevant environmental and resulting social effects which would result from a project" (Battelle 1978).

(iv) "... assessment consists in establishing quantitative values for selected parameters which indicate the quality of the environment before, during and after the action" (Heer and Hagerty 1977).

The diversity contained in this sample of definitions illustrate some of the inherent problems of the concept, and generate a number of questions which are critical when trying to understand not only the current role of EIA but alo its future potential:-

How and in what ways can EIA be of value to the decision-makers but at the same time help to maintain and enhance the quality of the environment?

Can EIA help to evaluate the costs and benefits of proposed actions to different sections of society?

Should EIA be narrowly restricted to physical environmental topics or should it encompass social, health, economic and psychological parameters?

Should EIA be selective (i.e. relevant impacts) or attempt to be all embracing (i.e. all impacts)?

What form of participation by government, scientists and the public is required in the EIA process?

Should EISs be produced in the technical language of science or in 'simple' language for politicians and decision-makers?

Can meaningful predictions be made about probable impacts?

Should techniques of assessment be qualitative or quantitative or attempt to be both?

In this paper environmental impact assessment (EIA) is

taken to mean the systematic examination of the likely environmental consequences of proposed projects, programmes, plans and policies. The results of the assessment - which are assembled in a document known as an environmental impact statement (EIS) - are intended to provide decision-makers with a balanced appraisal of the environmental, social and health implications of alternative courses of action. When an EIS has been prepared it is used by decision-makers as a contribution to the information base upon which a decision is made. In this way, EIA can assist in formation and evaluation of environmentally sound development proposals.

3. PRINCIPLES OF EIA

EIA aims at logical and rational decision-making. It is generally agreed that EIA should be concerned with the identification, measurement, interpretation and communication of environmental impacts of the proposed action. Attempts should be made to reduce potential adverse impacts and increase likely benefits through the identification and assessment of alternative sites and/or processes. The "no go" alternative should also be assessed. Public involvement has an important part to play in determining issues of significance, providing local information, and helping to identify alternatives. EIA is intended to be an "objective", technical and predictive exercise, with no decision-making component. The results of the assessment are presented in the EIS document as a discussion of the beneficial and adverse impacts considered to be relevant to the project, plan or policy. This report is one component of the information on which the decision-maker ultimately makes a choice. At this stage other factors - particularly economic and political factors - may influence the outcome of the decision. Ideally, a final decision can be made with due regard being paid to the likely consequences of a particular course of action. To be effective EIA should be implemented at an early stage of project planning and decision. It must be an integral component of the design of projects, rather than something utilised after the design phase is complete. Preferably, EIA should be part of an incremental decision-making process which has a number of decision points, and proposals when implemented should be subject to monitoring and auditing programmes. In this way there can be a continuous feedback between EIA findings, project design and locations.

Conceptually, it is useful to distinguish between EIA

methods and EIA techniques. Methods are structured mechanisms for the identification of impacts and the organisation of results. Generally, all methods share a common objective of ensuring that as many potential relevant impacts as possible are identified, measured and described, although some go further and incorporate means whereby impacts from different project designs can be evaluated and compared. These latter methods not only identify impacts, but also quantify, weight and aggregate them. Techniques are used to predict future states of specific environmental parameters. Techniques for EIA can be grouped into a number of subject areas such as hazard, noise, transport, air pollution, ecology, landscape character and visual impact. Information and data obtained using techniques can be organised, presented, and in some cases, evaluated, according to the guidelines of a particular method.

4. THE ADVANTAGES OF EIA

EIA is a mechanism which aids the efficient use of human and natural resources which has proved valuable to both those promoting developments and those responsible for their authorization. EIA may reduce costs and the time taken to reach a decision by ensuring that subjectivity and duplication of effort are minimzed, as well as identifying and attempting to quantify the primary and secondary consequences which might necessitate the introduction of expensive pollution control equipment, compensation or other costs at a later date.

There are a number of ways in which EIA can improve the efficiency of decision-making, but to be effective EIA should be implemented at an early stage of project planning and design. It must be an integral component in the design of projects, rather than something utilized after the design phase is complete. Preferably, EIAs should be part of an incremental decision-making process which has a number of decision-points in the project planning procedure. This means that there can be a continuous feedback between EIA findings, project design and locations. EIAs can be implemented to test alternative project designs at an early stage, to help choose the project design which emphasizes benefits and minimizes harmful effects. EIA therefore can be used not only to investigate and avoid harmful impacts but also to increase likely benefits.

The emergence of an optimum alternative in terms of

the objectives or goals relevant to a proposed project means that EIAs may have significant long-term financial advantages. If a potential problem is identified early in project planning it may allow considerable financial savings to be achieved. At the crudest level the abandonment of a project may be required if all alternative designs or locations are considered unsuitable in terms of likely detrimental effects. This would save capital costs. It is more likely, however, that design modification may reduce the need for expensive ameliorating action once a project becomes operational. If a development is not assessed for its likely impacts, it may cause serious social and health problems. For example, a proposed dam and reservoir may have health effects which may require an expensive health care programme. The wrong location for a resettled population may result in agricultural failure and the need for food supplies to be sent to the relocated people from other areas.

The incorporation of EIA into decision-making may create a number of benefits. If a forecast of the likely impacts of developments is available, allowances can be made and the infrastructure can be provided in a manner whereby impacts are minimized. Where uncertainty exists as to future development, EIA can identify those areas most susceptible to adverse impacts and so guide site selection. To be effective EIAs can only be used when the alternative sites are few in number; otherwise EIA can be time-consuming and expensive. EIA can, nevertheless, aid the identification of the most suitable site in terms of benefit maximization and reduction of harmful effects. Should no site be considered suitable, then the results of EIA aid the determination of broad environmental, social or health criteria to be used when a large number of sites are screened for the suitability. The relevance and importance of EIA for site selection has been recognized in a report published by UNEP entitled "Guidelines for Assessing Industrial Environmental Impact and Environmental Criteria for the Siting of Industry (UNEP 1980).

5. CONTEMPORARY DEVELOPMENTS IN EIA

The contribution of EIA depends not only on the availability of appropriate, effective methods, but also on the outcome of debate concerning its scope and application. The following issues are now of major importance.

5.1 The Use of EIA at Policy Level and in Forward Planning

In principle, EIA procedures should apply to all actions likely to have a significant environmental effect. A comprehensive EIA system would include the appraisal of policies, plans, programmes and projects. Lee and Wood (1978) have called this a "tiered" EIA structure. Higher order policy or plan evaluations would be conducted first, at national or regional level; lower order programme or project evaluations would then be implemented locally. There are a number of advantages to such a tiered approach. It enables major questions of need, safety, environmental protection and compensation to be decided in general terms so that subsequent proposals are more likely to be practicable, consistent, and not unnecessarily constrained. It facilitates optimum site selection and full consideration of alternatives, which cannot usually be achieved at project level. It also allows more time for collection and analysis of environmental data, and eliminates repetition, as a higher order EIA may obviate the need for numerous similar project EIAs.

Unfortunately comparatively few policy or plan level EIAs have yet been attempted, although the evaluation of higher order decisions is widely believed to be very important. O'Riordan, for example, has expressed the view that "unless an environmental instinct is implanted at the policy determining level ... EIA will tend to be a cosmetic exercise" (1981). Clark et al. (1981) have commented on the difficulties encountered when both project and policy issues are considered at the same time, as at the public inquiry into the nuclear fuel reprocessing plant at Windscale in the UK. Why should policy and plan level EIA be so poorly developed? Foster (1983) who has reviewed some practical and research approaches to policy and plan level EIA, has identified a number of difficulties in using EIA at these levels. They include the lack of a site-specific environment to study, imprecise knowledge of the future, freedom to establish goals and objectives, and lack of suitable methods. Another factor is undoubtedly the reluctance of governments to open the decision-making process to public scrutiny.

Although various mechanisms have been proposed for policy-level EIA, for instance by means of investigatory parliamentary committees backed by research secretariats (O'Riordan & Sewell 1981) or by independent assessment bodies (Clark et al. 1981) it appears unlikely that they will be implemented in the near future. In the meantime, limited progress is being made towards the use of EIA methods and techniques in strategic land use planning. For example, in Scotland the Scottish Development Department has compiled background information and generated and reviewed alternative strategies for siting of petrochemical development (S.D.D. 1977).

5.2 The Use of EIA to Identify Social Distibution of Costs and Benefits

Conflict between local and national or other interests in relation to a development proposal is a common problem. It cannot be resolved by EIA but EIA can sometimes help to clarify the issues at stake, before a decision is taken, usually on the basis of political factors. In addition to identifying, predicting and discussing individual impacts and how they affect particular environmental components and sectors of a human population, EIA can provide an overview of the distribution of impacts in a cumulative and spatial sense. This is especially important for impacts which affect the quality of human life, whether they affect people individually or collectively.

One of the best examples of an area in which EIA could encourage a greater degree of social justice is pollution control. A recent UK study (Miller & Wood 1983) criticises existing UK pollution control practice because it is based on the principle of "best practicable means" and does not specifically require consideration of the incremental effects of pollutants, nor permit application of quantitative environmental quality standards. With reference to a number of detailed case studies of air, water, land and noise pollution it concludes that had environmental impact assessment been required, serious pollution impacts which left local populations with a net detriment and no means of compensation might have been avoided.

5.3 Financial Aspects of EIA

A major criticism which has been directed at EIA is that it causes considerable costs and delays. Initially, EIAs may be expensive to implement, particularly in areas where little is known about existing environmental and social conditions. Design changes produced as a result of EIA findings may also increase capital costs, but it can be argued that savings to local, regional and national economies arising from the avoidance of deleterious impacts and from the maximisation of beneficial impacts will outweigh the costs of an EIA system in the long term. The cost of an EIA system will decline once procedures and techniques have been established.

The costs of EIA are commensurate with the complexity and significance of the problem and the level of detail required. In many countries, the cost is borne by the proponent of the development, whilst in others it is borne by the authorising agency. In those countries with experience in EIA, the costs vary between 0.5% and 2% of the project value. It is misleading, however, to regard the "actual costs" of EIAs as being saved if an EIA is not undertaken, because much of the information required will have to be collected by some means for submission of planning applications or other purposes. Moreover, it can be argued that thorough investigation of impacts at an early stage of project planning may save money by speeding up the process of implementing a proposal, because inflationary costs arise from delays. Project proponents may also benefit from improved project design and siting which may obviate the need for expensive ameliorative action such as the introduction of pollution control equipment or payment of compensation. British Gas, a UK industry which undertakes in-house environmental appraisal, claims an overall cost saving as a result (Dean and Graham, 1978).

The financial benefits to the public from the implementation of EIA have usually not been determined, because it is difficult to assign monetary values to such benefits. Many of the environmental amenities that would otherwise have been degraded or destroyed have a unique value, which over time will far outweigh EIA costs. Many cases show that the use of EIA has allowed the choice of an option that is both environmentally and economically superior to the original choice - the Alaskan Pipeline being a good example.

6. CONCLUSIONS

There is now a general recognition that environmental considerations ought to be integrated into the planning and decision-making framework, but differences exist as to the exact form that such integration should take. The administrative structures of the EIA process also vary. Some countries implement EIAs through legislative or administrative regulations whilst others will integrate it with planning or other authorization systems. EIA can ensure that environmental aspects are given equal status with economic, technical and social considerations during the evaluation of development proposals. Attention can also be given not only to immediate impacts, but also to indirect, secondary and long term effects. It is necessary to stress the importance of a proper framework for deciding which project activities ought to be subject to EIA and that project EIAs are limited by decisions made at a policy or plan level. The evaluation of higher order decisions is important and this is where EIA is now increasingly focusing attention.

REFERENCES

Battelle Institute: 1978, The Selection of Projects for EIA, Commission of the European Communities Environment and Consumer Protection Service, Brussels.

Clark, B.D., R. Bisset, P. Wathern: 1981, 'The British Experience', in T. O'Riordan and W.R.D. Sewell, (eds.) Project Appraisal and Policy Review, Wiley, Chichester.

Dean, F.E. and G. Graham: 1978, The Application of Environmental Impact Analysis in the British Gas Industry. Paper presented at the UNECE Symposium on the Gas Industry and the Environment, Minsk.

Foster, B.J.: 1983, Land-use Policy and Plan-making: the role of EIA in Forward Planning. Paper presented at the Symposium on Environmental Impact Assessment, Chania, Crete, 10-17 April 1983.

Garner, J.F. and T. O'Riordan: 1982, 'Environmental Impact Assessment in the context of Economic Recession', Geographical Journal Vol. 148(3) November, pp. 343-361.

Heer, J.E. and D.J. Hagerty: 1977, Environmental Assessment and Statements, Van Nostrand Reinhold, New York.

Lee, N. and C. Wood: 1978, 'EIA - a European perspective', in Built Environment 4(2) pp. 101-110.

Miller, C. and C. Wood: 1983, Planning and Pollution: An Examination of the Role of Land Use Planning in the Protection of Environmental Quality, Clarendon Press.

Munn, R.E. (ed.): 1979, Environmental Impact Assessment: Principles and Procedures, Scope 5 Report, 2nd edition, John Wiley, Chichester, England.

O'Riordan, T.: 1981, Environmentalism, 2nd edition, Pion, London.

O'Riordan, T. and W.R.D. Sewell: 1981, 'From Project Appraisal to Policy Review', in T.O'Riordan and W.R.D. Sewell (eds.) Project Appraisal and Policy Review, Wiley, Chichester.

Scottish Development Department: 1977, National Planning Guidelines for Petrochemical Developments, S.D.D., Edinburgh.
United National Environment Programme: 1978, Draft Guidelines for Assessing Industrial Environmental Impact and Environmental Criteria for the Siting of Industry, UNEP Industry and Environmental office, Paris.
United Nations Environment Programme: 1980, Guidelines for Assessing Industrial Environmental Impact and Environmental Criteria for the Siting of Industry, UNEP Industry and Environmental Guideline Series, Vol. 1, UNEP Industry and Environment Office, Paris.

AFFILIATION

Brian D. Clark is Executive Director of the Centre for Environmental Management and Planning, University of Aberdeen, Scotland.

Larry Canter

ENVIRONMENTAL IMPACT STUDIES IN THE UNITED STATES

Over 15,000 environmental impact statements (EIS's) have been prepared in the United States since the January 1, 1970 effective date of the National Environmental Policy Act (Public Law 91-190). Considerable discussion and debate has occurred over the value of environmental impact requirements in project planning and decision-making. Many EIS's have been controversial, and over 1500 legal actions have resulted from the requirements of PL 91-190. The purpose of this paper is to summarize U.S. experiences in environmental impact studies, and to delineate emerging issues and needs relative to these studies. Topics to be addressed include a summary of the National Environmental Policy Act and the Council on Environmental Quality regulations. A synopsis of U.S. experiences since the passage of NEPA will be presented. Finally, some key issues related to current and future studies are delineated.

1. NATIONAL ENVIRONMENTAL POLICY ACT

The basic legislation for environmental impact studies was the National Environmental Policy Act (NEPA) of 1969. It was the first law signed in the decade of the 1970's, and it gave significance to environmental issues and considerations (Canter, 1977). NEPA has two major sections, one delineating general environmental policy and the other establishing the Council on Environmental Quality (CEQ). The policy section contains broad national goals as well as specific requirements for preparing EIS's. The EIS requirement represents the "teeth" of the law in that Federal agencies are required to document the manner in which the environment is considered in project planning and decision-making along with traditional decision factors such as technical feasibility and economic analysis. Specifically, each Federal agency is required to include in every recommendation or report on proposals for legislation and other major actions significantly affecting the quality of the human environment, a detailed statement known as an EIS.

B. D. Clark et al. (eds.), Perspectives on Environmental Impact Assessment, 15–24.

An environmental assessment (EA) refers to a concise public document that serves to briefly provide sufficient evidence and analysis for determining whether to prepare an EIS or a finding of no significant impact (FONSI). Pending this determination, information from the EA can aid an agency's compliance with NEPA when no EIS is necessary, or facilitate preparation of an EIS when one is deemed to be necessary. The components of an EA should be similar to the components of an EIS. As originally listed in NEPA, the EIS was to address the following five points:

(i) the environmental impact of the proposed action,
(ii) any adverse environmental effects which cannot be avoided should the proposal be implemented,
(iii) alternatives to the proposed action,
(iv) the relationship between local short-term uses of man's environment and the maintenance and enhancement of long-term productivity, and
(v) any irreversible and irretrievable commitments of resources which would be involved in the proposed action should it be implemented.

The above-listed points were expanded in the CEQ quidelines issued in 1971 and 1973, and in the CEQ regulations which became effective in 1979. One of the definitions which has expanded since the passage of NEPA is for the term "environment". In the early 1970's studies focused on the natural environment with very little consideration given to the man-made features. The CEQ guidelines of 1973 gave emphasis to secondary or indirect impacts from projects, with most of the impacts reflected in human population growth and associated changes in community infrastructure. Accordingly, in U.S. studies the term human environment shall be interpreted comprehensively to include the natural and physical environment and the relationship of people with that environment.

2. COUNCIL ON ENVIRONMENTAL QUALITY REGULATIONS

The 1979 CEQ requlations were published as a result of a presidential executive order which required CEQ to conduct public hearings and develop regulations which would streamline the process and produce better environmental decisions related to programs and projects. It is beyond the scope of this presentation to completely review the 1979 regulations; however, the following key features will be highlighted: scoping, inclusion of the environment in planning, record of decision, mitigation monitoring, list of preparers, and scientific approach

(Council on Environmental Quality, 1978). Scoping refers to a process in which the range of actions, alternatives, and impacts to be addressed in an EIS are delimited from all possible actions, alternatives and impacts. The scoping procedure assists agencies in delineating the central issues and in assigning responsibility for the EIS among the lead agency and cooperating agencies. Scoping should begin early in the NEPA process; in most cases, it should begin shortly after the decision to prepare an EIS, and it should be integrated with other general planning considerations.

The basic purpose of NEPA was to insure that the environment is considered in project planning and decision-making along with traditional technical factors and economic analyses. The environment must be considered in conjunction with these factors rather than separately after other decisions have been made based on technical and economic grounds. Accordingly, the 1979 CEQ regulations give emphasis to integrating the NEPA process into early planning. If environmental review is tacked on to the end of the planning process, then the process is prolonged, or else the EIS is written to justify a decision that has already been made, and genuine consideration may not be given to environmental factors.

In the early years of preparing EIS's the general concept was that statements represented "disclosure documents" in that all of the impacts, both beneficial and detrimental, should be summarized. In fact, NEPA has been called a "full disclosure law". The 1979 CEQ regulations indicated that an EIS is more than a disclosure document since it should be used by Federal officials, in conjunction with other relevant materials, to plan actions and make decisions. In this sense, the EIS can be referred to as a "decision document". To ensure that the EIS is considered in the decision process, agencies are required to produce a concise public record, called a Record of Decision (ROD), indicating how the EIS was used in arriving at the decision. This ROD must indicate which alternative (or alternatives) considered in the EIS is preferable on environmental grounds. Agencies may also discuss preferences among alternatives based on other relevant factors including economic and technical considerations and agency statutory missions. Agencies should also identify in the ROD those essential considerations of national policy, including factors not related to environmental quality, which were balanced in making the decision.

The five EIS points listed in NEPA included one related to a description of those adverse environmental effects which cannot be avoided should the proposal be implemented. This suggests that some undesirable effects could be avoided through the application of mitigation measures. Mitigation measures may include avoiding the impact altogether by not taking a certain action or parts of an action; minimizing impacts by limiting the degree or magnitude of the action and its implementation; rectifying the impact by repairing, rehabilitating, or restoring the affected environment; reducing or eliminating the impact over time by preservation and maintenance operations during the life of the action; and compensating for the impact by replacing or providing substitute resources or environments (Council on Environmental Quality, 1978). During the first decade of environmental impact studies major emphasis was given to mitigation measures, and often, projects were approved based on the inclusion of mitigation measures. The 1979 CEQ regulations require that when an agency specifies environmentally-protective mitigation measures in its decisions, these measures should be implemented and monitored to be sure that they are achieving the anticipated reductions in undesirable impacts.

Many EIS's prepared during the decade of the 1970's can be characterized as lacking a scientific basis and approach. The 1979 CEQ regulations require accurate documents as the basis for sound decisions, with the lead agency being responsible for the professional integrity of environmental documents. Agencies shall insure the professional integrity, including scientific integrity, of the discussions and analyses in EIS's. To ensure more accurate professional documents, the 1979 CEQ regulations require a list of people who help prepare documents, and their professional qualifications, for inclusion in the EIS. This list of persons is required to encourage professional responsibility and to ensure that an interdisciplinary approach was followed in the conduction of the study.

The 1979 CEQ regulations gave considerable emphasis to using scientific approaches and techniques in impact prediction and analysis. Agencies should identify any methodologies used and make explicit reference by footnote to the scientific and other sources relied upon for conclusions in the EIS. This is important for ensuring better and more rational decisions, and it is also necessary since many EIS's end up in litigation. A key perspective is that an EA or EIS should be prepared like a technical report with appropriate documentation. The format for an EIS as delineated in the 1979 CEQ regulations is contained in Table 1, with brief descriptions of the contents of each of the major sections.

The most important technical element of an environmental impact study is associated with prediction of changes in environmental factors resulting from alternative plans, and the interpretation or assessment of the significance of those changes. Many EIS's contain qualitative approaches for impact prediction and assessment, and while this may be necessary for certain projects or impacts, every attempt should be made to quantify anticipated changes and to use scientific rationale for significance assessment. Table 2 contains a summary of considerations used for defining the significance of environmental changes in the United States (Council on Environmental Quality, 1978). Some of the items are specific while others represent general considerations.

3. UNITED STATES EXPERIENCES

As mentioned earlier, over 15,000 EIS's have been prepared since the January 1, 1970 effective date of NEPA. The number of EA's probably exceeds 200,000. Due to legislative procedures within the United States, it is possible for opponents of projects to file suit based on the lack of administrative or technical responses of proposing agencies to NEPA. Over 1500 court cases have resulted from the requirements of NEPA with the vast majority related to the proposing agency's fulfillment of the administrative requirements of NEPA. Taking one year as an example, during 1981 there were 1033 EIS's filed with the U.S. Environmental Protection Agency (Council on Environmental Quality, 1982). Filing of lawsuits in 1981 based in whole or in part upon NEPA grounds decreased moderately from filings during the previous 2 years. Statistics show that 114 such cases were filed during 1981, as opposed to 140 cases for 1980 and 139 cases for 1979. To date, the most NEPA lawsuits were filed in 1974 (189 cases) and the least were filed in 1977 (108 cases). Twelve injunctions based upon NEPA were issued during 1981. This compares to 17 such injunctions in 1980 and 12 injunctions in 1979.

The agencies which were challenged most frequently in court remain about the same, although the ranking varies slightly from year to year. In 1981, the Department of

Table 1: EIS Substantive Format from 1979 CEQ Regulations (Council on Environmental Quality, 1978)

Section	Description
Summary	Each environmental impact statement shall contain a summary which adequately and accurately summarizes the statement. The summary shall stress the major conclusions, areas of controversy (including issues raised by agencies and the public), and the issues to be resolved (including the choice among alternatives).
Purpose and Need	The statement shall briefly specify the underlying purpose and need to which the agency is responding in proposing the alternatives including the proposed action.
Alternatives Including the Proposed Action	This section is the heart of the environmental impact statement. Based on the information and analysis presented in the sections on the Affected Environment and the Environmental Consequences, it should present the environmental impacts of the proposal and the alternatives in comparative form, thus sharply defining the issues and providing a clear basis for choice among options by the decisionmaker and the public.
Affected Environment	The environmental impact statement shall succintly describe the environment of the area(s) to be affected or created by the alternatives under consideration. The descriptions shall be no longer than is necessary to understand the effects of the alternatives. Data and analyses in a statement shall be commensurate with the importance of the impact, with less important material summarized, consolidated, or simply referenced.
Environmental Consequences	This section forms the scientific and analytic basis for the comparisons of alternatives section. The discussion will include the environmental impacts of the alternatives including the proposed action, any adverse environmental effects which cannot be avoided should the proposal be implemented, the relationship between short-term uses of man's environment and the maintenance and enhancement of long-term productivity, and any irreversible or irretrievable commitments of resources which would be involved in the proposal should it be implemented.

Table 2: Considerations in Determining Impact Significance (Council on Environmental Quality, 1978)

"Significantly" as used in NEPA requires considerations of both context and intensity:

(a) Context. This means that the significance of an action must be analyzed in several contexts such as society as a whole (human, national), the affected region, the affected interests, and the locality. Significance varies with the setting of the proposed action. For instance, in the case of a site-specific action, significance would usually depend upon the effects in the locale rather than in the world as a whole. Both short- and long-term effects are relevant.

(b) Intensity. This refers to the severity of impact. Responsible officials must bear in mind that more than one agency may make decisions about partial aspects of a major action. The following should be considered in evaluating intensity:

(1) Impacts that may be both beneficial and adverse. A significant effect may exist even if the Federal agency believes that on balance the effect will be beneficial.

(2) The degree to which the proposed action affects public health or safety.

(3) Unique characteristics of the geographic area such as proximity to historic or cultural resources, park lands, prime farmlands, wetlands, wild and scenic rivers, or ecologically critical areas.

(4) The degree to which the effects on the quality of the human environment are likely to be highly controversial.

(5) The degree to which the possible effects on the human environment are highly uncertain or involve unique or unknown risks.

(6) The degree to which the action may establish a precedent for future actions with sigificant effects or represents a decision in principle about a future consideration.

(7) Whether the action is related to other actions with individually insignificant but cumulatively significant impacts. Significance exists if it is reasonable to anticipate a cumulatively significant impact on the environment. Significance cannot be avoided by terming an action temporary or by breaking it down into small component parts.

(8) The degree to which the action may adversely affect districts, sites, highways, structures, or objects listed in or eligible for listing in the National Register of Historic Places or may cause loss or destruction of significant scientific, cultural, or historical resources.

(9) The degree to which the action may adversely affect an endangered or threatened species or its habitat that has been determined to be critical under the Endangered Species Act of 1973.

(10) Whether the action threatens a violation of Federal, State, or local law or requirements imposed for the protection of the environment.

Defense was sued 22 times, comprising 19 percent of the total number of filings. The Department of Transportation was next with 20 lawsuits (18 percent), followed by the Department of the Interior with 18 lawsuits (15 percent), the Department of Agriculture with 15 lawsuits (13 percent), and the Department of Housing and Urban Development with 10 cases (8 percent).

Environmental groups became the most frequent plaintiffs in 1981, filing suit in 40 of the 114 cases. Individuals or citizen groups were responsible for filing suit in 32 cases; local governments were next, filing 24 lawsuits. Business groups filed 16 lawsuits; directly affected property owners and residents filed 13 lawsuits; state governments filed 9 lawsuits; and Indian tribes and labor associations were each responsible for filing 2 lawsuits. A legal foundation filed one case.

Since the enactment of NEPA, the most frequent allegation in lawsuits based upon the statute is that an agency failed to prepare an EIS for a major federal action which would significantly affect the environment. In 1981, 52 NEPA cases included this contention. Inadequacy of an EIS was raised in 50 cases, and inadequacy of an environmental assessment in 8 cases. Miscellaneous contentions were alleged in nine other cases (Council on Environmental Quality, 1982).

4. EMERGING ISSUES AND NEEDS

With over one decade of experience in the conduction of environmental impact studies in the United States, several emerging issues can be identified. These issues result from an evaluation of the quality and effectiveness of EIS's in the first decade, and the anticipation of needs to arise in the next decade. A listing of these issues is as follows (Canter, 1981):

(i) Use Impact Quantification
(ii) Use Systematic Approach
(iii) Have Scientific Basis for Studies
(iv) Apply Existing Methodologies and Techniques
(v) Conduct Post-Audit Analysis
(vi) Develop Value Judgment Approaches

An EIS should be based on as much quantitative information as possible in terms of describing the proposed action, alternative plans, environmental setting, and anticipated impacts of the alternative plans and proposed action. As a general rule, when a quantitative approach is not taken, anticipated impacts are often judged to be overly severe. Appropriate quantification provides a better basis for impact assessment.

Another emerging issue is the need for systematically considering the environmental effects of a project, and recognizing that the environment itself is a system. In many cases single impacts can be identified without recognition of subsequent consequences resulting from the single impacts. Therefore, it is necessary to utilize a systematic approach for conduction of environmental impact studies as well as to recognize interrelationships of environmental factors and the system within which the environmental factors exist.

Environmental impact studies must be conducted from a scientific perspective and basis. This means that the best scientific information and technology should be utilized in the conduction of any study. It may mean that new technologies and scientific approaches will have to be developed to address particular projects and their associated impacts. To consider EIS's as non-scientific documents represents a complete misunderstanding of the concept of environmental impact studies and their requirements and potential uses.

Another issue is that many environmental impact studies do not make use of the existing methodologies for impact identification and evaluation, nor existing techniques for impact prediction. A major need is to apply the technology which exists for the conduction of environmental impact studies. Application of this technology will enable the impacts to be quantified, and will ensure that both a systematic as well as scientific approach is utilized.

Another important issue is related to whether or not predicted impacts have actually occurred. This can be called post-audit analysis in that studies are needed to verify whether or not predicted impacts associated with projects actually occurred, and if their magnitudes matched those that were anticipated. If post-audit analyses can be made, this will enable better conduction of future environmental impact studies since the approaches and models can be calibrated or adjusted to more accurately describe anticipated impacts.

Throughout the environmental impact process is the need for value judgments in impact identification, selection of environmental factors, selection of techniques for predicting impacts, interpretation or assessment of the significance of changes, and preparation of written documentation. Many value judgments are necessary in deciding the relative importance of environmental factors and associated impacts, with this information being basic to the selection of a proposed action from a series of alternative plans. Value judgments can change over time and may vary with geographical location. Research is needed to better understand value judgments as a part of the process, and to better integrate value judgment within the process.

REFERENCES

Canter, L.W.: 1977, Environmental Impact Assessment, McGraw-Hill Book Company, New York, New York, pp. 1-19.

Canter, L.W.: 1981, "Benefits of Environmental Impact Studies in the United States", paper presented at EIA Seminar, Delft Hydraulics Laboratory, Delft, Holland.

Council on Environmental Quality: 1978, "National Environmental Policy Act - Regulations", Federal Register, Vol. 43, No. 230, Nov. 29, pp. 55978-56007.

Council on Environmental Quality: 1982, "Environmental Quality", Washington, D.C., pp. 234-235.

AFFILIATION

Professor Larry Canter is the Chairman of the School of Civil Engineering and Environmental Science and Co-Director of the National Centre for Ground Water Research, University of Oklahoma, USA.

Konrad von Moltke

IMPACT ASSESSMENT IN THE UNITED STATES AND EUROPE

1. INTRODUCTION

Environmental Impact Assessment (EIA) is an art as well as a science, in that not only must the techniques of EIA be thoroughly mastered, but they also need to be applied in a practical situation. While this is true of many activities, it is good to keep the distinction in mind when approaching the subject of EIA from either the research or practical perspective.

In quite a different field of environmental affairs, that of chemical control, there are three essential phases in the assessment and control that must be distinguished. Hazard assessment is a purely scientific exercise aimed at recognising the intrinsic qualities of a substance which might give rise to concern. Risk assessment marries the information obtained from the hazard assessment to the most scientific assessment possible of actual conditions of manufacture, sale, use and disposal of a substance. Obviously this is a scientific exercise which requires a substantial input from practitioners, in that inevitably a growing number of assumptions and judgements need to be incorporated into the process. It is hardly a surprising that risk assessment is a controversial topic although the general experience has been that agreement can be achieved on hazard assessment techniques between groups with highly divergent underlying interests. Finally, there is the whole range of control decisions, drawing practical conclusions from both hazard and risk assessments. This is, and will inevitably remain, a political process dependent upon an institutionalized weighing of costs and benefits, of acceptable and unacceptable risks in U.S. parlance. This is clearly not a science, but it still has the great virtue of producing decisions which people with no great knowledge of the merits of a decision can act upon. Scientific results can, in contrast, only be interpreted by the expert. In fact, there is an intimate relationship of mutual inter-dependence between all three phases. Decision-making is difficult without an adequate information base; and

B. D. Clark et al. (eds.), Perspectives on Environmental Impact Assessment, 25–34.

the development of scientifically verifiable data very quickly becomes a futile task if it does not link to some form of practical conclusion.

This tripartite distinction which has come to be accepted in the field of chemicals control, can also usefully be applied in the area of EIA. By this approach, some of the confusion which has surrounded EIA for some time may be clarified. This paper will focus upon the third topic, namely the circumstances under which results from what might be termed environmental hazard and environmental risk assessment are applied to policy decisions by public authorities in the United States and Europe.

From the above presentation, it will be evident that EIA, in practice, is significantly affected by the political and administrative culture of the country in which it is practised. Though almost trite, the statement is nevertheless of fundamental importance: the principles of risk assessment can readily be discussed between countries and important elements are ultimately capable of transferral; the principles of control decisions are the domain of each country; however, control decisions taken (or not taken) by one country often affect others directly, through exposure to or protection from environmental hazards, or indirectly through commerce and trade for example. Hence it is vital that the bases and substances of control decisions are understood and communicated effectively across national boundaries.

In examining the United States and Europe the risk is run of comparing apples and oranges. The United States is but a single country, with great diversity it is true, but yet with a remarkably homogeneous political and administrative culture. Hence a statute of the state of Vermont is inherently comparable to one of the state of New Mexico, and a federal regulation is part of a well established, some might say too well established, system of enforcement. Europe, on the other hand, is many things to many people: it is a political idea rather than a political reality. Indeed, "Europe" means quite diferent things to persons in the USSR, Poland, Sweden, the Federal Republic of Germany, Spain or the United Kingdom. Institutionally speaking, there are at least four "Europes". The most embracing is the Economic Commission for Europe, a United Nations "regional body" which includes all the states anyone would consider to be part of Europe, plus the United States and Canada. Then there are the socialist and western parts of Europe: CMEA or Comecon and the Council of Europe. Finally, there is the European Community. While the latter is the most limited, comprising only ten states of Western Europe, it is unquestionably the most significant legally, politically and economically. Institutionally, this paper will only address itself to the European Community (E.C.) because it alone has the ability to influence and change the policy context within which EIA is applied in practice. While numerous other international European bodies have contributed to our knowledge concerning EIA, only the E.C. is in a position to go beyond the exchange of information and the deepening of analysis to actually change the procedures of EIA. Thus this paper will discuss EIA procedures in the United States, selected countries of Western Europe and the European Community.

2. UNITED STATES

In the United States, the break-through for impact assessment was achieved by the introduction of a requirement for EIA through the National Environmental Policy Act (NEPA). The requirement was geared to the preparation of an Environmental Impact Statement (EIS) to accompany any action by a federal agency which is likely to have a significant impact on the environment. Over the last twelve years, this requirement has been interpreted by a number of court decisions. "Action" is taken to include not only direct action, as in road construction or the building of dams, but also indirect action as in licensing activities where others would be responsible for the execution of the action. Under certain circumstances, even programmes can be included in the requirement. An important court case led to a consent decree by the American aid agency (U.S. AID) accepting its obligation to undertake impact assessments when supporting projects in foreign countries. While NEPA does not make any specific provisions for taking into account the information generated by an EIS, there are ample means of recourse built into the general provisions governing American administration. Examples include the Freedom of Information Act and the Administrative Procedures Act. Sectoral laws also exist for the protection of nature and the environment, to ensure that an effective link can indeed be established.

Environmental Impact Statements under NEPA are essentially administrative documents, and they tend to be very bulky, quite difficult to understand and generally unattractive. The federal agencies concerned, and in particular the U.S. Army Corps of Engineers or the Department of the Interior, have acquired remarkable expertise in the preparation of these documents, even though the information they contain must frequently be obtained from other sources such as outside contractors and in particular from applicants in licensing procedures. NEPA also established an agency, the Council for Environmental Quality (CEQ), which is required to receive and evaluate EISs, and to provide guidance for their preparation. It was this function, mandated by law, which appears to have saved the CEQ from abolition by the Reagan administration.

The EIA requirement of NEPA must be seen in context. It is in many respects a typical piece of U.S. legislation, in that the Congress was not at all sure what it was creating at the time, and the EIA requirement has been the cause of much controversy over the last few years. NEPA was adopted at a time when few federal procedures existed in the area of land use planning or the integrated review of permitting procedures. NEPA in a sense filled a vacuum in the American administrative structure. The existence of the vacuum has contributed to the importance that the EIA requirement has acquired over the years. Moreover, the information produced by an EIS is subject to the characteristically competitive process of policy making in the U.S.. This competitive process results in a large number of important administrative decisions being ultimately subject to adjudication by the federal courts.

NEPA only applied to federal agencies as these agencies have no jurisdiction over actions which are purely the responsiblity of the states. Over the years, a number of states have adopted EIA requirements of their own, supplementing the federal requirements. Consequently the network of review created by EIA in the U.S. is very wide. It is accurate to say that no major project could now be planned without the preparation of an EIS. It is of course difficult to judge just what the effects have been. It is frequently claimed that the EIS lengthens the planning process, but this is no more readily ascertainable than the claim that it provides for a timely consideration of environmental issues and thus ultimately improves the quality of planning and may even expedite the process.

The introduction of an EIA requirement through NEPA was undoubtably a main breakthrough, as although there had been some discussion regarding techniques for impact identification and prediction within the U.S., the legislation went beyond the proven state of the art and forced the rapid development of EIA methods and techniques in the U.S. and beyond. The existence of the NEPA requirement, the insistent, almost world-wide search for effective techniques to prevent environmental damage rather than undertake its post facto repair, together started an international debate of quite remarkable dimensions. The literature is by now almost impossible to keep up with. Europe certainly was not immune to this development, and in fact had no intention or desire to be so.

3. EUROPE

The situation in European countries is different from that in the United States in a number of important respects. In most European countries a substantial body of regulations for the consideration of land use and environmental issues already exists, often long-established. In particular, zoning laws, which invariably require some kind of assessment of potential environmental impacts, have already existed for a long time, in some cases for decades, as have general laws to limit the creation of "nuisance". Moreover, these laws are executed in a much more integrated administrative structure than in the United States. The phenomenon of competing jurisdictions, between different levels of administration and agencies, is much less widespread in Europe than in the United States, largely because a sense of hierarchy underlies all administrative structures. Hence information generated in one part of the administrative system can pass to others, but very often it will not get into the public domain. As a consequence, the introduction of an EIA requirement is undoubtedly a very much more complex process in European countries, particularly at the level of the European Community. It is, therefore, hardly surprising that it is taking more time to materialize. Moreover, in most European countries, the reality of the legal and political situation means that major efforts must be made to get the legislation correct at the first attempt, since opportunities for subsequent adjustment either through the courts or through what is known in the United States as legislative review, simply do not exist for constitutional reasons.

The major exception in this regard is France which, in 1976 adopted legislation as part of its Nature Protection Act, mandating the preparation of environmental assessments for all major public or private development projects. The reasons for the much more rapid action in France, a phenomenon which has been repeated in several other areas of environmental concern, do not lie in a significantly more environmentally-minded parliament or administration. Comparisons are, of course, difficult but it is probably fair to say that the Netherlands and Sweden, and to a lesser extent Norway, are environmental innovators in Europe, while the major countries, the Federal Republic of Germany and the United Kingdom, are in no way inferior to France in terms of environmental commitment. Each of the major countries has particular areas of strength and occasional innovation, and patches of signal weakness, although the latter are not usually discussed internationally in public. France is able to legislate more rapidly and more unambiguously because the implementation of the legislation remains firmly in the hands, and to a remarkable degree at the discretion, of the administration (much as in the United Kingdom), and is not subject to judicial review except under the most genteel of circumstances. Hence it is administrative practice rather than legislation which will tend to define the scope of policy initiatives. In France, this implies not only analysing the implementing regulations very carefully, but also considering the way in which the new regulations meshes with established procedures. In this instance, declaration d'utilite publique is the most important element and the implementing regulation requires that an EIS be included in the public dossier on which official approval is to be based, without however, making this an absolute condition of approval. The public audition - I advisedly avoid the English word "hearing" with its connotations of formal procedure - is a rather informal process in which the appointed inspector meets any interested parties to receive their views. The public can examine the documents and express opinions on them. As a rule, all essential decisions about the project and its requirements have already been taken by then. It is consequently hardly surprising that the French EISs tend to be brief documents, often well illustrated, trying to make the best case for whatever environmental impact a project proposal may have.

Almost simultaneously with the passage of legislation in France, the Federal Republic of Germany adopted a cabinet level decision obligating all federal authorities to prepare environmental assessments of important projects they undertake. In most instances the actual responsibility for design and execution of specific projects does not lie with the federal authorities, so the cabinet decision has had no discernible impact. In the following years, some of the

Lander took the initiative and legislated impact assessment requirements. The Federal Republic of Germany also has a long established and very elaborate licensing and other procedures to control adverse effects of all manner of activities, and has consequently always maintained that the essential elements of EIA already existed. Recent studies have attempted to assess the validity of such claims, but have not proved conclusive. At any rate, there is no sense of an urgent need to reform practices in a fundamental manner in the light of experience with EIA in other countries. There is however, an attitude which welcomes a review of existing procedures in the light of the E.C. draft directive on EIA, and consequently the Federal Republic of Germany has a consistent policy position in favour of such Community legislation.

The Netherlands has been a driving force in the development of the state of the art of EIA in Europe. In the mid-seventies, the government announced its intention to submit legislation designed to create an EIA requirement. Since then it has been engaged in a protracted exercise of testing and analysis to define the best approach to the issue. Introduction of an EIA requirement is to become part of an important law containing administrative procedures for environmental protection. The typical legislative process in the Netherlands proceeds by a series of steps, slowly building a law which corresponds to practical requirements. The process is sometimes very slow, but the results often go beyond what can be achieved in other countries.

The Dutch government finally introduced its bill on EIA into parliament in May 1981, and a fairly protracted legislative process is still to be expected. The bill sets out the requirements for impact assessment in greater detail than has been undertaken elsewhere in Europe. The tricky issue of finding a definition to determine which projects and public plans are to be subject to an impact assessment does not, however, need to be settled through primary legislation. The operative general definition (Article 41b) will also require subsequent implementing legislation and the political debate will be of great significance in this respect. The bill goes further than any other legal instrument in Europe at this time, in terms of detailing the minimum content of an impact assessment. In all of these areas, action in the Netherlands is likely to provide important clarification for all countries in Europe.

In Luxembourg, the Act of 27 July 1978 on the Conservation of Nature and Natural Resources provides in

Article 1 that the Minister will cause an impact analysis to be prepared for all measures which might have an impact on the natural environment. Whereas the French law stated that the purpose of the act was also to assure a harmonious balance of populations living in urban and rural environments, the Luxembourg texts refers only and specifically to the natural environment. Hence it remains to be seen just what the effect of the EIA requirement in the context of this law will be.

In Ireland, the 1976 Planning and Development Act introduced the principle of impact analysis which was then subsequently defined in greater detail by a regulation published in 1977. A financial threshold for EIA has been provided, in that projects involving investments of 5 million pounds or more require an EIA.

In other countries of the Community, the policy debate on impact assessment has gone less far. In Belgium, since 1977 certain regulations have been in force within the Environment Ministry, but their effect is not known. The Minister announced in early 1979 that he intended to submit a bill on impact assessment to parliament during the year, but on account of changes in government this has not yet occurred.

Similarly, differences of opinion in the Environment Ministry have meant that in Denmark no agreement has been reached with respect to the desirability of introducing a law providing for environmental impact assessment.

Although no specific regulation concerning environmental impact assessment as such exists in the United Kingdom, there is still a highly rigorous administrative procedure taking into account environmental factors and instituting a very extensive system of public inquiries. In the specific area of land use planning for example, the Town and Country Planning Act of 1971, as amended in 1977, provides a procedure for the assessment of the environmental consequences of certain important projects.

4. THE EUROPEAN COMMUNITY

In its declaration of 22 November 1973 on the Community Action Programme, the Council of the European Community adopted one principle, among others, to guide its further action. The principle was "that effects on the environment should be taken into account at the earliest possible stage in the decision-making processes", and "to evaluate the effects on the quality of life and on the natural environment of any measure that is adopted or contemplated at national or Community level and which is liable to affect these factors". The introduction of a Community environmental impact assessment procedure would therefore be the logical next step from this declaration of principle. Several studies undertaken for the Commission with a view to determining the bases for the formulation of such a measure led to several drafts for a directive on EIA. A proposal for a directive, drafted after consultation with the various interested parties, has been submitted to the Commission. This text presents two particularly interesting aspects from a European environment point of view. The first is that although the directive will, like any other, leave the choice of means to the Member States, it will render obligatory the accomplishment of its aims; it will therefore make it possible to render more rational or equitable a procedure which risks leading to distortions between states by its very existence or non-existence. The second interesting aspect is that the existence of Community measures in this domain opens up the prospect that in future, not only will the national environment be taken into account, but account will also be taken of the environment of neighbouring states. Thus far no national provisions exist to cover this particular aspect.

The proposal for a directive states that for public or private projects listed in an annex, the applicant is required to include in the application dossier, information on the evaluation of the consequences for the environment of the project. Before authorizing a project, the competent authorities must publish the application together with this dossier. Public consultation must also be facilitated. The modalities of consultation are not, however, specifically defined.

Following publication of the proposed directive, some of the most important new elements in the debate have been contributed by the United Kingdom. In particular there is the question as to how and to what extent the existing national procedures are compatible with the potential requirements of the directive. This is of particular concern to the United Kingdom as existing procedures already provide for a complete and detailed scrutiny of the environmental effects of developments under certain circumstances. This is undoubtedly a central issue in the on-going debate about the directive. To what extent individual Member States will be required to amend existing procedures, and to what extent it will become easier to follow developments in another Member State on account of the directive remains to be seen.

Since publication, the proposed directive has been working its way through the Community institutions. The Economic and Social Committee has adopted an opinion which is favourable, while maintaining a careful balance between the argument that impact assessment improves environmental protection and may even expedite consideration of applications, and the attitude that EIA constitutes an additional cost and carries the risk of delaying approval procedures.

The Council of the European Community has also been working on the proposal, although in secret. No fundamental decisions have so far been taken, but it appears that a number of states are still negotiating under an overall reserve, that is they have not yet declared whether they accept the principle of such a directive or not. Manifestly, the United Kingdom has been hesitant about the directive, as have, on certain aspects, the French and Dutch governments.

4. CONCLUSION

What does all of this imply in terms of environmental impact assessment in "Europe"? Once more one can observe that policy-making in Europe is slower than in the United States. On the other hand the results which are achieved tend to be much more permanent in nature. Hence we are now at about the half way stage of a long and complex process which should, ultimately, lead to the establishment of an EIA requirement in all the countries of the European Community. As a result of the introduction of EIA, there should be greater transparency of decision-making and a basis for transnational comparability in the results of environmental policy such as exists nowhere else thus far.

AFFILIATION

Dr. Konrad von Moltke is the Director of the Institute for European Environmental Policy in Bonn.

Murray G. Jones

CANADIAN FEDERAL AND ONTARIO PROVINCIAL ENVIRONMENTAL ASSESSMENT PROCEDURES

1. INTRODUCTION

Canada is a country composed of ten provinces and a number of "territories". The provinces all have similar parliamentary systems of government to that at the national or federal level, although the division of responsibilities is such that the provinces have an autonomous role in many areas, the environment being one. This division of power in the Canadian situation illustrates why it is not appropriate to examine just the federal process. Each of the provincial processes provides an unique approach to Environmental Assessment (EA).

It is useful to note the evolution in environmental thinking that has taken place in Canada. In the 1960s and early 1970s, laws were passed at provincial and federal levels which were a reaction to increasing degradation of the environment. These laws were directed at the abatement of pollution from activities which were emitting pollutants to the air, water or land, and usually involved the use of licensing procedures of some sort. As time went on it became apparent that, while giving a certain amount of protection to those aspects of the environment just mentioned, there appeared to be many concerns which were not specifically covered and were therefore ignored. For instance, while it may have been possible to lower the air and water pollutant concentrations from a particular source, it was difficult to do anything about the fact that the activity was located on an environmentally unacceptable site in the first place. In addition, it was obvious that the installation of pollution control devices some years after the commencement of the industry was far more costly than incorporating such equipment from the start.

It was the opinion of the governments in Canada that developed such procedures, both federally and provincially, that consideration must not only be given to air, land and water, but also to man, nature, land and the interactions

B. D. Clark et al. (eds.), Perspectives on Environmental Impact Assessment, 35–50.

Table 1 Summary of the Main Features of the Federal and Ontario EIA Procedures

	Federal	Ontario	U.S.
System	Environmental Assessment and Review Process	Environmental Assessment Act	National Environmental Policy Act
Type of procedure	Cabinet policy	Legislated	Legislated
Relationship with existing decision-making process	Semi-integrated	Integrated	Integrated based on policies in law
Body controlling procedure	Federal Environmental Assessment and Review Office (FEARO)	Ontario Ministry of the Environment (MOE)	Council on Environmental Quality (CEQ)
Who is responsible for EIA preparation	Proposing federal department	Proponent	Lead agency or joint agencies
Field of application	Projects, federal departments and agencies and proponents requiring federal funds or lands.	Projects, plans, programme level. Provincial, municipal then private sectors (phased introduction).	Projects, plans, legislative level. Some exemptions in regulations.
How are activities identified	Self-assessment and precedent.	Exemptions for provincial projects and positive list for private sector.	Self assessment and negative declaration.
Definition of environment	Natural, physical, social and economic.	Natural, physical, social and economic.	Human environment, matural and physical and relationship of people to that environment.
Body for whom the EIS is prepared	FEARO	MOE	Lead agency.
Use of scoping	IEE identifies significant impacts also guidelines.	?	Yes.

Use of guidelines	Generic and project specific guidelines.	General guidelines, some generic but normally project specific.	General guidelines (CEQ) and generic guidelines (departments).
Public involvement	Early involvement with publication of project guidelines continues to submission of the EIS.	Suggested but only begins formally when EIS submitted.	Yes, incorporated into agency procedures.
Hearings or meetings	Always.	Ministers opinion, one hearing body, Environmental Assessment Board (EAB).	Possible, depends on individual agency procedures.
Decision-making body	Minister of the Environment plus minister of sponsoring department.	Minister of the Environment or EAB.	Lead agency plus others when necessary.
How is decision made	Based on recommendations by FEARO. Final decision by cabinet.	All minister comment on EIS, minister may alter EAB decision. Cabinet approval.	Review by federal agencies, public and others. Lead agency decides on basis of EIS and other information.
Appeal	To cabinet.	To cabinet.	To courts.
Monitoring	Recommendations from panel, Ministers can require a monitoring programme.	MOE with relevant ministries can require monitoring.	May be done by agencies.

between these components, and to cumulative, secondary and "off-site" effects. Included within the term "man" are the social, cultural and community aspects. The overall approach, which evolved in the mid 1970's can be called "anticipatory" as it provides for consideration of effects which could occur, and it attempts to prevent or mitigate those effects which are considered unnecessary or undesirable. It does this through the consideration of options or alternatives which could include siting, technology, design or, if necessary, abandoning the project.

The anticipatory approach serves as a partial definition of Environmental Assessment (EA), to which some additional points may be added:

(i) EA is only useful if it is really integrated within the decision-making processes of both the approving government departments and the organization which is seeking approval of the project;

(ii) EA should continue past the stage where a document or statement is prepared and submitted. It should proceed into the implementation and operational stages. This could be achieved by continuous monitoring of the consequences of a project during the construction and operational phases, as well as monitoring the EA process as a whole.

With this background on the EA framework, it is now possible to examine the federal EA procedure and the procedure used in the province of Ontario. The federal procedure was introduced in 1973, the Ontario procedure in 1975. While having a number of similarities, such as the definition of environment, there are many structural differences which provide for an interesting comparison. In addition the Ontario procedure was the first to be put into legislation in Canada and was perhaps the first legislated system to occur outside the U.S.. Both the federal and Ontario procedures are quite different from that used in the U.S.. A comparison of the three separate procedures is presented in table 1.

For further assistance, figures 1 and 2 provide an understanding of the two Canadian procedures.

2. FEDERAL PROCEDURE

In this section the main components of the federal EA procedure in Canada will be presented.

Figure 1 Basic Steps in the Canadian Environmental Assessment and Review Process

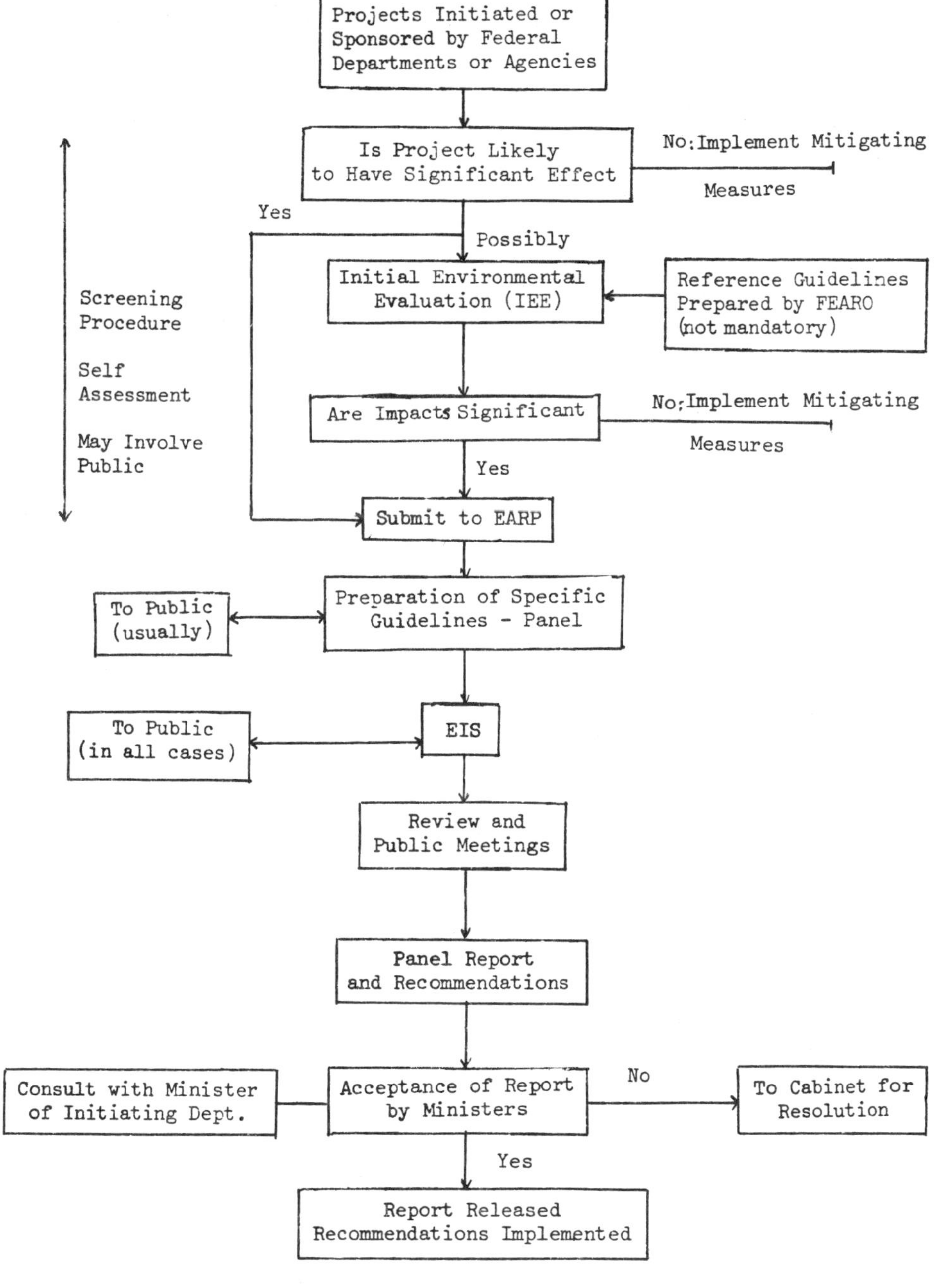

Figure 2 Ontario Environmental Assessment Process

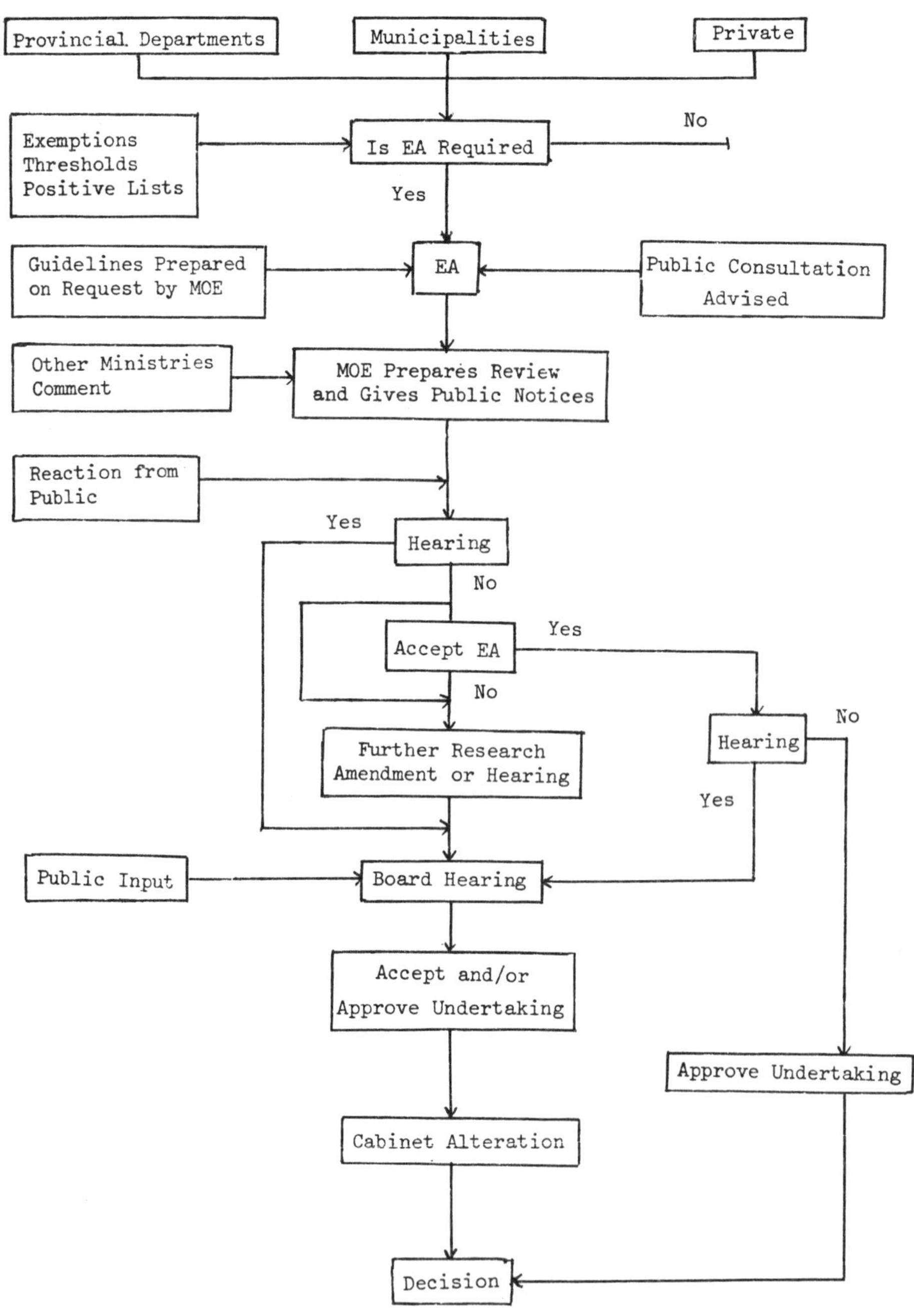

MOE: Ontario Ministry of Environment

2.1. Type of procedure

The federal procedure was initiated through a policy directive taken at the cabinet level. It consequently does not have the power of law, but it does provide an agreement of all cabinet ministers to apply EA to their departments. In addition, it has a degree of (political) flexibility, since it may be amended by the cabinet alone without the use of parliament. Since its introduction in 1973, the policy directive has been amended once, in 1977. In the view of the author, the system can be regarded as semi-integrated in that not all approvals are necessarily given at one time, and only the review carried out by the Department of the Environment is concerned with the environment. Indeed, approvals by other by regulatory agencies may come before or after the EA but the EA decision occurs at a time when options are available and the decision can be taken not to proceed.

2.2. Administration process

A special body has been established to oversee the EA procedure and plays an active role once a commitment to perform an EA has been made. Known as the Federal Environmental Assessment Review Office (FEARO), this body reports directly to the Minister of the Environment and Fisheries, through its executive chairman, on matters concerning the Environmental Assessment Review Process (EARP). FEARO determines policy and provides guidelines, procedures and administrative advice to the public and to participating agencies (FEARO, 1979). In addition, FEARO sets up "panels" for reviewing EAs and evaluates the process as a whole on a continuing basis.

2.3. Responsibility for EA preparation

The federal department or agency which initiates a project is responsible for any preliminary studies and if necessary the EA.

2.4. Scope of EA

The process applies to projects, programmes and activities which are initiated by federal departments or agencies. It also includes projects for which federal funds or lands are required. Some examples of completed EAs include a nuclear power station, a hydro-electric project in a national park, the expansion of a uranium refining capacity of a crown corporation, to name but a few.

2.5. Identification of impacts

A two step self-assessment of the likely impacts of the proposed project by the project proponent is the key aspect of the federal procedure. During the first step, environmental screening can be undertaken using criteria developed by FEARO, to determine whether there are likely to be any significant impacts. If the conclusion is that no significant impacts are to be expected then no further study is required. If significant impacts are expected, then the proponent should submit the project to FEARO for formal guidance on the EA to be undertaken. In situations where some doubt as to the significance of the impacts exists, an Initial Environmental Evaluation (IEE) can be undertaken in accordance with the general guidelines published by FEARO in 1978. The IEE is undertaken by the proponent (see figure 1). The self-assessment phase must involve consideration not only of technical data and information but also of the project s potential for causing public concern. A FEARO publication states that public reaction to a proposal is an important factor in determining significance. Thus a certain amount of interplay between the proponent and the public is expected to take place at an early stage in the assessment process.

One possible difficulty with such a process is that it relies on departments being open and straight-forward about the likelihood of significant impacts. This reliance is based on moral obligation, or perhaps the persuasive powers of the responsible minister in the cabinet. Over time public knowledge about the environment and the EARP process in particular has grown, and with it the expectation that EAs will be prepared. This knowledge and expectation provides some assurance that there will be a public review of major proposals.

2.6 Definition of the environment

The environment to be studied can vary from case to case depending on the requirements laid down. However, a consistently broad interpretation has been applied, including not only the natural (or biotic) environment, but also the physical (abiotic) environment and social-community aspects.

2.7. Organization for which EA is prepared

The EA is prepared for, and submitted to, a "panel" of experts. The panel is appointed by the executive chairman of FEARO and includes the chairman or his delegate and experts from inside or outside the federal public service (FEARO, 1980).

2.8. Use of scoping and guidelines

At present scoping and guidelines (two systems for focusing and directing an EA towards the appropriate issues and siginificant impacts), can be combined in the EARP. The combination occurs during the preparation of the project-specific guidelines by the panel. These have the objective of ensuring that the proponent is aware of the information required by the panel, technical reviewers and the public. During the preparation of the guidelines, the panel may seek whatever professional advice it considers necessary. There is also a move towards increasing public comment prior to issuing the guidelines. In addition, direction can be given to the main study through the use of any IEE completed. This IEE would identify those impacts significant enough to require an EA.

2.9. Public involvement and hearing

The involvement and participation of the public is a most crucial aspect of the EARP process. To summarize the public involvement (see figure 1), there is a strong suggestion that the public should be involved during the screening procedure. Secondly, when a project is submitted to the panel, the panel often receives comments on the project-specific guidelines. Thirdly, the proponent is encouraged to invite participation during the course of EA preparation. Finally, upon submission of the EA to the panel a very open review procedure takes place. This consists of series of public meetings, organized by the panel, in the immediate vicinity of the proposed location of the project. Extensive dissemination of information takes place before such meetings to the public, federal, provincial and private organizations. The meetings are not highly structured but have sufficient order that all views can be stated, questions asked and responses given.

Finally, the public has the right of appeal to their parliamentary representative and/or the cabinet should they disagree with a decision that has been made by the responsible minister(s).

2.10. Decision-making

Based on the environmental assessment and hearing, a report is prepared by the panel and sent to the Minister of the Environment. This report describes the background to the project, the project itself, the characteristics of the area, the environmental and related social implications, and the conclusions and recommendations of the panel. Decisions on the panel's recommendations are made by the Minister of the Environment and the minister of the initiating or sponsoring department. Should there be any disagreement the matter is then referred to the cabinet for resolution.

2.11. Appeals

If the public disagrees with the decision appeal can be made to the cabinet.

2.12. Monitoring

As a result of recommendations by the panel, or on their own initiatives, the Minister of the Environment and the minister of the initiating department can reach decisions relating to a monitoring programme. The implementation of the program could involve the project proponent and other ministries having an interest in the matter.

On a separate level, monitoring of the overall EARP is a continuing function of FEARO. It is claimed that this has allowed for constant adaption to changing circumstances and to lessons learned. In addition, in 1979 a study of EARP was carried out by an independent advisory council reporting directly to the Minister of the Environment. Mechanisms to strenghten the process were recommended (Canadian Environmental Advisory Council, 1979).

3. ONTARIO

In considering the EA procedure adopted in the province of Ontario, reference should be made to table 1 and to figure 2.

3.1. Type of procedure

The Ontario procedure was designed, from the earliest stages, to be expressed in law. Before the procedure was established, however, a green paper (Ontario Ministry of the Environment, 1973) was published as a consultation document. The green paper presented for comment, a series of procedural options. Industrial groups, government agencies, environmental groups and the public responded with suggestions and comments. In the author's opinion, this was a useful approach which should be considered as a means for seeking views and arriving at an appropriate procedure by any country/state interested in establishing environmental assessment. The results of the consultation, in Ontario, formed the basis for a law which was passed by parliament in 1975 and enacted in late 1976.

As with the federal system, this procedure is semi-integrated, but for different reasons. In Ontario the EA process should be carried out concurrent with other planning activities by the proponent at an early stage. Approval of the EA, however, must be obtained before other approvals can be acquired (Rudik and Young, 1976). This clause was made part of the legislation in order to avoid unnecessary commitments or expenditures on a project prior to approval to proceed. There is, therefore, what may be called a "front-end" approval.

The system has been adapted and steamlined first, so that information gathered during the EA and presented to the Ministry of the Environment, will be applicable to other agencies from which licenses must be obtained. Secondly, the choice of the alternative will not be disputed at later stages Finally, if hearings are held on the EA then subsequent hearings, which would normally be required in the approval stages, may be dispensed with. This streamlined approach has been favourably received by the government, industry and the public.

3.2. Administration process

In contrast to the federal procedure, no special body has been established to oversee EA in Ontario. Instead, responsibility has been given to the Ontario Ministry of the Environment which also has responsibility for other environmental matters. There is, however, one section within the Ministry, called the Environmental Assessment Branch, established in 1982, which co-ordinates the development of guidelines and the review procedure, advises the minister on the decisions to be made, participates in public hearings (if such are required) and monitors the overall application of the law.

3.3. Responsibility for EA preparation.

The proponent of an "undertaking" (the Ontario term for plan, programme, project or activity) is responsible for the EA and the resulting document. The proponent may be an individual, corporation, municipality, provincial department or agency.

3.4. Scope of EA and impact identification

The Ontario Environmental Assessment Act applies to all plans, programmes or projects undertaken by the province or by municipalities unless specifically exempted. This is a type of negative list approach (ie. everything is in unless exempted out). However each exemption must be requested from the Minister of the Environment and he must provide reasons for his decision when granting an exemption. Shortly after the introduction of the Act, many exemptions were requested because of the advanced state of development of numerous projects. After the first rush of exemptions, and an increase was observed in the number of EAs submitted. Exemptions, however, continue to be given to many projects, and this remains a contentious issue in the Province.

Private sector undertakings are dealt with in a different manner. The Act applies to "major commercial or business enterprises or activities or proposals, plans or programmes..." designated by regulations. Phasing of the introduction of the Act required that attention should be first given to the provincial sector, then the municipal sector and finally the private sector. Therefore, only a limited number of private projects have been designated for an EA. These include a proposal for a new pulp and paper mill, a private hydro-electric plant, and the development of a lignite deposit in the northern part of the province. Attention is now being directed towards the private sector, and a positive list is expected to be devloped in the future.

3.5. Definition of environment

A very comprehensive definition of the environment has been written into law (Ontario Ministry of the Environment, 1975). This definition includes aspects of air, land, water, plant and animal life and man. In addition, social, economic and cultural conditions, structures (such as buildings), the results of any activity (solid, liquid or gas) and any combinations or inter-relationships of the foregoing are included.

3.6. Organization for which the EA is prepared

The EA is prepared by the proponent and submitted to the Minister of the Environment. The minister then initiates a review of that document. This amounts to a review by all ministries co-ordinated by the Ministry of the Environment. The review and the EA are released to the public.

3.7. Use of scoping and guidelines

Scoping as a separate activity does not exist in the Ontario system; instead, at the request of the proponent and through communication with other appropriate ministries, guidelines can be produced on a case by case basis. In addition, there exist "General Process and Content Guidelines", and a few generic guidelines relating to broad categories of projects such as nuclear power stations, transmission lines, and hazardous liquid industrial waste disposal (Ontario Ministry of the Environment, 1978).

3.8. Public involvement and hearing

The ministry strongly encourages proponents to involve the public during the EA process prior to the submission of the EA. This is to allow for the exchange of information and to help identify those issues of greatest importance. If such involvement does not occur, then the first indication to the public that a project has been proposed usually occurs in a ministry publication on EA which states that an EA has been recieved and is now under review. On completion, the review, together with the statement are placed in accessible locations near the project site and an official notice is published. Comments can be directed towards the minister and requests made for a public hearing.

A separate body conducts the hearings on the EA when so directed by the minister. This body, the Environmental Assessment Board (EAB), has its own procedures for public notification and dissemination of information. Hearings are held in the vicinity of the project and are structured semi-formally. All persons are given an opportunity to question the evidence and to make their own presentations. The EAB is empowered to make a decision on the proposal. This decision and its justification are then published.

3.9. Decision-making

If no hearing is held, the minister makes the decision on the undertaking . This decision is in two parts, the first on the acceptability of the EA, and the second on whether the undertaking should proceed. The minister may decide that the report is acceptable and yet believe that a hearing on whether on not the development should proceed is required. In this case, the hearing is then restricted to a consideration of this issue alone.

In order that the decision-making process remains within the parliamentary system, the minister has the right to over-rule the decision of the EAB provided that he does so within 28 days and provides a reason for doing so.

3.10. Appeal

Any decision may be appealed to the cabinet. This is the ultimate level for decision-making (see figure 2).

3.11. Monitoring

The ministry is essentially responsible for monitoring and enforcing its approvals. This is usually achieved through regional divisions with the assistance of the head office staff. In addition, other ministries also co-operate, bringing their own expertise to particular problems.

The Ministry of the Environment has overall responsibility for the Act and, therefore, the procedure. Consequently, it can monitor the procedure for successful implementation.

4. CONCLUSIONS

To conclude this overview of the Canadian (federal) and Ontario (provincial) EA procedures, it is useful to highlight certain crucial differences between the two. These are:

(i) the procedure - policy versus law;
(ii) the initiation - self-assessment versus exemption/designation;
(iii) guidelines - required versus optional;
(iv) controlling procedure - separate versus integrated body;
(v) review - public versus combined government/public process;
(vi) hearing - compulsory versus optional; and
(vii) results - recommendations versus decisions.

In both systems the definition of the environment is wide and final decisions are taken at a ministerial level with appeal to cabinet.

The main advantages that can be seen in the Canadian procedure are the extensive public involvement and the use of specific guidelines to aid the proponent. On the negative side, the initial self-assessment is completely secret (unreviewed), the use of a policy statement provides for too much flexibility on the part of the various departments, and the time taken from beginning to end is too lengthy.

Disadvantages of the Ontario procedure include significant use of the ministerial exemptions power and the lack of any specific scoping or guidance. There have also been complaints on the time taken for the procedure, especially the presentation of the review. Advantages are that the system is legislated with an "up-front" approval. In addition, the broad definition of the environment clearly sets the scope of the assessment as a comprehensive review of all impacts.

REFERENCES

Canadian Environmental Advisory Council: 1979, Environmental Assessment and Review Process: Observations and Recommendations, Canadian Environmental Advisory Council, Ottawa.

Federal Environmental Assessment Review Office: 1978, Guide for Environmental Screening, Ministry of Supply and Services, Ottawa.

Federal Environmental Assessment Review Office: 1979, Revised Guide to the Federal Environmental Assessment and Review Process, Ministry of Supply and Services, Ottawa.

Federal Environmental Assessment Review Office: 1980, Environmental Assessment Panels: What they are and What They Do, Ministry of Supply and Services, Ottawa.

Ontario Ministry of the Environment: 1973, Green Paper on Environmental Assessment, Ontario Ministry of the Environment, Toronto.

Ontario Ministry of the Environment: 1975, The Environmental Assessment Act, Ontario Ministry of the Environment, Toronto.

Ontario Ministry of the Environment: 1978, EA Update, 2 (4), Ontario Ministry of the Environment, Toronto.

Rudik, V.W. and D.R. Young: 1976, Environmental Assessment in Ontario: Why, How, What and When, Paper presented to the Simcoe County Chapter of the Association of Professional Engineers of the Province of Ontario, April 22, 1976.

AFFILIATION

Murray G. Jones was formerly employed by the Ontario Ministry of the Environment within the Environmental Assessment Section, before becoming adviser to the Ministry of Health and Environmental Protection in the Netherlands. He now is a Staff Engineer in Environmental Affairs Department for Shell Canada Ltd (Toronto).

Xaver Monbailliu

EIA PROCEDURES IN FRANCE

1. THE BACKGROUND TO EIA IN FRANCE

EIA legislation in France dates from July 1976 when the Nature Protection Law was passed; it was enacted in October 1977. This legislation, designed to protect the environment, became operational in January 1978, and required EIAs to be prepared. Thus, EIA procedures are a recent innovation for France. Four years of experience may be regarded as a "running in" period, so that current legislation can be modified as appropriate. The law requires that the EISs (Etudes d'Impact) must examine the ecological, aesthetic and scientific aspects of a development. It also stipulates that some "minor" infrastructural developments are excluded from the EIA requirements. It is, however, surprising that the legislative body has excluded the following developments:

(i) transmission lines under 225 kV;

(ii) housing development zones to be constructed within the limits of local planning procedures (Plan D'Occupation Du Sol: POS), although local plans do not necessarily include the environmental repercussions of such housing schemes;

(iii) housing schemes occupying less than three thousand square metres;

(iv) areas for units of two hundred caravans which may be up to ten hectares; and

(v) all developments with a cost less than 60 million French francs.

Existing legislation is, therefore, not very clear and is sometimes evasive as to the contents of the EIA. For example, an EIA for a highway will include a study or statement as to its repercussions on forestry, wildlife and landscape, but will probably exclude the sociological effects on the rural population.

Some 4500 EISs were produced during 1978, of these, two-thirds were private projects and the remainder public projects. Deforestation projects accounted for 720 EIAs, while 445 related to low density housing schemes, 400 to

B. D. Clark et al. (eds.), Perspectives on Environmental Impact Assessment, 51–55.

agricultural consolidation projects, 95 to quarries, 50 to waste disposal sites, 30 to electricity transmission lines and 10 to reservoirs.

Some three hundred private firms, environmental associations and a few university teams are dealing with EIAs in addition to various governmental agencies. During 1978, professional fees and costs were estimated at 50 to 100 million French francs. It is the responsibility of the proponent of the development scheme to appoint the consultant or study teams and to finance the EIA.

The Ministry of the Environment, which combines the former Ministry of Equipment and the Ministry of Quality of Life, has produced over twelve sets of guidelines for the preparation of EIAs. Such guidelines are general descriptions of the repercussions of infrastructural developments as motorways, power-lines, housing schemes, sewage schemes and landfill projects (Ministere de l´Environnement et du Cadre de Vie, 1979).

2. PUBLIC INVOLVEMENT

According to the introductory statement of the 1976 law, every citizen has the right to protect his or her natural environment. Despite such statements, public participation in EIA procedures is very limited. It is restricted to a discussion of the published EIS. Up till 1979, only twelve EIAs have been questioned by administrative tribunals. The developer may, however, seek close ties with the local population during the preparation of the EIA. Such contacts are generally at a political level and often influence the final choice of the project location. These contacts are of great importance when proposals such as highways, pipelines or power-lines affect various departements (local government units), villages or towns. The geographical position of these linear developments may be determined by the group exerting the greatest influence. The position regarding public participation has been highlighted by the consultation and confrontation that occurred at Plogoff in Brittany, the site of a proposed nuclear power station. The consultation was held in so-called "annex town halls", which were Citroen vans, as the local population had decided not to allow the public inquiry to take place in their town hall. It is also important to note that a lawyer appointed by the local Plogoff people was not allowed to plead in court for the local protestor held in custody.

3. HOW EIA PROCEDURES COULD BE IMPROVED IN FRANCE

The following proposals to make EIA more efficient in France could also be of interest to other countries where EIA legislation is proposed.

3.1. EIA procedures

One of the major obstacles to establishing a more objective and improved EIA, is that the EIA is the responsibility of the developer. Clearly, attempts will be made to submit EISs at the lowest cost, possibly omitting some important environmental information. To overcome this problem, a regional environmental impact fund to finance EIAs should be established. Such a fund should be jointly controlled by the regional authority, local or regional public organizations or associations, and the developer through the Chamber of Commerce for example. The independent nature of this fund would increase the likelihood of achieving a comprehensive approach to EIA. The fund could also sponsor the research and editing of alternative EIAs in crucial cases when there is disagreement on vital issues. The administrators of the fund could form an ad hoc committee to provide advice on how to undertake an EIA. The committee could also determine if an EIA is necessary in those cases where one is not officially required, but may be demanded due to special circumstances.

The publication of the final observations on an EIA by the Ministry of the Environment, is not a legal requirement in France. Such observations, as well as a report of the public participation, should be made available. A more open attitude by the authorities would contribute greatly to a democratization of the EIA procedures.

A final suggestion is that the authority responsible for administering EIA procedures should not be directly associated with a department which is responsible for proposing or financing projects. Ministries responsible for both projects and the EIA procedures are likely to be viewed by the public as lacking the necessary impartiality in an EIA. Instead, responsibility for EIA should be transferred to the Office of the Prime Minister or to a panel of independent ombudsmen.

3.2. Information requirements for EIA

Impact studies rely on information about local resources such as agriculture, forestry, ecology and landscape, as well as on the technological assessment of the proposed project. As a consequence, agencies will be seeking information on the current state of the environment repeatedly for different developments within the same region. Data banks containing such information would eliminate unnecessary duplication of effort. Such data banks could contain maps (soil, ecological and geological maps), air photographs, bibliographic references and other relevant information. A list of local environmental experts could also be included, so that local associations defending the environment, or the project manager, could seek advice. The first data bank has been installed in the Departement de l´Ain, east of Lyon, while others are soon to be installed in the Jura, Saone et Loire, L´Isere, Nievre and Loir-et-Cher.

A second phase of data handling could result in up-to-date mapping and interpretation of stressed and sensitive areas of the regions. Such maps should be regularly updated and modified where necessary. A multi-disciplinary team at the county level could meet in order to evaluate recently published EISs and then to update the information on environmentally sensitive areas.

It is important to identify constraints and repercussions of the proposed project, for example the visual effect of a power-line, therefore, it may be useful to include information concerning EIA methods and techniques in the data banks. The use of such data banks as a source of guidance for those involved in EIAs would be an importance benefit, since most EIAs are limited to a general description of the area and the possible impact. Very few attempt to justify the location of the development project; even fewer attempt to identify the site of least impact.

3.3. Improvements to public participation in France

Various measures to improve public participation have already been suggested earlier. Another aid would be the publication of a regional newsletter. Such a newsletter detailing the various published EISs would be welcomed by local citizens and associations. A summary of EIAs, as well as the public inquiry dates, should be included.

A final suggested mechanism for improving public participation is that of increasing the number of public hearings (one has been held at Carry-le-Rouet, near Marseille), and public meetings during the preparation of an EIA. These meetings could lead to a modified or re-directed

EIS which included the concerns of the local population, and hence, provide citizens with a greater say in their regional development.

3.4. Impact prevention

Efforts should perhaps be aimed at developing methods which would allow, at a regional level, the identification of those areas in which development would or would not cause environmental problems. A creative approach to EIA procedures is required rather than the negative aspects of planning control, which determines where a development cannot take place. Development projects should be designed to be more environmentally "acceptable", in that alternatives to projects both in concept and design should be considered, for example material recycling as an alternative to resource extraction.

4. CONCLUSIONS

EIA in France, as indicated above, has many failings. However, this is partly due to the relatively recent nature of the legislation, and the attitudes of the French towards the environment which have been influenced by the relatively low population densities. This paper has highlighted one particular area of procedural weakness, that is the issue of public participation in EIA, and attempted to suggest a few remedies for the situation. One question remains that is: are EIAs, as they are presently undertaken in France, the most appropriate measure to halt the present deterioration in environmental quality?

REFERENCES

Ministere de l'Environnement et Cadre de Vie: 1979, _L'Impact sur l'Environnement d'Une Decharge Controlee-Guide Methodologique, Direction de la Prevention des Pollutions et Nuisances des Problemes de Dechets_, Ministere de l'Environnement et Cadre de Vie, Paris.

AFFILIATION

Xaver Monbailliu is a land-use planning consultant and has prepared many EISs particularly on power-lines.

Murray G. Jones

THE EVOLVING EIA PROCEDURE IN THE NETHERLANDS

1. INTRODUCTION

An examination of the EIA procedure being proposed in the Netherlands is a useful introduction to EIA procedures for individuals and countries who, like the Netherlands, are just becoming involved in this subject. This paper will examine, firstly, the background to the proposed EIA procedure, including definitions, a justification for EIA, the uses of EIA, and the key criteria which have governed the development programme; secondly, the development programme itself and the research which provided recommendations for a procedure and methods useful to that procedure; and finally, an analysis of the principal elements of the procedure. Where appropriate, cross reference will be made to other systems which have aided the Netherlands choice of procedures and methods.

As might be expected, the procedure proposed is suited to the needs of the Netherlands and is different from those used elsewhere. One of the objectives of the procedure is the elevation of environmental considerations to their "full and proper place" in the decision making process. In this regard, the definition of EIA put forward in a government position paper is of importance (Ministry of Health and Environmental Protection, 1979). It is as follows: "Environmental Impact Assessment must be seen as an aid to planning and decision-making. This aid consists of the compilation, review and application of an EIA, and the subsequent monitoring of the environmental consequences of the implementation of a decision, taking into consideration a number of procedural rules".

Although there are a number of existing laws in the Netherlands expressly concerned with the environment, they are oriented towards specific sectors: for instance air, noise and water. In analysing these and other laws which have environmental implications, (such as some of the planning, transport and conservation laws) in relation to the principles of EIA, a number of deficiencies in the

B. D. Clark et al. (eds.), Perspectives on Environmental Impact Assessment, 57–68.

system which restrict an adequate and integrated consideration of the physical environment were identified. The government position paper (Ministry of Health and Environmental Protection, 1979), lists these deficiencies as:

(i) at present, often no information is requested about the environmental effects of a proposed action;
(ii) in some cases no environmental information is required at all;
(iii) the required information does not lead to a comprehensive integrated picture of the environmental consequences of a proposed action;
(iv) at present there are still large differences in the current legislation regarding the requirements for the contents of the environmental information to be submitted concerning a proposed action, regarding its presentation, and also the consultation, public participation, review and application of the information; and
(v) it is not always required to mention the purpose of the proposed action or to submit alternatives.

Turning from these weaknesses in the system to the benefits of EIA, EIA was seen in the Dutch context as having a number of positive attributes. These were presented in a memorandum to Parliament in 1978 (Ministry of Health and Environmental Protection, 1978) as:

(i) the possibility is created for all environmental consequences of importance to planning and decision-making, and their inter-dependencies, to be systematically reviewed;
(ii) the integrated review of the environmental consequences, made possible by EIA, could lead to a harmonized consideration of the requests for licences and permits within a framework of relevant environmental legislation and could also lead to a streamlining of the licensing procedures;
(iii) making the relevant information public at an early stage provides for more meaningful public participation and can prevent revocation of a decision arising from objections made after investments have begun;
(iv) the formulation of alternative solutions during the early stages is stimulated; and
(v) it provides an insight into the gaps in available knowledge or information which must be taken into account during planning and decision-making.

While EIA was perceived to hold great promise, a number of problems had to be overcome before it could be incorporated into the planning system. Two principal criteria were used to guide the development of solutions to such problems. The first was that EIA should not cause any delay in the planning and decision-making process. This concern arose from the early U.S. experience with EIA, where costs and delays were viewed as issues of importance. The second criterion was the limitations caused by the standards and rules of the existing environmental policy. Since the environmental interests would need to be balanced against other interests, application of these criteria meant that EIA would need to be integrated to a great degree with the existing planning system.

Before describing the chosen system, it is appropriate to examine the type of research that has been undertaken into both the development of a procedure and the methods which can form part of that procedure.

2. THE DEVELOPMENT PROGRAMME

Upon recognition of the growing attention being given to EIA during the early 1970s in a number of countries, the Minister of Health and Environmental Protection asked an independent advisory council for advice on whether such a procedure would be of use in the Netherlands and, if so, what the components of the system might be. In 1976, the Council provided the Minister with a unanimous opinion that EIA was useful and desirable (Provisional Central Council for Environmental Hygiene, 1976). Its opinion was important for others interested in EIA to consider, especially as the Dutch legal and planning system was already very extensive. The main difficulties with the existing system were identified as the sectoral approach and a lack of an integrated consideration of the environment. Such problems were also quoted in the U.S. and in Canada as some of the reasons for the introduction of EIA. The Provisional Central Council for Environmental Hygiene suggested that further study should focus upon a series of trial runs to allow for some experimentation with methods and procedures for a variety of activities. The Minister agreed and the first research programme was established. Two consultants were retained to recommend how trial runs should best be carried out, and which additional subjects should undergo investigation in a complementary research programme.

After two months the types of activities to be the subject of trial runs were identified (Twijnstra Gudde and DHV, 1977). These activities included a broad cross-section of plans and projects, for example, a highway, the expansion of a large paint factory, a sewage treatment plant, the disposal of dredged silt, an industrial island in the North Sea, and a land use zoning plan. In total nine specific studies were initiated and completed within twelve months. Due to the preliminary nature of the studies, interest in the procedures and methods, and the relatively short time available (given that the planning of these activities was already underway), it was decided not to gather field data. Only existing data sources were used. Since the studies were intended to present facts, the assessment of the presented information was left mainly for the decision-makers. This approach has now been changed and environmental impact statements (EISs) are required to contain a certain amount of analysis, if this is needed for the decision-making process.

A key procedural point was the use of panels of experts in the trials. The panels of experts were used to analyse the technical merits of most of the statements and to examine, from a broad perspective, their quality and comprehensiveness.

Meanwhile, the complementary research programme was examining the systems adopted in other countries, in particular the U.S. and Canada. These studies resulted in a preliminary EIA manual which outlined a procedure for EIA (including links with the existing planning and decision-making system). The manual also contained information on data availability, standard techniques, and suggested contents for an EIA. A procedure for introducing EIA including the use of an interim policy and the drafting of legislation was suggested (Twijnstra Gudde and DHV, 1979a and 1979b).

Throughout these preparatory stages, the Ministry of Health and Environmental Protection appointed staff to participate in the steering committees. Consequently, when all the information was collected and internally analysed, it was possible to begin drafting a position paper on the subject. There were two Dutch ministries interested in the development of EIA. One of these was the Ministry of Health and Environmental Protection, which had responsibility for most of the laws relating to the environment and had departments concerned with aspects of air, water and land.

The other was the Ministry of Culture, Recreation and Social Welfare which was responsible for nature conservation, national parks, landscape, recreation, monuments etc. A position paper was, therefore jointly prepared.

In 1979, after discussion and agreement within cabinet, the position paper became formalized as the 'Governmental Standpoint on Environmental Impact Assessment' (Ministry of Health and Environmental Protection, 1979) and was passed to parliament for consideration. Based on the Standpoint and the motions of amendment, an Environmental Impact Assessment Bill was drafted and presented to parliament in May 1981. Research into EIA had not ceased however, since the Standpoint identified a number of areas requiring further intensive effort. The research had to that date provided only the framework for future EIA procedures, since there was little point in working out the detail until the fundamentals had been agreed by the cabinet. Briefly, the further research programmes included:

(i) the field of application of EIA (a list of legislation, plans and projects to which EIA would apply);
(ii) the organization and costs (the administrative system and manpower requirements, with particular attention being paid to the independent panel);
(iii) a central bank (a system to allow the proponent to easily identify what information is available and where);
(iv) methods for evaluating the significance of impacts and for choosing between alternatives; and
(v) scoping and guidelines (systems for focusing and guiding EIA towards the most significant impacts and issues relating to the most appropriate alternatives).

As the results from these studies became available so they were incorporated into the law or explanatory documents or guidelines.

To briefly summarize, the programme for the introduction of EIA has comprised:

(i) advice on necessity from an independent council;
(ii) establishment of trial runs and complementary research to provide some detail;
(iii) internal analysis of recommendations received;
(iv) drafting of the Standpoint and an agreement with cabinet;

Table 1: Summary of the Main Features of the Dutch EIA Procedure

Type of Procedure	Legislative and integrated.
Body controlling process or procedure.	Ministry of Housing, Physical Planning and Environmental Management is responsible for the law and regulations only.
Who is responsible for EIA preparation?	Proponent.
Field of application.	All levels; national, provincial, municipal and Private. Phased introduction.
How are activities identified?	Positive lists.
Definition of environment.	Physical, natural, cultural, historic and health.
Body for whom the EIS is prepared.	The competent authority.
Use of scoping.	Possible.
Use of guidelines.	General, generic and specific guidelines will be used.
Public involvement.	Proposed.
Hearings or meetings.	Dependent upon competent authority procedures.
Decision-making body.	Competent authority.
How is decision made?	As part of overall approval on project.
Appeal.	Proposed, where existing decision-making permits.
Monitoring.	Individually by regional inspectors.

(v) discussions with parliament;
(vi) drafting the law; and
(vi) continued research programme.

3. PRINCIPAL ELEMENTS OF THE EIA PROCEDURE

In this section the main elements of the procedure which is considered to be best for the Netherlands will be outlined. Reference should be made made to Table 1 and Figure 1 which provide the framework for this analysis.

3.1. Type of procedure

The Dutch system will be initiated through a single law which will lay down the basic procedure and content. By opting for an integrated system, no new decision points will be created. Instead existing decisions will be examined to determine which are the most important and relevant for the particular activities which will undergo an EIA. These decisions and the related procedures will be amended as appropriate to require the consideration of the EIS along with other information normally of relevance.

3.2. Administrative process

The law was drafted jointly by two ministries and it was expected that they would have to monitor the implementation of the law, as well as provide technical assistance where requested. The basic enforcement responsibilities, however, come under the aegis of the competent authorities, those responsible for decision-making, whether they are state level authorities, provincial or local authorities. In 1982, a new cabinet was formed in which new ministries were created. The Ministry of Housing, Physical Planning and Environmental Management is now primarily responsible for EIA. The Ministry of Agriculture and Fisheries also has some responsibility for EIA as nature conservation comes under its aegis.

3.3 Responsibility for EIA preparation

The initiator of an activity is responsible for ensuring that an EIA is performed either directly or through the use of consultants.

Figure 1 Proposed Environmental Impact Assessment Process for The Netherlands

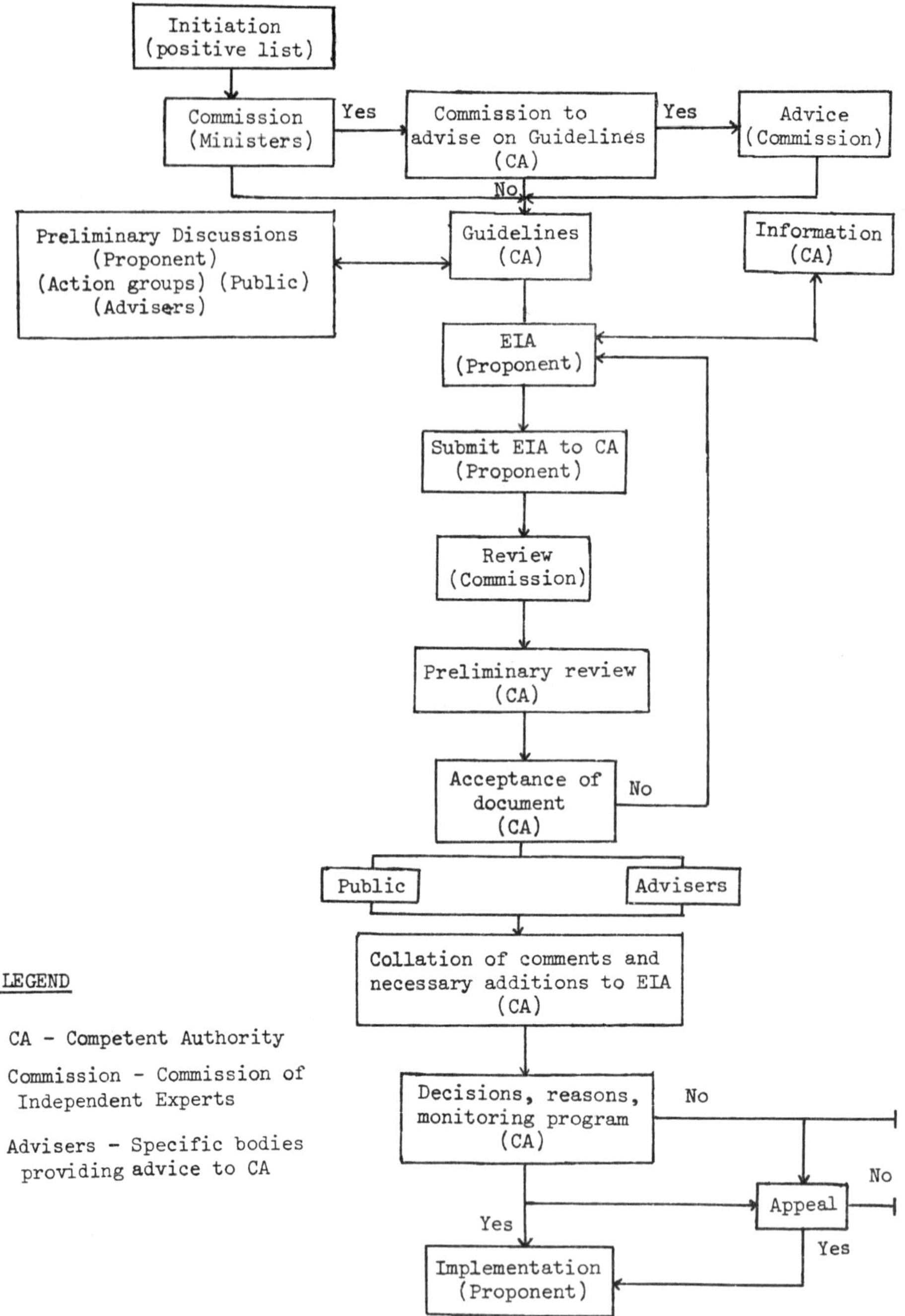

3.4. Scope of EIA

EIA is to apply to legislation, plans and projects at all levels: that is, national, provincial and municipal. In order to avoid difficulties during the introduction of EIA, due to the lack of experienced personnel and developed procedures, a phased approach has been taken. This started with an interim policy, under which the government has applied the EIA procedures to its own projects since the bill was introduced to parliament in May 1981. As a result EIAs have been applied to several projects, one of which was a proposed artificial island for dredging material storage.

The government has adopted a positive list approach to deciding whether an action requires an EIA. This list allows the proponent to know if and when an EIA will be required.

3.5. Definition of environment

It is important to note that the definition of environment is based upon the physical environment and includes water, soil, air, animal, plants and inanimate objects. The interrelationships of these elements, as well as aesthetic, natural-historic, cultural-historic and psychological aspects must be considered. In addition, primary and secondary, reversible and irreversible, direct and indirect, cumulative and synergistic effects are all considered to be within the definition of the environment.

3.6. Use of guidelines

Guidelines for EIA are to be used by the competent authority. Two types of guidelines are possible: project specific guidelines, and general process and content guidelines. The competent authority will discuss the guidelines with the proponent and with its own advisers who are semi-independent persons employed by the national government. These advisers include the usual advisers to the competent authority and the 'regionale inspecteurs en natuurbeschermings - consulentschap' (regional inspectors of the Ministry of Health and Environmental Protection, and nature conservation advisors from the Ministry of Culture, Recreation and Social Welfare). The local knowledge of the advisors will still play a valuable role in the preparation of the guidelines before they are given to the project proponent.

The general process and content guidelines are now being published in a preliminary form and are expected to be coupled with technical appendices to provide broad terms of reference on the use and application of EIA as whole. Consideration is also being given to the preparation of generic guidelines on a topic by topic basis. Such guidelines could then provide initial guidance to a proponent prior to the preparation of project specific guidelines.

3.7. Public involvement

It is proposed that the opinions of the public should be solicited during the preparation of guidelines or scoping activities. In addition proponents will be encouraged to communicate with the public and interest groups. Once the EIS has been accepted as fulfilling its terms of reference, the public will be notified of this acceptance before the EIS becomes available for public examination (see Figure 1). Comments can then be provided in writing and/or at meetings, depending on the procedure adopted for the assessment of the project. Finally, there is the possibility of appeal over the decision.

3.8. Independent review

The concept of an independent review of EISs via a panel procedure originated in Canada, but the Dutch have considerably adapted the concept for their use. The commission will be made up of experts having a wide range of backgrounds. An independent review will be required in all cases. The involvement of the commission will commence at an early stage in the EIA process, although its main involvement will be when the EIS is submitted.

Advice will be prepared on the acceptability of the EIS solely on the judgement of the members of the commission. This advice will then be given to the competent authority before it is released to the public.

3.9. Decision-making

The competent authority is the organization that will decide on the acceptability of the EIA. Once the authority decides to accept the EIA, it becomes responsible for the content of the EIS. This unique feature dictates that, on the basis of comments received from the advisers and the public, the competent authority must amend the EIS as necessary and then finally make a decision on the proposed activity. This decision is not solely based on the EIS, but also on other relevant studies and reports. When the decision is announced, it must be accompanied by a justification of the decision. In addition, conditions of approval and a description of the monitoring programme must be announced if the proposed action has been approved.

3.10. Appeal

Appeals will be linked to the existing appeal structure for major decisions, although appeals will not be allowed in all cases. Both proponents and objectors have the right to appeal if the decision is not in their favour. This includes the right of proponents to appeal against the conditions of approval.

3.11. Monitoring

Monitoring is viewed as an important element of the overall EIA process. Not only will projects have specific monitoring programmes, as identified in the approval document, but the Ministry of Housing, Physical Planning and Environmental Management will also monitor the EIA law and the process as a whole.

4. CONCLUSION

This paper has outlined the intensive step-by-step approach that has been undertaken in the Netherlands during the development and introduction of their EIA procedure. The Dutch approach has involved general policy research at the first level, before becoming progressively more detailed as the process was defined and approved. As a result of this approach, the Dutch should have a well-defined set of procedures together with a series of technical aids, such as guidelines, handbooks and case studies when full implementation occurs.

The most noteable elements of the Dutch system are the positive list approach, the use of an independent panel, and public review.

REFERENCES

Ministry of Health and Environmental Protection: 1978, Explanatory Memorandum to the 1978 Budget, Lower House, Session 1977-1978, 14800,17,2, p.77.

Ministry of Health and Environmental Protection: 1979, Governmental Standpoint on Environmental Impact Assessment, Ministry of Health and Environmental Protection, Leidschendam.

Provisional Central Council for Environmental Hygiene: 1976, Environmental Impact Statements, Provisional Central Council for Environmental Hygiene, Voorburg.

Twijnstra Gudde N.V. and DHV Consulting Engineers BV: 1977, Preliminary Investigation Concerning Experiments with Environmental Impact Assessment and Complementary Research, Ministry of Health and Environmental Protection, Leidschendam.

Twijnstra Gudde N.V. and DHV Consulting Engineers BV: 1979a, Environmental Impact Assessment Recommendations, Ministry of Health and Environmental Protection, Leidschendam.

Twijnstra Gudde N.V. and DHV Consulting Engineers BV: 1979b, Summary of Recommendations on the Basis of Trial Runs and Complementary Research, Ministry of Health and Environmental Protection, Leidschendam.

AFFILIATION

Murray G. Jones was an adviser to the Ministry of Health and Environmental Protection before becoming Staff Engineer in Environmental Affairs for Shell Canada Ltd (Toronto).

Robert G.H. Turnbull

EIA IN THE PLANNING PROCESS: A SCOTTISH OVERVIEW

1. INTRODUCTION

This paper outlines the United Kingdom (U.K.) approach to environmental planning and protection, and from the approach adopted in Scotland to the environmental impact of oil and gas developments, suggests a number of issues that must be taken into account in any realistic environmental planning system. This approach has also provided a means by which a positive vision of the future can become a reality.

The interest in Environmental Impact Assessment underlies the dual nature of the approach directed towards the improvement of the environment. There is a need first, for direct measures designed to identify, protect and enhance general and specific features of the environment, such as the media of air and water, flora and fauna, land and the built environment. Such measures are not sufficient by themselves. They must be accompanied by others which indirectly protect and enhance the quality of the environment by preventing or regulating actions which may give rise to adverse effects. To make such a system of control effective, there must be adequate procedures and methods for identifying, predicting and evaluating environmental effects of proposed actions in order to provide a basis for decision-making.

In the U.K., this duality in the approach to environmental policy has long been recognized. Measures for the protection and conservation of the physical environment include the designation of nature reserves, protected species, conservation areas, smoke control areas and noise abatement zones. These are generally directed towards the maintenance of the existing situation and provide partial protection for the most valuable or vulnerable environmental assets. To complement these measures, other procedures are concerned with actions or developments likely to have significant environmental effects. These procedures are embodied in the planning and pollution control systems which are more comprehensive and have broad environmental objectives. The aim of environmental impact assessment (EIA) has been to

B. D. Clark et al. (eds.), Perspectives on Environmental Impact Assessment, 69–79.

increase the effectiveness of this second approach. This has led to considerations of the range of actions to be covered by EIA, the levels of government at which it should operate and the way it should relate to existing control procedures.

The approach adopted in the U.K. has been to concentrate on developments covered by the planning system, since it is in the provision of additional information on such projects that EIA can most readily be applied. Although new statutory procedures have not been established for the assessment of major developments, existing procedures and methods have been reviewed in the light of EIA methods and additional environmental considerations have been introduced in the decision-making process. Since the concept of EIA was first developed in the United States, the U.K. has been engaged in a continuing programme of environmental research.

2. PROJECT AUTHORIZATION PROCEDURES

In the U.K., development projects require planning permission from public authorities before they can commence. Within the authorization procedure a number of permissions may be required depending on the scale of the project, the intended activity or use, detailed design, location and other factors. Not all of these authorization procedures are concerned with the environmental effects of the proposed development. Two forms of control are of prime significance: planning control, and pollution control including hazard and safety. Other kinds of control are of lesser or occasional importance.

Other procedures in which environmental considerations have a role are usually specific to particular types of development. Energy developments such as new power stations, transmission lines, pipelines and opencast coal mines all require authorization from the Department of Energy which normally carries with it planning permission. Oil refineries also require authorization from the Department of Energy, and this must be obtained before an application to the planning authority is made. Major inter-urban highways are authorized by the Department of Transport, while new towns and water supply schemes are authorized by the Department of the Environment and the Scottish Office.

The essential characteristics of the present U.K. system for the authorization of development projects are as follows:

(i) existing procedures are comprehensive; no major development with significant environmental effects can escape some form of control;

(ii) private and public developments are subject to broadly similar considerations;
(iii) the majority of decisions on major developments will be taken by local authorities, although contentious proposals will generally be decided at central government level;
(iv) in reaching decisions to authorize developments, local discretion is important in the choice of criteria taken into account and their precise values: there is a minimal application of national standards; and
(v) at some stage, the public are informed and have the right to object.

3. OIL-RELATED ONSHORE DEVELOPMENTS AND ENVIRONMENTAL RESEARCH

So far, the reference has generally been made to developments which have many precedents and have a fairly predictable range of likely impacts. In contrast, some of the proposals for onshore developments in Scotland associated with North Sea oil involved a high degree of uncertainty and thus posed problems for the planning authorities. This gave impetus to the development of thorough and positive methods of project appraisal within the planning system.

In the early 1970's, sites were being sought on the west coast of Scotland for the construction of concrete oil production platforms. This was a totally unfamiliar form of devlopment which would result in major site works, a large labour force and the transport of large quantities of raw materials. In addition, several of the proposed sites were located in remote and beautiful areas with small resident populations, limited infrastructure and no industrial tradition. Proposals for concrete platform construction sites resulted in the Scottish Office commissioning the first project assessment. Some of the unusually difficult problems that the planning authorities encountered were:
(i) the sheer size of the projects;
(ii) the unusual nature of the activity;
(iii) the small resident population in most areas;
(iv) the inadequacy of the existing infrastructure;
(v) the scenic significance of the host area;
(vi) the temporary nature of most projects;
(vii) questions of national interest; and
(viii) changing site requirements.
In the rural areas, where many of the early applications were made, planning authorities were not accustomed to handling

projects of this complexity. It was clear that a special effort would be needed to ensure that the proposals were adequately understood and that the planning staff would not be overloaded in dealing with them. To this end, a technical advice note was published in 1974 which provided outline guidance on how to approach the appraisal of the impact of oil related developments (Scottish Development Department, 1974a). The essential steps in posing the right questions concerning a proposal were set out and the possibility of carrying out an EIA was discussed. The main question at issue was: is an EIA necessary? If so, how should it be carried out, what should it cover and who should prepare it? Since that time some thirty EIAs have been carried out for oil and gas related developments in Scotland. A number of others have examined electricity pumped storage schemes, overhead transmission lines and reservoir schemes. These have been undertaken in various ways. Some were commissioned by the Scottish Office from consultants; others were carried out by the local authorities concerned, sometimes employing consultants to study specialist aspects such as noise, pollution or hazard. In the case of the Flotta oil terminal and the Beatrice oilfield, the studies were commissioned by the developer, while in others the local authority commissioned consultants in partnership with the Scottish Office. Clearly, a wide range of different approaches have been adopted.

The techniques and methods used in the earlier studies drew upon the experience gained in the U.S., but attempted to avoid over-complication by identifying the key environmental issues. This concern prompted the U.K. government to sponsor research on project appraisal at Aberdeen University, concerned primarily with methods (Clark et al., 1981) and the Catlow and Thirlwall study which considered the need for EIA in the U.K. (Catlow and Thirlwall, 1976).

There has been a slow growth of interest in the U.K. towards EIA. In 1976 the Royal Commission on Environmental Pollution endorsed the view that developers should provide an assessment of air, water, waste and noise pollution arising from certain major developments. By 1978, the Commission on Energy and the Environment was formed to consider the implications of the production and use of coal, oil, nuclear power, gas, electricity and renewable resources in the U.K. (a report was published in 1981 by HMSO).

4. THE APPLICATION OF EIA

Development projects in the U.K. are undertaken by both private and public developers, local authorities, nationalized industries and statutory undertakers. Some forms of development, such as highways and reservoirs, are exclusively undertaken by public agencies, while others are largely the preserve of private developers, such as petro-chemical plants. There is no distinction in terms of environmental impact between private and public development. The concern is to ensure that the most comprehensive assessment of possible environmental impacts should take place whenever it is warranted by circumstances.

5. NATIONAL PLANNING GUIDELINES AND EIA

The evolving approach to EIA in Scotland must be seen against the background of the national planning, development planning and the development control system. Following the introduction of new planning legislation in 1972 and the reorganization of local government in 1975, the Secretary of State for Scotland issued a circular in 1977 (Scottish Development Department, 1977a). In this he stated that he wished to only be notified of proposals which raise national issues, and those which are not in accordance with an approved structure plan or a local plan called in and approved by him. The circular was accompanied by a series of guidelines covering sites for large-scale industry, petro-chemicals, agricultural land, aggregates, landscape, nature conservation and recreation. Reference was also made to the Coastal Planning Guidelines issued in 1974, which divided the Scottish coast into preferred development and conservation zones (Scottish Development Department, 1974b). These were to form the basis for policy, and the existing arrangements for notification of oil-related developments were to continue to apply.

In the National Planning Guidelines for Large Industrial Sites and Rural Conservation (Scottish Development Department, 1977b), site selection criteria were set out together with an indication of the areas most likely to meet the criteria. The National Planning Guidelines for Petro-chemical Developments (Scottish Development Department, 1977c), suggested that in advance of specific proposals, it would be prudent for planning authorities to establish the potential for petro-chemical developments in their area, and to design their subsequent structure plans, local plans and development control policies so that sites could be readily identified, if and when required. Those authorities most involved with oil developments have pursued the suggestions and recommendations in these guidelines thereby developing an understanding of the potential for oil, gas and petro-chemical developments in their respective areas.

In carrying out this work, the authorities have commissioned consultants to undertake special studies ranging from the economics of feedstock and product transport, to site selection criteria, environmental capacity and safety aspects. One authority undertook a detailed assessment of environmental costs and problems of a possible petro-chemical development at a number of sites. It was found that development on almost all of the sites identified would cause environmental damage which, combined with the potential hazard to existing development and population, was considered to be an unacceptably high price to pay. The conclusions drawn from these initial EISs led to a further review of alternative petro-chemical sites acceptable in environmental terms. Experience suggests that the preliminary application of EIAs is clearly of value as, at a very early stage, it is possible to identify the main environmental impediments to the use of the identified sites for petro-chemical development.

From work done under the guideline initiative, including the issue of Planning Information Notes (PINs) on the type and characteristics of petro-chemical developments, both the Scottish Office and local authorities now have a better understanding of the environmental issues and operational factors involved in the advanced selection of industrial sites. Further work involved the identification of deficiencies in the information and what further investigations should be undertaken. The deficiences include environmental, infrastructure and operational considerations as outlined below.

5.1. Environmental considerations

The main environmental considerations identified are as follows:

(i) Safety: Consultants were commissioned by a number of authorities to define the dimensions and layout of a petro-chemical plant; to formulate general guidelines concerning community safety zones and to apply these hazard guidelines to the identified sites.

Consideration was also given to the potential impact which hazardous events at the potential petro-chemical sites would have on neighbouring industry.

(ii) Pollution: Studies were commissioned into marine, air, visual and aural aspects of pollution. Evidence obtained indicated that in some areas there are no difficulties which cannot be overcome at acceptable levels of cost.

(iii) Land acquisition: While no formal approaches were made to the landowners of the identified sites, an effort was made to establish ownership patterns and to form a view about the problems which might arise over the purchase of the land, if required.

(iv) Utilities and communications: Through the co-operation of the service departments of various local authorities, assessments were made of the existing patterns of availability of water, electricity, effluent disposal facilities and road, rail and air services. Attempts were made to identify the engineering costs and environmental implications of extensions of these utilities should this prove necessary.

5.2. Industrial infrastructure

Assessments were made of the ancillary industrial requirements of a petro-chemical complex, including such services as civil engineering, metal fabrication, plant cleaning, mechanical and electrical engineering, pipe insulation, painting and stockholding. Where these facilities did not exist in a fully developed form, particularly in relatively remote locations, it was appreciated and accepted that the expansion and diversification required for such activities could give rise to secondary environmental problems. Similarly, assessments were made of the labour supply and the consequent need for housing, education and community facilities.

5.3. Operational factors

From the studies undertaken, it was clear that there are a number of operational factors which limit the range of potential sites. These factors include the need for an adequate area of relatively flat land with good load bearing characteristics, proximity to both suitable pipeline landfall conditions and sheltered deep water for marine facitities, the availability of adequate infrastructure, and in particular the desirability and feasibility of providing an adequate safety zone around the plant or complex. This clearly implies the need for preliminary sieving of search areas to establish that operational factors can be met prior to undertaking an EIA.

6. CONTRIBUTION OF THE GUIDELINES AT THE NATIONAL LEVEL

In issuing the guidelines for larger-scale industry, and in particular for petro-chemical developments, it was appreciated that major chemical developments inevitably raise pollution and other environmental issues. It was, therefore, thought important that these should be evaluated by the planning authorities at an early stage, and preferably in advance of receiving the planning application. Furthermore, it was recognised that resistence from the environmental lobby to certain types of industrial developments was likely to increase, not only in the U.K., but throughout Europe. The causes of this resistence deserve to be treated with respect. Great care is necessary in the preparation and presentation of counter-arguments. The case for development proposals must cover the economic and environmental issues and give a general analysis of hazard and risk. In addition to the need for recognition of the environmental concerns, there is a need to safeguard specific areas of land of importance to industry. These were identified at a national level through the issue of the guidelines covering agriculture, forestry, nature conservation, landscape, recreation and the coast. The guidelines provided a broad indication of the national policies which the planning authorities should take into consideration when preparing development plans and exercising their development control functions. The guidelines thus provided a broad backcloth which assisted in the identification of suitable industrial sites.

7. CONCLUSIONS ON THE SCOTTISH EXPERIENCE WITH EIA

A number of conclusions can be drawn from Scottish experience of EIA research and application:

(i) there are many different situations in which EIA is relevant;

(ii) to compile a satisfactory report, adequate information and expertise, a methodical approach, and an awareness of the context of national, regional and local policies and attitudes is necessary;

(iii) there are significant variations in the type of information that is relevant according to the stage of the development process that has been reached;
(iv) techniques for the evaluation of the significance of EIA require further refinement;
(v) there is a wide variety of techniques that can be used to examine impacts; and
(vi) commonsense will be required to limit the assessment to issues of real importance.

A few further ideas on EIA are offered for consideration. It is generally believed that the phrase "environmental impact assessment" comes from the U.S. National Environmental Policy Act of 1969. It may be said, however, that the idea of assessing environmental impact and making plans to accommodate the impacts began with Noah's ark - therefore it has a long tradition in the history of superstition and prophecy. The recent trend has been to replace prophecy by extrapolation from present scientific knowledge and to assess the probable consequences of some human intervention on nature.

In carrying out an EIA, development must be considered in three distinct stages. Impacts must be identified, measured and their significance both singly and in total must be assessed. Each of these stages has attendant difficulties:
(i) Since knowledge of the environment and the effects of activities is incomplete, impacts may fail to be identified. Effects on human health for example, may not become apparent for many years.
(ii) There may be no acceptable means of measuring impacts which are believed to exist. For example, subtle changes in values, attitudes and culture may prove impossible to predict or describe.
(ii) The most difficult area is assessing the significance of impacts identified and if possible measured. Hence in making an overall assessment of a project, decision-makers may be faced with a range of impacts, some quantifiable and some not, whose significance can in many cases be assessed only in an intuitive way.

When such difficulties arise in relation to a single project, it is easy to imagine how they can be compounded when appraisal methods are extended to the comparative assessment of alternative sites. Various techniques of ranking, weighting and normalizing have been used to simplify EIAs, but these procedures often contain hidden subjective assumptions about the relative values of the impacts, which ought to be made explicit rather than concealed in a numerical weighting.

In these circumstances, the question may be asked whether the trend towards making explicit the basis of decision-making by means of EIA has been worthwhile, or whether the result has simply been to complicate and lengthen the process of consideration. The answer must be that public concern for the environment has grown to such an extent that it is essential to ensure that major issues are seen to have been identified and adequately examined by the appropriate authorities. The complexity of such decisions is real, not imaginary, and if the public are aware of this complexity, then the level of debate can only be improved. Herein lies the challenge.

8. SUMMARY

In this paper, an attempt has been made to draw attention to the status of EIA in the U.K., the approach and initiatives taken by the Scottish Office in issuing National Planning Guidelines, and their relationship to EIA. It has highlighted the relationship of EIA with development planning and control procedures, how the preparation of an outline approach can assist in asking the right questions concerning a new development, and how it has been extended into research on methods and forward contingency planning for industrial developments.

REFERENCES

Catlow, J. and C.G. Thirlwall: 1976, Environmental Impact Assessment, Research Report 11, Department of Environment, London.

Clark, B.D., K. Chapman, R. Bisset and P. Wathern: 1981, A Manual for the Assessment of Major Development Proposals, HMSO, London.

Commission on Energy and the Environment: 1981, Coal and the Environment, HMSO, London.

Scottish Development Department: 1974a, Appraisal of the Impact of Oil Related Development, DP/TAN/16, Scottish Development Department, Edinburgh.

Scottish Development Department: 1974b, Coastal Planning Guidelines, Scottish Development Department, Edinburgh.

Scottish Development Department: 1977a, National Planning Guidelines, SDD Circular 19/1977, Scottish Development Department, Edinburgh.

Scottish Development Department: 1977b, National Planning Guidelines for Large Industrial Sites and Rural Conservation, Scottish Development Department, Edinburgh.

Scottish Development Department: 1977c, National Planning Guidelines for Petro-chemical Developments, Scottish Development Department, Edinburgh.

AFFILIATION

Robert Turnbull was formerly Deputy Chief Planning Officer for the Scottish Development Department and was instrumental in the promotion of EIA in Scotland. Mr Turnbull now holds the position of Honorary Research Fellow at the University of Aberdeen.

Suraphol Sudara

EIA PROCEDURES IN DEVELOPING COUNTRIES

Since the Stockholm Conference on the Human Environment in 1972, concern for the environment has reached almost every corner of the world. While developed countries were involved in environmental protection long before the Conference, most developing countries began to focus their attention on environmental quality after 1972. They often responded by establishing national agencies for the protection of the environment. Such agencies tended to be styled upon the format used in the U.S.. The adoption of the U.S. approach was frequently accompanied by only minor changes aimed at making the process suit their needs. EIA accompanied the creation of such agencies as it was considered to provide a tool for the protection of the environment against the possible adverse effects of development projects.

This paper will briefly present some of the problems encountered by developing countries with EIA, before examining the procedures adopted in six countries, although only a cursory presentation of four systems will be made.

1. PROBLEMS ENCOUNTERED WITH EIA

In many developing countries, the problems of introducing EIA originate from a shortage of qualified persons together with a lack of any real understanding of the EIA procedure. The multi-disciplinary requirements of EIA can give rise to additional personnel problems. In some countries however, this problem has been overcome by establishing a review panel comprising experts from various agencies. The second problem is perhaps more intractable since there is often a lack of commitment to EIA among the various agencies, even though EIA is often required by law in some developing countries. This situation might be due to a lack of understanding and acceptance of the value of EIA. National agencies responsible for EIA are, unfortunately, not making sufficient effort to improve the situation, as many agencies simply consider EIA to be another stage in the regular process.

B. D. Clark et al. (eds.), Perspectives on Environmental Impact Assessment, 81–90.

As a consequence of the limited understanding of EIA, the actual implementation and performance of EIA in most countries has been rather poor. While it is generally agreed in developed countries that EIA ought to be initiated during the project formulation stage, in developing countries an EIA is often undertaken in the later stages of a project, EIA being regarded as a process which will produce evidence as to why the project should be allowed. Such EIAs may contain data which is biased to reflect such objectives. Nevertheless, such EIAs frequently attract much criticism, on occasions leading to project delay or rejection.

In order to encourage the decision-makers, at all levels of government, to take environmental impacts into consideration, an EIA procedure ought to involve all implementing government agencies in charge of both government and private development projects. The typical approach in developing countries is to establish a new agency which is usually only responsible for the final approval of the EIS. As a result the procedure is often not as effective as it might otherwise be.

In developing countries, EIA has served as a mechanism for increasing public participation in the project authorization process. The results have not always been beneficial, since public participation can be a sensitive issue in some countries, particularly those countries not accustomed to direct public debates on controversial issues. Great care is therefore needed, in order to introduce a type of public involvement in accordance with the traditional practice of each country.

Some developing countries have made an attempt to implement EIA; for example the Asian and Pacific countries, Thailand and the Philippines have been involved with EIA since 1975 when laws were passed. Other countries, such as Malaysia, India, Indonesia and Papua New Guinea, though not possessing EIA legislation, have a requirement for some form of EIA, although the approaches are often different. These differences reflect the legal and administrative systems, as well as, social and political systems, and will be examined in the remainder of this paper.

2. THAILAND

The National Environmental Quality Act (1975) gave Thailand a new agency with responsibility for the environment, known as the National Environment Board (NEB). Attempts were made to introduce EIA into the 1975 Act and to provide the NEB with authority over EIA. These attempts failed and EIA was a voluntary activity. This situation continued until 1978 when a revised version of the Act entitled: "Improvement and Conservation of National Environmental Quality Act (No.2)" was passed. This Act required that the proponents of projects or activities of specified categories or characteristics must submit a report presenting an assessment of the environmental consequences of their proposals together with an analysis of any mitigating measures. The categories and magnitudes of projects or activities requiring an EIA were announced in 1981. Table 1 details these projects and activities.

In order to assist in the EIA procedure, the NEB published the "Manual of Guidelines for the Preparation of Environmental Impact Evaluation". This manual provides an explanation of the procedure and is supplemented by guidelines for 17 project categories, for example, agro-industries, highways, human settlements, industries, industrial estates, mining, ports and harbours. The EIA procedure (outlined in figure 1), commences with the submission of a proposed project brief to the Division of Environmental Impact Evaluation (DEIE) within the NEB. The project brief provides sufficient information to give a brief but clear understanding of the type of project, location, size or magnitude of operation, and proposed schedule for implementation.

The environmental assessment procedure commences with an Initial Environmental Examination (IEE) which may be undertaken by the proponent of the DEIE. Basically, the IEE was designed as a means of reviewing the environmental integrity of projects in order to determine whether the second level, the EIA, must be performed. The IEE assesses the potential environmental effects of a proposed project within a very limited budget, and is based upon information which is easily available, or upon the judgement of an expert on such projects. An IEE must be carried out for any project within any of the specified categories. If the IEE results indicate that a fullscale EIA is not required, then any necessary environmental management measures, such as mitigation or monitoring activities, can be prescribed. This completes the environmental assessment for such a project.

Table 1 List of those Projects or Activities for Which an EIA is required

Types of Projects or Activities	Sizes
Dam or Reservoir	storage volume greater than 100,000,000 cubic metres or storage surface area greater than 15 square kilometers
Irrigation	irrigated area greater than 80,000 rais (12,800 hectares)
Commercial Airport	all sizes
Hotel or Resort Facilities in environmentally sensitive areas such as areas adjacent to rivers, coastal areas, lakes or beaches or in the vicinity of national parks	greater than 80 rooms
Mass Transit System and Expressway	all sizes
Mining	all sizes
Industrial Estate	all sizes
Commercial Port and Harbour	with capacity for vessels of greater than 500 ton gross
Thermal Power Plant	capacity greater than 10 MW
Petrochemical Industry	greater than 100 tons/day of raw materials required in production processes of oil refinery and/or gas separation
Oil Refinery	all sizes
Natural Gas Separation or Processing	all sizes
Chlor-Alkaline Industry requiring NaCl as raw material for production of Na_2CO_3, NaOH, HCl, Cl_2, NaOCl and Bleaching Power	production capacity of each or combined product greater than 100 tons/day
Irons and/or Steel Industry	requiring iron ore and/or scrap iron as raw materials for production greater than 100 tons/day or using furnaces with combined capacity greater than 5 tons/batch
Cement Industry	all sizes
Smelting Industry other than Iron and Steel	production capacity greater than 50 tons/day
Pulp Industry	production capacity greater than 50 tons/day

(Adapted from Boriboon, 1982)

Figure 1 : Environmental Impact Assessment Procedure in Thailand

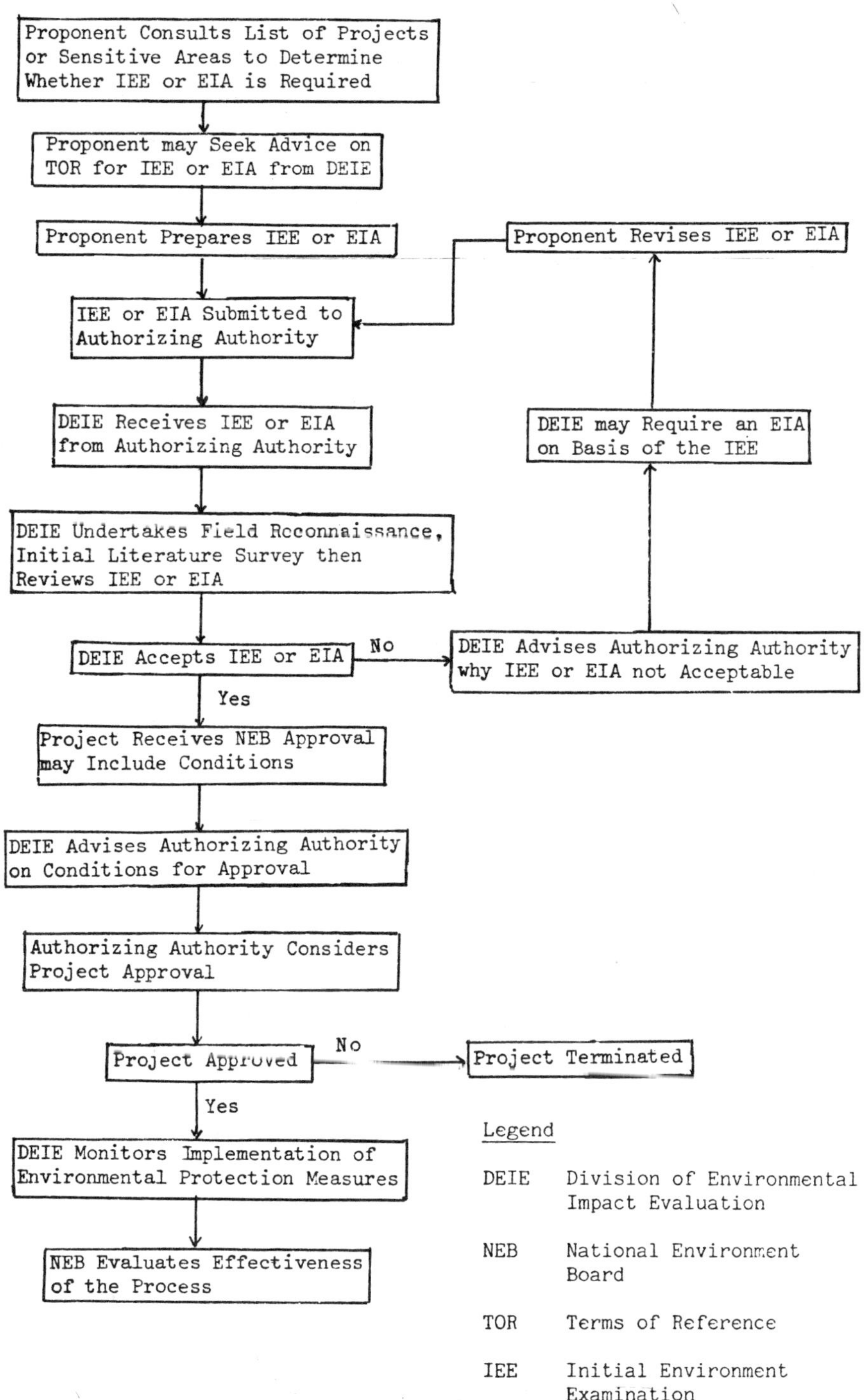

In the event of a fullscale EIA being required, the IEE can be of great value as an aid to the identification of the key issues requiring detailed examination, as well as those needing only a cursory study. The DEIE then formulates the terms of reference for the EIA which is to be prepared by the project proponent. The quality of the EIS that is produced by the proponent is largely dependent on the capabilities of the researchers, their financial budget and their terms of reference. The EIS is reviewed by the DEIE and if it is accepted the NEB provides authorization for the proposal. Following project initiation, the implementation and performance of the environmental protection measures are monitored to ensure compliance.

3. PHILIPPINES

EIA became a legal requirement in 1977 as a result of Presidential Decree No. 1151, which promulgated the Philippine Environmental Policy and Presidential Decree No. 1586, which provided the necessary rules and regulations for the implementation of EIA. A few months earlier the National Environmental Protection Council (NEPC) had been established (Presidential Decree No. 1121), in order to "rationalise the functions of government agencies charged with environmental protection and with the enforcement of environment related laws" (NEPC, no date). The NEPC prepared guidelines for the EIA procedure, which were published in 1979. NEPC also identified a number of Environmentally Critical Areas, such as National Parks and mangroves, and Environmentally Critical Projects, for example, resources extractive industries. These designations were designed to aid the determination of whether an EIA is required for specific projects.

In the Philippines the proponent of the project initially determines whether the project is specified within the EIA system. In cases of uncertainty, the assistance of the lead agency, for example the Ministry of Agriculture, may be sought. In the case of a negative determination, the project proponent may proceed with the project. Should the project fall within the EIA procedure, then a Project Description must be filed with the NEPC and be reviewed by the EIS Review Committee. If the project proposal is not in the category of Environmentally Critical Projects, but is to be located within an Environmentally Critical Area, then the Committee will determine whether negligible or significant environmental effects are to be expected. If no significant effects are perceived then the Committee may recommend to the President or his duly authorized representative the granting of an Environmental Compliance Certificate.

When significant effects are expected or the project is found to be within the category of Environmentally Critical Projects, it is then the responsiblity of the proponent to submit a draft EIS to the lead agency. After review by the lead agency, a final EIS is submitted to the NEPC. A period of consultation with relevant government agencies and other interested parties then takes place. The NEPC may, if a substantial impact is expected, call a public hearing to examine the proposal. The EIS Review Committee, after consideration of the EIS, may recommend one of three courses of action, namely:

(i) recommend to the President that an Environmental Compliance Certificate should be granted;

(ii) advise the project proponent that modifications to the project are required together with a revised EIS before a certificate may be granted; or

(iii) advise the proponent that the project is unacceptable, stating the reasons for this decision.

By the end of 1980, the NEPC had processed more than 281 IEEs and 13 EISs. Among the 13 EISs, four were infrastructure projects and six for resource extraction, with the remaining three being industrial projects. Most of the IEEs, about 70 percent, were related to mining concessions, the remaining 30 percent being for commercial fishing, seaweed culture, infrastructure projects and heavy industry.

To complement the EIA procedure, which is essentially directed at project planning and environmental management, the Ministry of Human Settlements has the responsiblity for the preparation of environmentally sound land and water use patterns for environmentally critical projects and areas. The Ministry is also responsible for the establishment of ambient environmental quality standards.

3. OTHER DEVELOPING COUNTRIES

This section will briefly summarise the status of EIA in four countries, namely; Malaysia, India, Korea and Indonesia.

3.1. Malaysia

The Environmental Quality Act of 1974 provided Malaysia with a comprehensive legal instrument for environmental control, although it placed greater emphasis on pollution control than on prevention. Certain activities are excluded from this act, specifically those of non-point pollution, for example timber logging. Against this background the Division of the Environment, established in 1975, introduced non-statutory controls in the form of guidelines which can be adopted by the local authorities for environmental management. The guidelines that have been introduced cover the following subjects:

(i) the control and prevention of erosion and siltation;
(ii) the siting and zoning of industries;
(iii) Environmental Impact Assessment; and
(iv) the selection and management of sites for the disposal of solid and hazardous wastes (Suhaimi, 1980).

Concern for the environment was incorporated into the Third Malaysia Plan (1976-1980) which dealt mainly with the balance of development and the preservation of environmental quality. Such concern has been expressed by calls for EIA. A system is thus being proposed for new development projects and the expansion of existing projects. The proposed procedure calls for an EIS if the project is described on a mandatory list or is located in an environmentally sensitive area. It is the responsiblity of the proponent to prepare a preliminary environmental assessment which is then reviewed by an independent panel and the authorizing agency. If the proposal is acceptable at this stage then it may proceed for authorization; otherwise a detailed EIA is required. The EIS is then submitted to the authorizing agency, the review panel, and other environment related agencies. The public may be invited to submit written comments on the EIS as part of the public participation procedure. Finally, a decision is made on the project proposal. At present the proposed EIA procedure has yet to be approved.

3.2. India

The use of EIA in India remains very limited, although the recently established Department of Environment is responsible for the promotion of EIA. The Department prepared a series of questionnaire type guidelines on environmental information in order to create environmental awareness among developers. Guidelines have been prepared for industrial devlopment, power plants including hydro-electric power stations, river valley development and mining. As developers are requested to provide the government with information on the environmental consequences of their activities, these guidelines may go some way to improving the quality of such information.

3.3. Korea

In 1977, Korea enacted the Environmental Preservation Law which required the application of EIA to major government projects. With the establishment of the Office of Environment in 1981, the efforts to encourage the application of EIA procedures were strenghtened (Kim, 1981). EIA was made compulsory for ten major governmental activities, for example dam construction, energy projects and industrial projects.

3.4. Indonesia

Based upon the general principles of state policy, the Ministry of Development, Supervision and Environment has responsiblity to ensure that EIAs are prepared in the planning and implementation of projects, programmes and activities. Those projects which require an EIA have been listed, although there is no legal requirement for such projects to undergo an EIA. Water resource development projects, which are internationally financed, generally have an EIA undertaken, as do some private projects such as the Kujang fertilizer plant at Cikampek, West Java, despite the fact that it is not legally required. In order to further the application of EIA, the Ministry of Development, Supervision and Environment has distributed limited sectoral guidelines of a similar nature to those of Thailand.

4. CONCLUSION

In most of the developing countries which are attempting to incorporate EIA into their system, the procedures adopted are more or less similar to those of the U.S.. Many issues need to be resolved in order to maximize the benefit of EIA. The principal drawback has been a lack of understanding amongst the project proponents, who regard EIA as another impediment to their project. In many developing countries, the lack of qualified people for EIA is also a problem. It is therefore vital that personnel are educated and trained in the techniques of EIA.

REFERENCES

Boriboon, S.: 1982, An Implementation of Environmental Impact Assessment in Thailand, Paper presented at the Inter-country workshop on Rapid Techniques for Environmental Assessment in Developing Countries, Bangkok.
Kim, H.C.: 1981, EIA in Korea, Paper presented at a seminar on EIA, Seoul.
NEPC: No Date, The Philippine Environmental Statement System, Ministry of Human Settlements, National Environmental Protection Council, Manila.
Suhaimi, A.: 1980, Environmental Control by Legislation: Malaysia as a Case Study, paper presented at the Asian-American Conference on Environmental Protection, Jakarta.

AFFILIATION

Dr. Suraphol Sudara is President of the Faculty of Science, Institute of Environmental Research, Chulalongkorn University, Bangkok.

PART B:

THE ROLE OF ENVIRONMENTAL HEALTH ASSESSMENT

A. Gilad

THE HEALTH COMPONENT OF THE ENVIRONMENTAL IMPACT ASSESSMENT PROCESS

1. INTRODUCTION

Economic development activities usually result in alteration or modification of the natural, manmade or social environment, thus producing environmental and/or social stresses. These stresses may take the form of pollution of air, water and soil, disturbance of the ecological balance, overloading of infrastructures and alteration of behavioural and/or social patterns.

Historically, homo sapiens has shown a remarkable degree of tolerance for environmental stress. Judging from the continuing mass migration from the countryside to the overcrowded cities with their slums and shanty towns, noise and pollution, man seems to possess an immense capacity for adapting to adverse conditions. Yet, as a species, we have been fully developed long before the stone age and there is no evidence that we have acquired, since that time, any genetic adaptation to the drastically changed conditions in which most of us live today. Since our cultural adaptation has not been accompanied by structural changes, we are ill-equipped to deal with many of the environmental insults to which we are subjected in modern society, and the resulting physiological or psychological stresses produce a range of various adverse health effects.

2. HEALTH EFFECTS OF ENVIRONMENTAL POLLUTION

Environmental pollution may affect health in any of the following ways:

(i) Contact with agents of disease: exposure to agents of communicable diseases; exposure to toxic substances causing immediate acute disease; exposure to toxic materials or agents which may cause acute or chronic disease long after exposure.

(ii) Exposure to substances which may cause genetic changes.

(iii) Lowering of resistance to infection.

B. D. Clark et al. (eds.), Perspectives on Environmental Impact Assessment, 93–103.

(iv) Producing sub-clinical irritation, nuisance and discomfort.
(v) Contributing to aggravation of existing disease.
(vi) Creating conditions incompatible with, or derogatory to, the achievement of physical, mental and social wellbeing.

There seems to be a distinct correlation between the degree of economic development on the one hand and the type of prevalent environmental health effects on the other, the trend being that with the increase of economic development the relative importance of the various environmental health effects changes; the problems of prevention and control of communicable diseases becoming less common and more manageable, while the problems connected with toxic, carcinogenic and mutagenic materials in the environment assuming an ever-increasing importance.

This does not mean, however, that the developed countries have finally solved the problems of communicable disease. The sudden outbreak of cholera around the Mediterranean several years ago, the persistence of infectious hepatitis in some European countries, the resurgence of malaria in south-eastern Europe, the epidemic of the so-called "Legionnaire's Disease" in the United States, the sudden advent of AIDS, all point to the fact that continuing vigilance is needed.

3. CHEMICALS IN THE ENVIRONMENT AND THEIR HEALTH IMPACT

The emergence of highly industrialized societies has brought new chemical dangers which we as a species are biologically ill-equipped to handle.

According to some recent estimates, about 60000 chemicals are used in daily life. Of these, about 1500 are the active ingredients of pesticide formulations, 4000 are used in drugs and 5500 are food additives of various kinds. The remainder are industrial chemicals, agricultural chemicals (other than pesticides), fuels for power production, and chemical consumer products. Many of these chemicals appear in the working environment, or in air, water, food and soil as pollutants resulting from the wastes of production and consumption. The number of chemicals in use is increasing rapidly. It is estimated that about 200, and possibly as many as 1000, new chemicals are put on the market every year.

Chemicals are a public concern because of the possible

adverse effects they may have on human health and on the environment; on the other hand, there are many benefits derived from their use, both in industrialized and developing countries. Banning or substantially reducing the application of some chemicals could have very serious repercussions on industrial development, agriculture, disease control, food production, power production and transport.

Man can be exposed to chemicals directly by skin contact and ingestion or through environmental media such as air, water and food. Some chemicals persist in the environment for a long time and may accumulate in the food chain. Most of them, however, are subject to environmental transformations, chemical or biochemical, and new porducts are formed that may be less or more toxic. Accidental high exposures of epidemic proportions have occurred during the last two decades in several countries. Examples are mass poisonings arising from organomercury compounds and hexachlorobenzene used in seed treatment, from polychlorinated biphenyls used as heat exchange liquids in rice oil production; and environmental pollution by chlorinated dibenzodioxins that escaped during the production of trichlorophenol. The major concern, however, centers on the possible effects of low-level, long-term exposure to chemicals that may occur at work or in the general environment. The prevention of chronic adverse effects requires long-term policies because they may be irreversible, and the latency period may be as long as 30 years or even more for some carcinogenic chemicals. There is a widely accepted opinion that 60 to 90% of all cancers are of environmental origin and among the environmental carcinogens, chemicals certainly play a very important role.

Future generations may also be adversely affected. Concern that some chemicals may pose a mutagenic threat to our offspring is based on sound theoretical reasoning and endorsed by a large amount of biological information. Information concerning the teratogenicity of chemicals in humans is, however, at present inadequate. Estimates of the frequencies of malformations resulting from teratogenic agents, including chemicals, vary from 3 to 5% up to about 10%.

The joint action of chemicals and/or agents occurring simultaneously in the environment has not yet received sufficient attention. Such interaction may be additive, synergistic or antagonistic. Synergism is generally important in toxicology and may play a significant role in

Figure 1: Environmental Health Impact Assessment Process

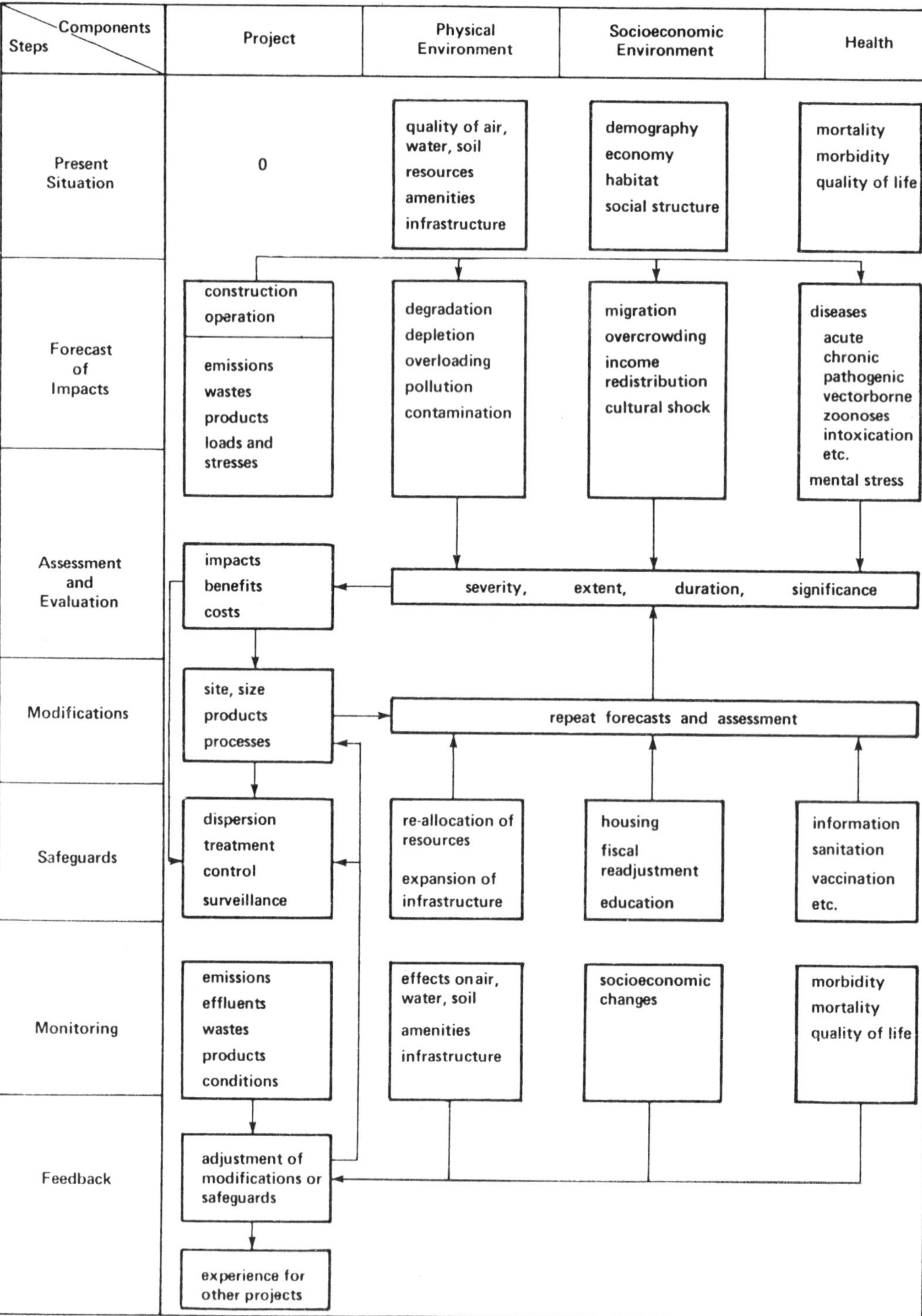

the contribution of chemical exposure to the premature onset or aggravation of chronic diseases, such as cancer or chronic respiratory diseases.

4. THE HEALTH COMPONENT OF EIA

Environmental impact analysis is assuming a major role in the efforts for achieving rational solutions to the conflicts between the societal goals of economic development on the one hand, and preservation of the quality of life and of the environment on the other. It is, therefore, important to assure that the health component of EIA which, up to now, has been either completely neglected or only superficially treated, should receive attention commensurate with its importance.

Various sectors of the population may be affected by developments in different ways, depending on the nature of the development, its stage (construction/operation) and the nature of the environment in which the development takes place.

Certain industrial developments may have a direct health impact on workers employed in the facility itself by exposing them to potentially toxic substances without providing adequate protection. Similarly, introduction of new agricultural practices may expose the workers to pesticides or to waterborne agents of disease, producing direct detrimental health effects. On the other hand, most of the health impacts on the community in which the development takes place are likely to be indirect, i.e. the development may affect the environment by polluting it or by upsetting the ecological balance, thus creating conditions favourable to the biological vectors of diseases. Certain types of development are likely to have significant effects on socioeconomic environment, with the resulting stresses exerting secondary impacts on the health of affected population.

A flowchart of the environmental impact assessment process against the various types of primary and secondary impacts, including the health impacts is shown in figure 1. It is clear that health impact assessment is intimately connected with, and to a large extent depends on the assessment of impacts on the physical and socioeconomic environment. The first stage of the EIA process, as shown in Figure 1, viz. analysis of the situation in the project area before the commencement of the project, depends on data which can be collected, given adequate time and resources. The difficulty at this stage lies in deciding which data are relevant for the subsequent assessment process so that the adequate data base is produced with a minimum of waste. The second stage, forecast of impacts, depends on extrapolations from other projects or other situations.

The two main approaches to the evaluation of health effects are epidemiology and laboratory experimentation. The experimental approach involves the use of animal models for estimation of effects on humans. One of the major difficulties in this approach is the number of extrapolations that are inherent in the process, such as extrapolation from animals to humans or from large doses and high levels over a short time to low doses over prolonged periods. The epidemiological approach studies effects on humans and thus obviates the necessity for extrapolating from one species to another. Nevertheless, in drawing conclusions from epidemiological evidence, some important and difficult extrapolations have to be made, e.g. from occupational exposure to community exposure or to multiple exposure. The difficulties in design, evaluation and interpretation of epidemiological studies are well known and need not be enumerated here. When attempting to use the results of these investigations for the purpose of risk assessment and the subsequent decision-making, the various explicit or implicit assumptions and uncertainties which are inherent in the extrapolation process are important to keep in mind.

Having defined the hazards to the best of our ability, we are faced with an equally difficult task of risk estimation, which we define as estimating the extent of harm and the probability of its occurrence. In the context of EIA, this approach will usually mean prediction of additional incidence of various diseases, disabilities or defects caused by the predicted exposure insults, acting either singly or in combination with other chemicals which are likely to affect given individuals or population groups. Up to this stage of the process, we are dealing with data (imperfect as they may be), extrapolations (admittedly difficult) and predictions based on the best available evidence. The next step of the process involves value judgements. In trying to assess and evaluate the impacts, we are in essence attempting to answer the question: how significant or important are the expected effects? This question raises, of course, a host of other equally difficult questions. Important for whom? According to what criteria? From whose point of view? Who is to judge? The decision-makers must answer these questions either explicitly, by formulating formal decision criteria, or implicitly, by making decisions based on intuition or on their assessment of the balance between the various private and public pressure groups.

The decision-making process must also consider the public propensity for taking various kinds of risk, which varies widely from society to society and from individual to individual depending on a host of cultural, socioeconomic and pyschological factors, as well as on the type and nature of the risk in question. Risk predictions are made in terms of statistical probabilities, which most people find difficult to visualize and understand. We seemingly face a situation in which there is an almost total lack of meaningful communication on this subject between the scientists who generate the information, the public for whose benefit it is supposedly generated, and the decision-makers, who must make far-reaching decisions based on imperfect, incomprehensible and frequently scant information. At the same time it is necessary to respond to the pressures and demands from, on the one hand, an insufficiently informed public and, on the other, frequently biased interest groups. Meaningful public participation in this decision-making process requires a conscious effort to improve communication between the scientific community, the developers, the public and the decision-makers.

In an effort to develop a method of effective communication, the WHO Regional Office for Europe, in collaboration with the Technical University of Copenhagen, has developed computerized techniques for translating emission inventories, environmental dispersion models and health effect information into coloured three-dimensional maps. This process facilitates rapid manipulation of data and visual demonstration of probable health effects of the various alternatives of the proposed action, under a range of probable circumstances.

Figure 2A represents a contour map of SO_2 levels prevalent in a metropolitan coastal area. The levels are such that at present no significant health effects from the exposure to SO_2 can be expected.

Figure 2B shows what would happen if two large point sources of SO_2, located on the axis parallel to the prevailing wind would be added. The ambient levels of SO_2 would cause significant morbidity and mortality due to

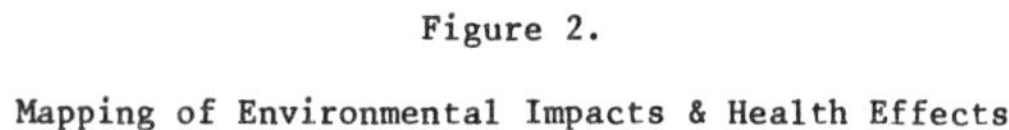

Figure 2.

Mapping of Environmental Impacts & Health Effects

Section of a coastal
metropolitan area

A.

B.

C.

Contours represent average yearly levels of SO_2 from a variety of area, line and point sources in different configurations. (Explanation in text)

Increased mortality rates among most susceptible sectors of population

Increased morbidity: reduction of pulmonary function, increased incidence of chronic bronchitis and other pulmonary diseases

No noticeable health effect

exposure.

Figure 2C shows what would happen if the same two sources were located on an axis perpendicular to the prevailing wind direction. While the health effects would be spread over a wider area, the excess mortality would be practically eliminated.

The process is in its initial stages of development, but the results are quite promising. The WHO Regional Office for Europe plans to continue this effort to develop the methodology needed for effective communication of the health component of EIA results to the decision-makers and/or to the public.

Health impacts of developments can be frequently minimized by institution of adequate safeguards. Institution of health information and education programmes, provision of adequate sanitation both during the construction and operation phases, pest, parasite and vector control programmes, vaccination campaigns, etc. have been found effective in mitigating or eliminating the risks to health of exposed populations and workers. The environmental impact assessment process is most useful in alerting the health authorities ahead of time to the need for such safeguards so that they can be planned and budgeted for and instituted before the health situation gets out of hand.

Forecast of environmental and health impacts are based, to a large extent, on assumptions, estimates and extrapolations made during the planning stage. It is therefore necessary to institute monitoring, feedback and adjustment mechanisms to test the validity and accuracy of the assumptions and to provide a mechanism for rapid adjustment of control strategies to changing or unexpected conditions. The precise content of the environmental health impact assessment study will of course depend on the nature of the proposed development and on the characteristics of the environment and of the population in the area. The minimal contents of the health component of EIA related to an industrial development should contain at least the following elements:

(i) Inventory of pollutants likely to be released into the environment as the result of the proposed action.

(ii) Quantitative description of the health situation in the region likely to be affected by the proposed action or project, including morbidity statistics on diseases related to pollutants likely to be released into the environment as the result of the proposed action.

(iii) Quantitative description of the dispersive, absorptive and adsorptive mechanisms in air, water and soil.

(iv) Based on (iii) above, prediction of concentrations of the various pollutants over the various periods of time to which the various sectors of the affected population will be exposed.

(v) Information on environmental, social and economic factors which are likely to influence the susceptibility of the affected population (or sectors thereof) to the pollutants identified as above.

(vi) Forecast of health effects including their intensity, duration, and the time likely to elapse between exposure and appearance of the effect.

(vii) Assessment of the health effects, as to their importance and short- and long-term significance.

(viii) Means by which the health effects could be eliminated or reduced and the associated cost.

(ix) Assessment of the health impacts of alternatives developed as the result of the general environmental impact assessment process.

(x) Definition of the health monitoring system to be instituted together with the activities of the proposed project or action to measure the actual health impacts in order to check the accuracy of prior predictions and to permit introduction of remedial measures if and when necessary.

The information inputs identified in sub-paragraphs (i), (iii) and (iv) above should be available as the result of the general environmental impact analysis process of which environmental <u>health</u> impact assessment forms an integral part.

A planning guide containing items to be considered in the preparation of environmental health impact assessment for a major water resource development project is contained in the World Bank publication, "Environmental Health and Human Ecologic Considerations in Economic Development Projects" (1979).

5. THE ROLE OF HEALTH AGENCIES IN EIA

The extent to which national health agencies participate in the process of environmental health impact assessment and the mechanisms by which their contributions can be made inevitable depend upon the structure of the health administration and the type of the EIA process in a given country. Since both vary greatly from country to country, it is not possible to establish widely applicable models for their interaction. It is clear, however, that health agencies at the local and national levels, as appropriate, must be fully involved in the process of environmental impact assessment at the various stages of the process. Health professionals must be included in the interdisciplinary teams developing environmental impact analyses. To enable the health professionals to make a contribution commensurate with the importance of the health component of EIA, special training courses dealing with assessment health effects within the context of EIA, should be introduced into the curricula of schools of public health.

The World Health Organization, through its Regional Office for Europe, has been cooperating with Member States on a number of country and intercountry projects containing significant components of environmental impact assessment. A number of country projects containing establishment of environmental impact assessment systems as one of their main objectives have been implemented with the assistance of the United Nations Development Programme. Formulation, institution and implementation of environmental impact assessment systems are now explicitly incorporated in the technical cooperative projects, for which the WHO Regional Office for Europe acts as the executing agency.

The World Health Organization is ready, subject to the availability of resources, to contribute its accumulated experience and expertise to national and international efforts aimed at development and implementation of environmental impact assessment systems in general, and of the health impacts components of EIA in particular.

AFFILIATION

Dr A. Gilad is a Consultant for Environmental Systems Management (formerly the Regional Officer for Environmental Systems Management) at the Regional Office for Europe, the World Health Organization, Copenhagen, Denmark.

R.J. Donaldson

MEDICAL EFFECTS OF ATMOSPHERIC POLLUTION AND NOISE

1. INTRODUCTION

The effects on health of atmospheric pollution and noise are discussed in the light of present knowledge. The problems of measuring environmental health impacts are outlined. A case study is included which concerns the attempt to predict the environmental impact of constructing a large steel works in Teesside in the early 1970's.

Britain is an industrialised nation and as such has a problem to strike a balance between the benefits of a rising standard of living and its costs in terms of the environment and the quality of life. This was one of the issues taken up by the Royal Commission on Environmental Pollution (1) and is a thread that runs through this paper.

A broad definition of environmental health is when an organism, both psyche and soma, is in balance with its total environment. So far as the physical environment is concerned the purpose is to maintain and improve the human ecosystem by promoting the benign environmental components and reducing or eliminating the malignant ones. Unlike most other species, man has the power to manipulate certain aspects of his environment, and although his intention has been generally to produce a beneficial effect this is frequently accompanied by adverse changes.

If the size of a country's population outstrips the capacity of the land to feed it then the choices are starvation or industrialisation. Britain, since the 13th century, has not been able to support its population from its own resources, and as the population increased this became more evident. Industry, necessary for the health and well-being of a population through the provision of resources to support a better diet, living standards and services such as health and education, can give with one hand but can take away with the other. A population which abandons the

B. D. Clark et al. (eds.), Perspectives on Environmental Impact Assessment, 105–119.

rural life associated with an agricultural economy will find itself largely in crowded noisy towns and cities breathing air of indifferent quality.

2. ATMOSPHERIC POLLUTION AND HEALTH

In England the concern about atmospheric pollution has a long history. A Royal Proclamation by King Edward I in 1306 prohibited artificers from using sea coal in their furnaces on the pain of stringent penalties (the execution of at least one offender has been recorded). In the 16th century Queen Elizabeth "findeth herself greatly grieved and annoyed with the taste and smoke of sea coals". In the 18th and 19th century the Industrial Revolution brought increasing problems of atmospheric pollution from the chimneys of factories and houses in the new industrial towns. Legislation to control pollution at that time was directed mainly at industry. The zeal of the sanitary reformers more than 100 years ago in achieving safe drinking water and proper disposal of sewage was not matched by an attack on the other environmental evil, air pollution, which had effectively turned the atmosphere over the large towns into an air sewer. Over the years the public showed little interest: indeed a major contributor to air pollution - the open coal fire - was, and sometimes is, stoutly defended.

A dramatic turning point in attitudes to atmospheric pollution occurred in December, 1952. A London "smog" (the word "smog" was coined to describe fog filled with smoke) resulted in over 4,000 deaths and although excess deaths had been noted in other smog episodes, nothing quite as sensational had occurred before. The matter received wide media coverage and a curious coincidence made the story even more sensational. A number of prime young cattle at a show in London also succumbed from the effects of pollution. Some cynical observers attributed the swift action that followed to the British love of animals. A committee (under the chairmanship of Sir Hugh Beaver) was set up immediately following the episode which examined, in a comprehensive way, pollution arising from the combustion of various fuels. The Beaver Report (2) identified the importance of pollution from domestic sources which it considered to account for 60% of pollution, but this was later updated to about 80%. The link between pollution and respiratory diseases was spelt out. The recommendations formed the basis of the Clean Air

Act, 1956. Thus the main concern of those involved in environmental control is with man-made pollutants from the combustion of fossil fuels.

It is difficult to be precise about the importance of atmospheric pollution on health because of problems in estimating the degree of exposure of individuals and the influence of other confounding variables such as cigarette smoking, nutrition, occupation, and the effects of climate. In addition, the exact way in which pollutants in the atmosphere affect the lungs is not completely clear. Experimental work on both man and animals using sulphur dioxide and other pollutants has produced results that are somewhat inconclusive. The pollutants in smog may influence the growth of certain bacteria and there is evidence that a change occurs in the bacteria in sputum following exposure to fog. It has also been suggested that the effects of sulphur dioxide are enhanced by the presence of other pollutants (synergism). Many experts regard the oxides of sulphur as the most potent cause of symptoms from pollution, although the precise mechanism is poorly understood (3). It may be that pollutants act as irritants, vasoconstrictors or stimulators of hypersecretion of mucus in the lungs and bronchi.

2.1 Acute health effects

There is wide agreement that patients with well-established respiratory or cardiac disease, particularly the elderly, suffer from the effects of atmospheric pollution when it reaches peak levels. This was amply demonstrated in the London fog of December 1952 and since then studies on daily mortality and on hospital admissions via the emergency bed service in London have been carried out. In periods of major fog the death rate and hospital admission rate increased until the winter of 1962/63. Since that time this relationship has not reappeared. The fogs that have occurred have held relatively much less pollution from smoke and sulphur dioxide. Morbidity has been studied in London and elsewhere by asking people with chronic bronchitis to keep a daily diary of their symptoms. The London studies (4) suggest that deterioration in symptoms occurs at smoke levels up to 250 $\mu g/m^3$ and with sulphur dioxide levels over 500 $\mu g/m^3$. However, a study in Teesside (5) with a small sample of 73 chronic bronchitis patients using a similar diary method showed they were affected at much lower levels of 150 μg/m³ of smoke and sulphur dioxide. Thus those who had diseased lungs were adversely affected by relatively moderate rises in the level of pollutants.

2.2 Chronic health effects

Population studies provide the main source of information about the long-term effects of atmospheric pollution but unfortunately many of these have proved difficult to interpret. Of the diseases said to be associated with atmospheric pollution the two most commonly mentioned are chronic bronchitis and lung cancer.

Chronic bronchitis has been referred to as the "Englishman's Disease" and accounted for 6% of all deaths in England and Wales in the early 1970's. this was approximately four times higher than most industrial regions in Western Europe. However, there has been a sharp decline in deaths from this cause since 1968 for both men and women (Standardised Mortality Ratios (SMR) = 100 in 1968, SMR = 69 in 1976). The disease has the following characteristics: more than twice as common in men than women, a steep increase in the gradient from Social Class I (top professionals and managerial) to Social Class V (unskilled labourers), more common in smokers, and usually occurs in people over the age of 50 years. In Britain it is usually less frequent in rural than urban industrial areas. Studies of sickness rates due to chronic bronchitis in civil servants and postmen revealed links with exposure to atmospheric pollution. A survey of children in Sheffield (6) in the 1960's showed significant correlation between air pollution and lower respiratory tract infection. A similar survey was carried out in Teesside (5) in the 1970's comparing the health of 7-year-olds living in high and low pollution areas which broadly confirmed the Sheffield results. However, there are other factors, particularly cigarette smoking, although not all cigarette smokers get chronic bronchitis, so there is probably some individual predisposition involved. Also linked with chronic bronchitis are occupation, social class, standard of housing and nutrition as well as climatic conditions. It is thought unlikely that atmospheric pollution causes chronic bronchitis, but it does exacerbate symptoms in the established disease.

Year of death	Approx. Year of birth	Deaths per 100,000 Greater London	Rural Districts
1946-50	1886	290	140
1951-55	1891	310	150
1956-60	1896	280	155
1961-65	1901	255	175
1966-70	1906	215	175
1973	1911	160	135

Table 1 : Bronchitis: age-specific death rates, males aged 60 to 64 years, Greater London and Rural Districts of England and Wales

Source: R.E. Waller, Control of Air Pollution: Present Success and Future Prospect. In "Recent Advances in Community Medicine". Edited by A.E. Bennett, Chp 4, 1978.

Table 1 shows the improvement in age-specific death rates from bronchitis in Greater London. This of course may be explained by other reasons such as the less fit moving out of London and the fitter moving in or changes in smoking habits, but it equally could be the change in urban disadvantage for bronchitis.

Statements are often made that atmospheric pollution is the cause of lung cancer. Certainly lung cancer rates are higher in urban than in rural areas. It is also true that smoke from coal contains well-known carcinogens. However, over a period when coal smoke has been substantially reduced, lung cancer rates have continued to increase. Furthermore, in some European countries where there is little pollution in urban areas, there is still a marked rural-urban gradient for lung cancer. This has given rise to the hypothesis that there is an "urban factor" which is a combination of adverse features. Most studies carried out during the last 20 years point to the prime cause of lung cancer as cigarette smoking and the part played by atmospheric pollution is probably quite small.

2.3 Smoke control areas

Control of pollution from domestic sources was achieved by smoke control areas which were first set out nationally under the Clean Air Act, 1956. In general, it is an offence to emit smoke from chimneys of a building in such areas. However, these are not smokeless zones because controlled amounts of smoke from specified buildings are permitted. For example, exemption from prosecution may be given for causing smoke as a result of lighting up a furnace. The owner-occupiers receive grants for conversion equal to 70% of the total cost which is shared by central and local government. There is no doubt that smoke control areas have resulted in a dramatic improvement in the quality of air but other events have also helped, particularly the trend towards building houses with central heating and a switch to natural gas. Mainly as a result of the policy of establishing smoke controlled areas, smoke concentration in Britain has fallen very markedly over the last 20 years and in large towns the amount of smoke is 1/5th or less than its previous level.

3. NOISE AND HEALTH

Noise is an unwanted sound which causes discomfiture to the listener. Sound is a transmitted form of energy. Man-made sources have made the modern environment much more noisy. Although convincing evidence of the adverse effect of environmental noise on health is difficult to come by, there is little doubt that a noisy environment affects the quality of life for many people. On the other hand, noise has a beneficial effect in providing arousal, stimulation and alerting to danger.

Noise may be either natural such as thunder or man-made such as road traffic. There has been a substantial increase in noise from man-made sources during the last hundred years. Annoyance may be caused either indoors or outdoors to some members of the population. Noises from the following sources commonly cause annoyance: aircraft, road traffic, industry and construction, trains, musical instruments, radios, barking dogs, shouting, singing, neighbours (especially their children) and parties.

A survey of member countries of the European Economic Community found that 45% of the population were exposed to high environmental noise levels from road traffic, 30% from intermittent industrial noises and 2% from aircraft noise (7).

3.1 Effects on health

Studies have examined the effect of noise on: hearing, sleep, communication, work and leisure as well as other general physiological and psychological parameters. It is difficult to generalise from the wide range of published studies. Indeed, there is a lack of really convincing evidence of hazards to health from environmental noise. On the other hand there is ample indication of very high degrees of annoyance caused to sections of the population.

There is a substantial subjective element in response to noise: broadly, 20% of people appear to tolerate high levels of noise without complaint but about 10% express great annoyance with even low levels of noise. The concept of annoyance reflects the attitude of the individual, the source, duration, and intensity of the noise as well as the time of day or night in which it occurs. The problem is enhanced where the noise is perceived as dangerous, for example, the fear of an aircraft crashing.

When measuring the effects of noise on hearing it is important to remember that loss of hearing is a common accompaniment of ageing (presbyacusis). Occupational deafness occurs in noisy industries only after fairly lengthy periods of exposure, although more rarely permanent deafness may follow a single loud explosive sound. Transient deafness (temporary threshold shift) may also be the sequel of exposure to a sudden loud noise or to prolonged intense noise. There is general agreement that hearing loss is not caused by environmental noise, although some concern is expressed about possible damage to hearing by prolonged exposure to the level of noise in discotheques.

From studies on sleep it appears that a continuous equivalent noise level of about 35 dBA inside the room does not disturb sleep. Above this level, particularly with a rise of a full 10 dBA, disturbance of the pattern of sleep appears, although there is considerable individual variation. Poor task performance may follow sleep in a noisy environment.

Although it is difficult to quantify, noise appears to have an adverse effect during rest, relaxation and study periods. Ordinary conversation, listening to the radio, television viewing or talking on the telephone begin to be affected when a continuous equivalent sound level above 50 dBA is recorded indoors. Conversation outdoors is affected with equivalent sound levels above 65 dBA.

In the experimental situation, noise affects humans physically in many ways including changes induced in the circulatory system and alteration of skin resistance, but the long term effects are inconclusive. Environmental noise which produces a high degree of annoyance in a section of the population can interfere with periods of relaxation and produce stress. Some studies around airports in Britain and elsewhere have claimed a higher rate than normal of mental illness and an increase in cardiovascular disease. One study in Japan (8) suggested that aircraft noise affected foetal development, though in many such studies the findings are questionable because of the nature of the study populations and the choice of control groups. More hard evidence is needed but there is little doubt that the quality of life for a section of the population would be considerably improved if noise levels were reduced.

3.2 Control of noise

There is a variety of approaches to the control of noise. In industry, the operation of machinery may be modified or a less noisy type installed, or the ears of workers may be shielded by protectors. Screening of the source will reduce transmission through the air or it may be moved in its entirety to a site away from residential properties.

Sometimes the solution lies in removing highly sensitive individuals from the noise location or installing double glazing.

Many lessons can be drawn from the 'quiet town' experiment operated in Darlington (population 97,500) in the North-East of England between 1976 and 1978. Noise abatement zones were introduced, heavy vehicles excluded from some central areas and legal measures enforced to control nuisance. However, the main thrust of the scheme was education. The population were made aware of the 'quiet town' by newspaper articles, leaflets, road signs, lectures, radio and television. The Darlington experience showed that much of the source of annoyance and complaints regarding noise related to standards of public behaviour: shouting, singing, banging of car doors, noise from children. At the end of the period, a quarter of those interviewed had been encouraged themselves and thought that others had been similarly encouraged to do things more quietly. Other surveys also show that people are often the chief culprits rather than traffic or aircraft noise.

Source of Sound	dBA
Low flying jet aircraft	150
Heavy industrial noise	100
Loud shout	100
Car engine	80
Normal conversation	50
Whisper	30

Table 2: Sound pressure values in Decibels

4. CASE STUDY: REDCAR IRON AND STEEL WORKS

From 1968 to 1974 I was Medical Officer of Health for Teesside County Borough in the North of England. This local government authority had a population of approximately 400,000 and it was a heavily industrialised urban area with large steel and chemical works on both banks of the River Tees. As one of the chief officers of the local authority I had responsibilities for the health of the population.

In August 1971 outline application was made by the British Steel Corporation for planning permission for the construction of the first stage of an integrated steelworks for the manufacture of steel products at Redcar which was a small coastal resort and part of the County Borough. The plan accompanying the application also indicated a substantial further phase and the Steel Corporation stressed that they would like the overall situation to be taken into account.

The scale of the development proposed was considerable and would have had far-reaching implications for the immediate area and beyond. Information accompanying the application was minimal, thus it was considered vitally important that the impact of the development should be studied in depth and objectively assessed. Much of the information to set up the studies was not available and had to be collected, so that over a year elapsed before a comprehensive report was submitted to the council from the chief officers concerned.

Unemployment in Teesside at the time was about twice the national average. Understandably, the County Borough Council were anxious to have the development, particularly because if it went elsewhere there would be a further loss of jobs since small steelworks in the area would close as part of the rationalisation process.

Thus, in November 1971, before they received the report of their officers, the council at a special meeting resolved that: "Believing that the problems of protection of the environment can be satisfactorily resolved by negotiation, the Government is requested to make an early decision permitting the British Steel Corporation to establish the whole of the proposed steel complex in Teesside". The effect of this council resolution could be interpreted as a message to the chief officers that the council wanted the steel complex and the officers should negotiate the best possible deal for the protection of the environment. In the event, however, the independence of the officers to produce an objective report was not affected. (9)

One of the chief grounds for opposing the development was concern about the effect it would have on the health of the population. This was expressed by some members of the council and more particularly by environmental groups as well as members of the public living in the proximity of the proposed site. The perceived dangers to human health by those concerned were never clearly defined but reference was made to diseases of the respiratory system including cancer. Apart from public pressure there were other compelling reasons for attempting to predict the impact of the proposed steel works on the health of the people. The overall mortality rate in 1971 for Teesside was 20% above that of England and Wales as a whole. Bronchitis and emphysema, which is not a significant cause of death under 45 years, was over 20% higher for those over 55 years than in the rest of the country. In the case of lung cancer there was a persistently higher mortality rate for men but a markedly lower rate for women than the national average. This may suggest that air pollution as a single factor may not be so important.

The mathematical model to predict the effects of the steel complex on the environment was based on technical data supplied by the steel corporation, pollution data from the local authority and meteorological data. Not taken into consideration however was the situation arising from plant breakdown or poor plant performance. It was also important to analyse trends over time of the level of atmospheric pollution in order to make future predictions more confidently. This was carried out for four pollutants (in soluble deposits, ferric oxide, smoke and sulphur dioxide) over the ten years 1962-1971 based on readings from 35 sites in Teesside. Computerised calculations were done for each site and each pollutant, and the average pollution for each month of the year was calculated. A seasonal adjustment factor for each month was computed to give an indication of seasonal variation in pollution. Then seasonal adjustment factors were subtracted from the original readings to give a series of "readings" free from seasonal effects. Regression analysis was used to examine the seasonally adjusted series of readings to determine whether there was any long term trend for pollution to decrease or increase and to estimate the magnitude of this trend (assumed to be linear).

4.1 Predicted results

From analyses of the meteorological data it was predicted that about 50% of grit and dust and sulphur dioxide would be directed over the sea and a substantial proportion of grit and dust would fall within the perimeter of the works. For only ten per cent of the time would the wind blow over the town of Redcar. It was not expected that the steel works would produce any smoke and at the time a smoke control programme was proceeding apace in Redcar which should reduce the major source of ground level smoke - the domestic chimney. The predicted increase in levels of undissolved solids (grit and dust) was that it was unlikely to exceed 3%.

Reliable predictions of sulphur dioxide levels were more difficult. It was estimated that the overall winter daily averages might increase from about 80 μg/m³ to 100 μg/m³. Short term peaks could increase from 300 to 400 μg/m³ to as high as 600 μg/m³. However, the prediction to process was handicapped because of lack of extensive meteorological and topographical data. Furthermore, the model did not allow for plant breakdowns or for the closure of other plants when the new steel works came into operation. Subsequent to providing the data for the mathematical model the Steel Corporation found it possible to use a desulphurising process in the coke ovens which could reduce the level of sulphur dioxide emissions by 10%. Further reductions could also be brought about by using fuel oil with low sulphur content. It was apparent that the population of Redcar would only be minimally affected by increases in grit and dust and sulphur dioxide when the plant was working normally.

Increased noise levels from the manufacturing processes could be substantially reduced if the necessary measures were taken and took into account the terrain and wind direction. It was thought possible to achieve noise levels below 65 dBA at the site perimeter giving a level of 48 dBA at the nearest house. Only a limited number of houses were in the vicinity of the site. Applying the information from epidemiological studies discussed in the previous sections, there did not seem to be any evidence for opposing the establishment of the steel works on the grounds of increasing the health hazards to the population.

However, the steel works are now in operation, though not to full specifications, and thus a splendid opportunity presents itself to assess the validity of the predictions made in the report not only about the health aspects but other environmental impacts. It is encouraging to know that work in this field has been undertaken by P.A.D.C. (in press).

5. PROBLEMS IN MEASURING ENVIRONMENTAL HEALTH IMPACTS

In assessing the impact on health of air pollution and environmental noise it is important to keep in mind the limitations of the present state of medical knowledge.

The effect of large doses of many pollutants on health is well-known: the physical damage caused by heavy smog and intense noise has been well documented. What remains unclear is the effect of low doses over a long period of time, and whilst an apparent correlation may be found between an agent and ill-health, the population is exposed to so many potential hazards that it is often impossible to prove a causal relationship. Epidemiological methods in these studies have severe limitations. For example, many agents in the environment, although carcinogenic, are relatively weak and give rise to only a few cases. In contrast an infective agent such as the measles virus can result in a large number of cases of the disease. Epidemiological methods to study diseases in the population work well when there is a high proportion of people affected who have been exposed to a single agent, but it would require huge study populations when only a small number are affected and especially, as in the case of cancer, when other carcinogenic agents are also present.

Mortality data are frequently used to measure outcome of an adverse factor and are useful because they are relatively accurate and record a single event. As already mentioned they have great limitations with small doses of the agent acting over a long time. Morbidity (sickness) data are usually not collected routinely in a form that can be used, so often have to be gathered specifically for the particular project in question. There are also problems in defining the criteria for the particular illness.

Furthermore, if mortality and morbidity are the only measures of outcome taken, then an essential feature will be missing - quality of life. For example, as has already been discussed, the small proportion of the population who become highly annoyed by relatively low levels of environmental noise would not show up in statistics of death or sickness, yet these individuals undergo considerable suffering. Although quality of life is a more nebulous concept it fortunately also frequently coincides with many of the amenity values which can be quantified.

Much discussion also involves whether or not there are acceptable levels for given pollutants, particularly some of the agents which have been more recently introduced. However, as discussed, some noise is beneficial. If on the other hand the aim is to remove all pollutants from the atmosphere adopting the so-called zero option, the cost would be prohibitive.

6. CONCLUSIONS

This chapter outlines all the difficulties in making predictions from health data. Frequent claims are made of a causal link between environmental conditions and health. Yet it is difficult to provide conclusive proof. It does not seem sensible to await such conclusive proof. It is better to accept that quiet, clean, attractive surroundings are more conducive to contentment than a drab, noisy, dreary environment. If industrial development is necessary to provide employment its effects should be minimised by providing additional amenities, thus reducing the adverse physical, psychological and emotional effects of unfavourable environmental conditions.

ACKNOWLEDGEMENT

The material in this article draws heavily on Dr. R.J. Donaldson's book "Essential Community Medicine (including relevant social services)" published by MTP Press Ltd. (1983), Falcon House, Lancaster, and we are most grateful for the publisher's permission to use this material.

REFERENCES

1. The Royal Commission on Environmental Pollution: 1970.
2. Committee on Air Pollution: 1954, The Final Report on the Beaver Committee, HMSO.
3. Royal College of Physicians: Air Pollution and Health: 1970.
4. Waller, R.E.: 1978, Control of Air Pollution: Present Success and Future Prospect, in A.E. Bennett (ed.), Recent Advances in Community Medicine, Chp. 4.
5. Donaldson, R.J. OBE: 1974, Air Pollution and Health: A Teesside Report.
6. Lunn, J.E., Knoweldon, J. and Handyside, A.J.: 1967, "Patterns of Respiratory Illness in Sheffield Infant Schoolchildren", Brit. J. Prev. Soc. Med., 21, 7-16.
7. Large, J.B.: 1977, Methods of Assessing Community Response to Environmental Noise, R.S.H. J. 4, 147-152.
8. Ando, Y. and Hattori, H.: 1971, Statistical Study on the effects of Aircraft Noise during Foetal Life. Proceedings of the 7th International Congress on Acoustics, Budapest. Paper 24H, 13.
9. Fairbank, W., Donaldson, R.J. and Radcliffe, J.W.: 1972, British Steel Corporation Proposed Complex - Redcar Phases II & III.
10. P.A.D.C. E.I.A. and Planning Unit (now Centre for Environmental Management and Planning): (in press), Post-Development Audits to Test the Effectiveness of Environmental Impact Prediction Methods and Techniques,

to be published by Oklahoma/Aberdeen Universities Worldletter as an Occasional publication.

AFFILIATION

Dr. R. J. Donaldson was formerly Director of Studies at the Centre for Extension Training in Community Medicine, London School of Hygiene and Tropical Medicine, England and is now Honorary Specialist in Community Medicine at the Royal Free Hospital, School of Medicine, Hampstead Health Authority, England.

L. K. A. Derban

HEALTH IMPACTS OF THE VOLTA DAM, GHANA

"The most valuable capital of the state and of the society is man, and to protect him is not merely the dictate of humanitarian sentiment but a duty imposed by self-interest on every community". Prince Rudolf.

Ghana has no known coal deposits although small quantities of oil have recently been discovered off-shore. Hence the principal energy assets for sometime to come will be hydro-electric power.

Of the 6 principal hydro resources, the Volta and its tributaries are the main sources of hydro power.

SUMMARY OF PRINCIPAL HYDRO RESOURCES

VOLTA RIVER		Installed Capacity (MW)	Annual Energy GWH	Remarks
Akosombo	-	882	5625	Commissioned 1966
Kpong	-	160	940	" 1982
Bui	-	300	1175	Preliminary studies
Oti	-	200	710	
Pwalugu	-	36	133	
PRA RIVER				
(4 plants)	-	230	640	
TANO RIVER				
(4 plants)	-	137	356	

Water development projects whether for generating hydro-electric power for irrigation or for water supply, are important in socio-economic development. However, such water development projects often result in impacts on the health and well-being of people including the well-being of the ecosystems on which the people's survival depends.

B. D. Clark et al. (eds.), Perspectives on Environmental Impact Assessment, 121–132.

1. THE VOLTA RIVER PROJECT (1961-1964)

The Volta River Project was the first major element in developing hydro potential of Ghana. The main aim was to generate electric power for industrial expansions in Ghana especially to assist in the establishment of an aluminium industry.

Towards this end, a Preparatory Commission conducted detailed feasibility studies between 1952 and 1956 and later, in 1959, a re-assessment was made by Kaiser Engineers. Even though in the final plans some consideration had been given to the impact of the project on human environment to be affected and on the ecosystem both up-stream and down-stream of the dam, it appears the need at the time for the country to have a cheap source of electricity was so paramount that the production of electricity had over-riding priority over and above any other consideration.

It was when dam construction was under way and the lake began to form did a more conserted effort start on the resettlement programme. The Volta River Authority was established and assigned the responsibility of seeing to the construction of the Volta River Dam at Akosombo and the eventual generation of power.

The construction of the dam started in 1961 and completed in 1964 and power generation started in 1966. Kaisers were the engineers and Impregillo the contractors.

The construction of the dam resulted in the formation of the new Volta Lake (the largest man-made lake in the world) with a surface are of about 3275 miles^2 (8500 km^2), 3.6% of the surface of Ghana and a shoreline exceeding 3000 miles (4800 km). It also changed the existing physical, biological and socio-economic environment of some 80,000 people living in 739 villages displaced from the flooded area.

The Volta River Authority was to plan, organise and implement by itself, a resettlement programme. The human problem of resettlement was one of the earliest and by far the most serious impact of the Volta River Project which the Authority had to face.

2. RESETTLEMENT PROGRAMME

At the time of the construction of the Volta Dam, the only activity in the socio-economic field was the resettlement of the 80,000 people who were living in the area to be flooded.

There were no comprehensive pre-impoundment studies made. However, before the closure of the dam in 1964, some basic studies on various aspects of aquatic biology were carried out by the Ghana Institute of Aquatic Biology and a census-type of social survey by the Volta River Authority (VRA) with the assistance of the University of Science and Technology, Kumasi, Ghana. Work on existing fishing gear was also undertaken by the Department of Fisheries and basic epidemiological data were collected by the Ministry of Health.

The planning of the resettlement and the actual implementation of the resettlement exercise was carried out by the VRA with the assistance of a number of Government Organisations.

The main resettlement objectives were:-

(i) To provide dwelling, work and recreation places for the settlers.

(ii) To offer the settlers an opportunity for productive work.

(iii) To improve the physical appearance of their surrounding landscape.

(iv) To help the settlers to develop a community spirit.

About 85% of the people (67,000) had to be settled in 52 Resettlement Sites scattered around the lake. Some 12,000 people (15%) chose to settle themselves. The people who were resettled came from four regions of the country and had lived in the river basin for many years. A total of 739 villages were destroyed. About 14,657 households were disrupted, 12,799 households with 4,200 sheep and goats, 2,954 cattle and 16,000 chicken had to be moved. The people were mostly subsistence farmers. Only 2% of them were river fishermen.

The resettlement construction programme consisted of 52 new villages containing 13,000 houses, 82 schools, 46 markets, water system for 146 pipes and over 500 miles of new roads. APproximately 100,000 ha of new farmlands were established.

2.1 Socio-Economic Problems

One of the major problems was the organization of the settlements into institutions with budgets and personnel to make them work. The challenge was therefore to support the settlers so that they could develop their farm and build their towns into living and orderly communities which satisfied their needs for health, safety and well-being.

Houses were provided, but the settlers had to make them into homes. The resettlements were bleak and featureless with no familiar market days for buying, selling and social contact. The settlers had exchanged a place of comfort for a place of insecurity. This initial stage was a period of social and economic constraint and stress. It took a long time to develop reasonable efficiency in fishing and farming. It was therefore necessary to supply food through the World Food Programme.

Another serious problem which the Volta River Authority had to face during the early years of resettlement was that of administration. The Volta River Authority, set up primarily to generate electric power, found itself not fully equipped to handle the administrative problems of the resettlement. Town Managers were created to manage the affairs of the resettlement through Town Development Committees. This arrangement made it very difficult to integrate the statuses and roles of the different traditional authorities brought together in the exercise. The social cohesion and order were seriously affected.

Lack of community life and the slow improvement in the physical and economic environment led to the drift out of the resettlement towns of some of the able-bodied people, leaving behind the very young and the old and infirm. Thus in 1968, 60% of the original settlers had moved out, some because of the difficulties encountered in the acquisition of land for cultivation, a trend which has recently been reversed by a positive and more effective approach to land acquisition and distribution. Others being fishermen, have moved closer to the lake to take advantage of the new economic environment.

On the other hand, a substantial number of people have settled on their own in unauthorised new fishing villages mostly in very isolated areas. These were migrant fishing communities coming from the lower Volta and spending part of their time on the lake and the other part in their home villages. For effective planning, information had to be obtained about their numbers, their social and economic organisation and their problems.

2.2 'Self Help' Resettlements

A survey carried out of the approximately 14,000 people who

decided to resettle themselves provided contrasting results. These people were paid compensation for both crops destroyed and for their own houses. Most of them were scattered but four villages decided en bloc to resettle themselves. The survey showed that there was more social cohesion amongst these four settlements but the quality of their housing and the level of social amenities were lower than those in the resettlements.

The transfer of responsibility to the local authorities and Technical departments of the Government had its initial problems, but has improved the situation.

3. ENVIRONMENTAL SANITATION AND HEALTH

Ignorance, apathy, local customs and habits led to marked indifference to public health problems in the settlements. The communal latrines provided (of aqua-privy variety) soon ceased to function as a result of lack of proper use and maintenance. Frequent breakdowns of the sewage collection and disposal system led to indiscriminate dumping of refuse and defaecation. Pipe-water systems operated by diesel pumps soon failed. The pumps broke down and there were no funds for fuel. The public water standpipes were not properly looked after and were soon in need of repair. With the breakdown of the water supply and sewage disposal in some settlement areas, conditions became favourable for the possible increase of water-borne diseases such as typhoid fever, dysentery and other diarrheal diseases, parasitic diseases such as hook-worm diseases and insect-borne diseases.

Yellow fever and dengue were the most important of the arthropod-borne viral infections found in the resettlement villages. These received serious attention because of the high fatality rates associated with them.

The vector of these diseases is Aedes aegypti which breeds in small collections of water in tins, coconut shells and discarded car tyres. The breeding and the distribution of this mosquito was encouraged during the early part of the resettlement programme by the breakdown of the refuse collecting system and the water supply pumps in the resettlement villages, which meant that water had to be stored in water jars and drums.

The Yellow fever has been successfully controlled by mass innoculation and by attack on the artificial breeding sites.

Local councils had great difficulty in maintaining

these misused health facilities. Deep pit latrines were found to be more suitable and pipe-water systems operated by hand pumps gave satisfactory results. An intensive health education programme had to be launched to help people who were then facing conditions which required a higher standard of health than they were used to.

3.1 Mental health

The psychological trauma caused by the sudden removal of the settlers from their familiar surroundings to a strange environment made their cooperation in the move difficult. The settlers needed a lot of encouragement and support to raise their economic status in order to boost their morale.

3.2 Urinary Schistosomiasis

Urinary Schistosomiasis (Bilharzia) is a snail-borne infection which might be expected in water development schemes in tropical countries. An epidemiological survey was made before the lake formed in 1960-61, along the Volta River. The prevalence in school children was 5%.

The creation of the lake, the consequent biological explosion of aquatic weeds associated with the vector snail and the mass migration into the area of fishing communities from areas where urinary schistosomiasis was endemic, led to great increases in the prevalence of the disease in many localities around the lake. Surveys in 1964-67 after the construction of the dam, showed a 90% rate in school children with a ratio of two boys to one girl.

Transmission occurred in the lake itself. Exposure to infection was high especially for fishermen who dive to get fish traps as well as for people, particularly children, who draw water from the lake or bathe in it.

The present situation of the disease in the Volta Lake cannot be viewed with optimism. It is realised that to prevent continuing transmission of the disease, a long term education programme combined with effective sanitary measures, including an efficient disposal system for excreta, a safe water supply and adequate bathing and washing facilities, is necessary.

3.3 Onchocerciasis (River blindness)

The disease is carried by the small black biting fly

(Simulium damnosum). Breeding is limited to well-oxygenated water, rapidly flowing streams and rivers. The fly and disease were previously widespread along the Volta River. When the lake was formed many of the major breeding sites in the river were eliminated.

Even though major areas of the lake were free of the disease some of the streams and rivers flowing into the lake are still breeding the flies. Some resettlement villages close to the breeding sites still have prevalence rates as high as 80%.

On the whole the prevalence of the disease has been drastically reduced along the Volta Lake by the formation of the lake. The eradication of this disease in Ghana now forms part of the WHO Onchocerciasis control programme in the entire Volta River Basin area in West Africa. It started in March 1974. This inter-regional control programme is a combined effort of the seven West African countries - Ghana, Togo, Niger, Mali, Upper Volta, Ivory Coast and Benin - and the Food and Agricultural Organisations, the World Health Organisation, the World Bank, and the UN Development Project. It is a twenty-year comprehensive and imaginative, well planned, control programme.

3.4 Trypanosomiasis (African Sleeping Sickness)

African sleeping sickness is carried by the tsetse fly (Glossina palpalis) its principal vector. The fly is associated with the lakeside and riverside, more especially in the light forest fringing the edge of the water.

The formation of the lake has been beneficial in that it drowned large areas of forest which harboured the tsetse flies. The prevalence of the disease has been low. In 1971 six cases of sleeping sickness were diagnosed and treated at the health centres. In 1972 only two cases were reported.

With the vegetation pattern around the lake changing and evolving during the years, and population movements and river transportation now on the increase, attention is being paid to the possible development of new breeding foci and the appearance of clinical cases of the disease.

4. THE VOLTA LAKE RESEARCH PROJECT

The formation of the Volta Lake created a number of development possibilities in fisheries, agriculture, transportation, wildlife and tourism. It was realised very early that there was a need for a comprehensive study to be made in order to assess the full potentiality of the scheme and the extent of the problems to be solved.

As early as 1962, the Government of Ghana raised in a draft request, the possibility of obtaining assistance from the then United Nations Special Fund for such research work. In 1964, the matter was studied by a Unesco Mission and again in 1966 by UNDP which approved it in June 1967.

The plan of operations for the first phase of the project (1968-1971) was signed in January 1968 by the Ghana Government, the UNDP and FAO which acted as the executing agency for the UNDP. The VRA was designated Government co-operating agency.

In the second phase of the project (1972-77) the VRA was designated the Government Responsible Agency and FAO, the participating Specializing Agency. Thus the responsibility for implementation was vested on the VRA which appointed a Ghanaian national to head the project.

The Government Ministries, Departments and other national agencies which participated in supplementing the projects in both phases were:-

Ministry of Health
Fisheries Department
Irrigation Department
Department of Games and Wildlife
Medical School, University of Ghana
Volta Basin Research Project, University of Ghana
Forest Products Research Institute (CSIR)
Institute of Aquatic Biology, Council for Scientific and Industrial Research (CSIR)

4.1 Objectives

The purpose of the first phase of the project was to assist the Ghana Government, through the VRA, to strengthen research in fisheries and hydrobiology, public health and the resettlement of the people displaced by the Volta Lake. Thus specifically, emphasis was laid on; the limnology and fish biology of the lake and the means of exploiting the fish stocks; the public health problem of water-borne diseases connected with the lake and their means of control; new systems of agriculture and other incentives for the resettled people.

The main aim of the second phase of the project, on the other hand was to strengthen selected research programmes for the development of the resources of the Volta Lake and the implementation of the recommendations arising from the first phase project.

4.2 Operation of the Project

The policy-making forum of the project was the National Co-ordinating Committee, which had as members all the cooperating agencies and participating agencies, and was headed by the Chief Executive of the VRA. The Committee had 3 specialised sub-Committees charged with the responsibility of advising the Project Administrator in their areas of specialized activity. The Sub-Committees were:-

(1) Fisheries and Hydrobiology
(2) Social, Agricultural and Economic questions
(3) Public Health

4.3 Activities

Activities were sub-divided under three sectors:-

- Fishery and Hydrobiology
- Public Health
- Social, Agricultural and Economics

4.3.1 Fishery and Hydrobiology

Identification of the fishing villages and camps and collection of data on their populations.
Fishery methods used by the fishermen and the type of fishing gear in use.
Experiments on improved techniques in fishing processing and marketing.
Fish catch surveys.
Fish biology and population studies.
Experiments on new prototype fishing canoe.
Fishery development plans - pilot fishery complex included such facilities as:

(i) Concrete landing ramp.
(ii) Weighing area for fish.
(iii) Fish processing area.
(iv) Wholesale fish market.
(v) Storage facility for fish and other agricultural commodities.
(vi) A boat yard.
(vii) A fishing school.

(viii) A fitting shop.
(ix) A general office.
Limnology - physical and chemical.
Plankton production and Benthic organism studies.
Aquatic weeds studies.

4.3.2 Social, Agricultural and Economics

Studies into the physical nature and extent of the draw-down area.
Crop trials in the draw-down Irrigation pilot trials.
Post-resettlement social survey.
Re-settlement farms experiments.

4.3.3 Public Health

Establishment of baseline epidemiological data on water-borne diseases - Urinary Schistosomiasis, Onchocerciasis etc. and drawing up control programmes.
Studies on vectors of disease -
snail host of schistosomiasis,
simulium fly related to Onchocerciasis,
tsetse fly related to African sleeping sickness,
mosquitoes related to malaria, yellow fever, filarias.

4.4 Results

Results of the research project have made it possible for recommendation to be made for exploiting the benefits that could be derived from the formation of the new lake and finding solution to some of the many problems encountered. Emphasis now is to be placed on the development of the fishery and agriculture sectors supported by research, public health and socio-economic activities.

Feasibility studies on the various projects could now be undertaken. Private entrepreneurs and public corporations could also be provided with the necessary data to assist them in the establishment of enterprises in fisheries and agriculture.

The lakeside health unit of the Health and Safety Department of the VRA continues to update health data relating to the lake, to evaluate and keep under surveillance the major health problems, to make field investigations into incidence and to forecast disease trends. It makes recommendations on health facilities and plans, and carries out and evaluates vector control programmes in cooperation with the Ministry of Health.

5. LESSONS LEARNT

The Volta Resettlement operation has been described as a brave and imaginative attempt with limited resources to tackle a challenge and an urgent crisis. The problems encountered in the exercise were enormous and were due mainly to inadequate time span and administrative capacity, insufficient political will and finance and absence of prior commitment and meaningful local involvement.

The chief lessons that could be learnt from the Volta Resettlement experience are:-

(a) The importance of assessing, at the feasibility stage, preferably by law, the environmental, social and health impact of the project at the same time as the engineering, economic and socio-political assessment requirements.

(b) Sufficient time must be allowed for detailed outline programme for ensuring adequate consultations with and proper care of the people to be affected including the well-being of the ecosystem on which their survival depends.

(c) Resettlement lands and social requirements must take the same time priority as the dam construction recognising that it is more difficult to adhere to tight schedules where resettlement is concerned.

(d) Sufficient finance must be provided for social planning and implementation as well as the engineering construction.

(e) The importance of inter-disciplinary approach to water resource development and management.

6. SUMMARY

Development of the Volta River Basin began in 1961 with the start of construction of the Akosombo Dam to generate hydro-electric power for industrial expansion in Ghana.

Even with the benefit of a comprehensive environmental study made under the Preparatory Commission in the 50's, attention was focused mainly on building the dam when the project started in 1961.

The VRA initially established and assigned the responsibility of seeing to the construction of the Dam and the generation of power had eventually to undertake the Volta Lake resettlement programme.

It was when the dam construction was well on the way and the lake began to form, did a more conserted effort start on the resettlement programme which must be regarded the most serious impact of the Volta River Project.

The importance of a multi-disciplinary team in the developmental process is emphasized not merely in writing reports but also implementing an integrated programme and avoiding splitting the project into two camps, Engineering and Social.

The economic development must be in the best interest of the people, the environment and the totality of the economy.

To quote Dr Johnson, 'To preserve health is a moral and religious duty; for health is the basis of all social virtues - we can no longer be useful when not well'.

REFERENCES

Ackermann, W.C. et al.: 1973, 'Man-made lakes, their problems and environmental effects'. Geophysical Monograph No. 17, American Geophysical Union, Washington D.C.

Chambers, R.: 1970 (ed.), The Volta Resettlement Experience, Pall Mall Press, London.

James, L.P., and R.R. Lee: 1971, Economics of Water Resources Planning, M. Graw-Hill, New York.

Jones, C.R.: 1973, Health Component in the Volta Lake Research Project - Report on project results, conclusions and recommendations, World Health Organisation.

Scott, D. et al.: 1980, Epidemiology of Human Schistosomiasis haematobium Infection on Volta Lake, World Health Organisation.

Volta Lake Research: 1971, UN Development Programme, Food and Agricultural Organization, Rome.

Volta Lake Research: 1979, Project Findings and Recommendations, UN Development and Agriculture Organization, Rome.

Volta River Authority: 1965, Volta Resettlement Symposium Papers, Volta River Authority, Accra.

Volta River Project Commission: 1956, The Volta River Project, Report of the preparatory Commission. H.M.S.O. London.

AFFILIATION

Dr L.K.A. Derban is the Chief Medical Officer of the Volta River Authority, Accra, Ghana.

Cameron G. Ramsay

ASSESSMENT OF HAZARD AND RISK

1. INTRODUCTION

A hazard can be defined as the inherent property of a system that could cause injury or damage. Risk is the chance that an event happens which demonstrates these harmful properties. In other words, risk is a measure of probability of harm occurring in a certain period of time. This paper considers the assessment of hazard and risk with particular examples from the oil, gas and petrochemical industries. The following topics are discussed:

- The need for hazard and risk assessment;
- The aims of the assessment;
- Suitable risk criteria;
- A systematic, quantitative assessment method.

Finally, examples of quantitative hazard and risk assessments are given, drawing particularly from the author's experience of applying such assessment in the field of Environmental Impact Analysis (EIA) for planning purposes.

2. THE NEED FOR HAZARD AND RISK ASSESSMENT

In recent years there have been a number of major accidents, both in the UK and abroad, that have served to heighten public awareness of the risks posed by certain types of industrial installation, and to bring about change in the way such installations are viewed by government and planning authorities. Examples of such accidents which are having far reaching effects on the attitudes of regulatory and legislative bodies include:

(i) The explosion at the chemical plant at Flixborough, England in 1974;
(ii) The release of toxic dust from the factory at Seveso in Italy in 1977;
(iii) The explosion of the oil tanker "Betelgeuse" at Bantry Bay in Ireland in 1979;
(iv) The accident at the nuclear power plant at Three-Mile Island (Pennsylvania) in the USA in 1979;

B. D. Clark et al. (eds.), Perspectives on Environmental Impact Assessment, 133–160.

(v) The capsizing in the North Sea of the accommodation rig Alexander Keilland in 1980.

In this environment of increased public awareness and increased involvement by regulatory authorities, proposals to site major oil, gas and petrochemical plants in the vicinity of dwellings or other industries inevitably arouse interest and initial concern. Both the planning authority and the responsible industrialist will need to make some assessment of the risks imposed on the communities by potentially hazardous developments before the design, layout and site location are finalized and planning permission is granted. (Health and Safety Commission, 1979; Health and Safety Executive, 1978; Cremer and Warner, 1980; Slater, 1979).

3. THE AIMS OF THE ASSESSMENT

The assessment must provide information which can answer the following major questions:

(i) Does the design and layout of the hazardous industrial plant permit the effects of minor hazardous events to be localized, without serious risk of spreading to the rest of the plant?

(ii) Is the plant sufficiently far away from neighbouring industrial installations to prevent "domino effects" where a failure on one plant leads to a more serious failure on the neighbouring plant?

(iii) Is the plant sufficiently far away from the surrounding community (houses, schools, hospitals, public roads etc.) to decrease the risk to an acceptable level?

If the assessed risk is unacceptably high, then it can be decreased either by modifying the plant design, or by changing the layout and chosen site, or by a combination of both.

4. SUITABLE RISK CRITERIA

In order to judge whether the risk is appropriately low or unacceptably high, it is necessary to derive suitable criteria and thereby put the risk into perspective.

4.1. Criteria to protect the public

It is an inescapable, but sometime forgotten or ignored,

fact that everyone is living in an environment where risk is always present such as: risk of injury, of illness, of death. While sensible precautions are taken to minimize the risks from some causes, other risks are often ignored or blithely accepted. Thus, for example, we immunize children against most serious diseases, yet furnish our houses with materials which are flammable and give off highly toxic smoke as they burn.

The general risk background is illustrated in Figure 1, where the risk of death faced by any individual in one year is plotted against the age of that individual (using mortality data from western European countries). It can be seen that between birth and the age of ten, the risk of death decreases from approximately 10^{-2} per year (a probability of 1 in 100 of dying in the first year of life) to the minimum of about $3x10^{-4}$ per year (1 in 3333 per year) at age ten. Thereafter the risk increases, particularly rapidly for the 15-20 age group owing to adolescents taking greater voluntary risks in activities such as motorcycling and hazardous sports. By the time people reach the age of 100, they face a probability of death of about one in any year, reflecting the very small number of people who live longer than 100 years. Against this risk background, then, it would be unrealistic to insist upon zero risk to the public from hazardous oil, gas or petrochemical installations.

For the population living near to a hazardous installation, the risk is perceived as one which is essentially unavoidable (involuntary) and from which they do not receive a direct benefit. It is, therefore, comparable with risks of an everyday kind without any compensating benefits, which are accepted principally because they are fundamentally unavoidable, such as falling, fire, natural and environmental factors (including lightning strikes), and poisoning. The difference is that an oil, gas or petrochemical installation (or other hazardous plant) is a man-made hazard rather than a natural one. It is, therefore, reasonable to expect the man-made risk to be sufficiently low that it does not make a significant difference to the total pre-existing comparable natural risk.

From mortality statistics for western European countries and the USA, the percentage increase in risk brought about by the presence of the hazardous installation can be assessed. In the context of the fundamentally unavoidable risks, excluding falls (which affect mainly the elderly and are, therefore, not comparable), a man-made risk of death of

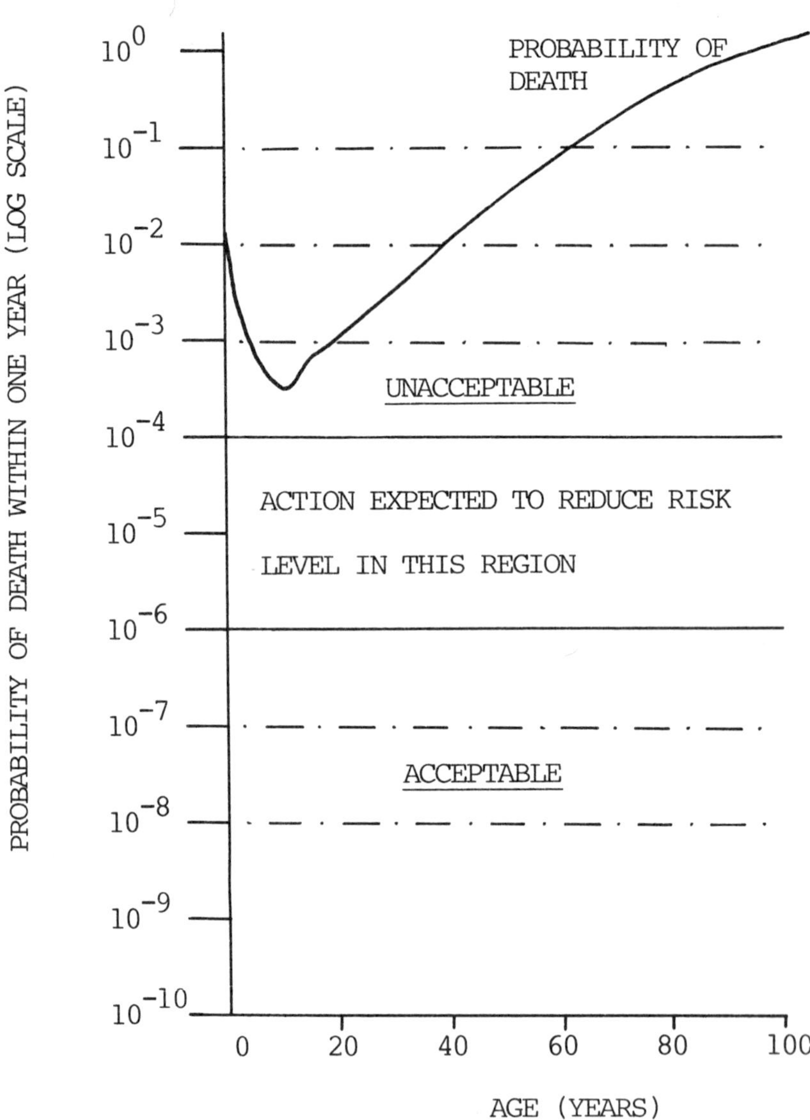

Figure 1: Risk Background

1 in 1 million years ($1x10^{-6}$ per year) would make about 3% difference to the total involuntary risk, about 0.2% difference to the total risk of accidental death of all kinds, and about 0.01% difference to the total risk of death from all causes. This would, therefore, appear to represent an appropriate level of acceptable risk. In other words, if the expected frequency of death for any individual in the external population due to the hazardous installation does not exceed a value of one in one million per year, then the risk is clearly acceptable.

At higher levels of risk of death, of the order of 10 in 1 million per year, the risk is not necessarily unacceptable, but consideration should be given to expending effort and money to reduce that risk. At levels of the order of 100 in 1 million per year, the imposed risk of death is certainly significant, and is therefore generally considered unacceptable (see Figure 1).

4.2 Criteria to protect neighbouring workforces

Since people spend only a relatively small part of their lives at work, and since workforces are generally considered to be capable of much more rapid and orderly evacuation in an emergency than the public, a higher level of risk is usually considered acceptable in the workplace than is allowable at dwelling houses, schools and shops.

In considering the risk imposed on the workforces in the neighbouring plants, it is helpful to make comparisons with the occupational risk of death. The generally recognized form of presentation of occupational fatal accident statistics is the Fatal Accident Frequency Rate (FAFR), defined as the average number of fatalities per 100 million (10^8) working hours, which is approximately 1150 working life-times. Values of the FAFR for various industries in the UK are:

	FAFR
Chemical industry	4
General manufacturing	4
Steelmaking	8
Fishing	35
Coal mining	40
Construction	67
Air crew	250

From these it may be concluded that the existing FAFR for employees on plants or factories (before neighbouring oil, gas or petrochemical plants are developed) would be expected to be about 4 or greater. The total FAFR includes a wide variety of accidents of a "conventional" kind which are common to all industries, such as falling or being hit by a dropped object. Indeed, studies have shown that only about half the FAFR for the chemical industry is attributable to hazardous chemicals or processes.

Consider a situation where there is an expected accident frequency of 25 in 1 million years for process malfunctions on a hazardous plant, which are sufficiently serious to cause death to employee(s) on a neighbouring plant who happen to be present when the malfunction occurs. This risk of a process accident (of fatal severity) of 25 in 1 million years may be expressed as a risk of 25 in 8760 million hours. In each year, a person will work on average about 2000 hours, and so someone working in this risk environment faces a risk of death of:

$$25 \text{ in } \frac{8760}{2000} \text{ million calendar years}$$

that is 5.7 in 1 million calendar years.

Expressed as an FAFR, this risk is equivalent to 5.7 in 2000 million working hours, or an FAFR of 0.29, which represents an increase of only 7% in the pre-existing occupational risk exposure. Thus a reasonable target for the maximum allowable imposed risk of death would appear to be 25 per million years, since this gives only a slight increase in the occupational risk to the employees on the neighbouring plant.

4.3 Criteria to protect plant and workforce

A hazard and risk assessment undertaken as part of an EIA for planning control purposes tends to concentrate on the plant's potential to cause damage beyond its boundaries. The risk outside the plant boundaries, however, is ultimately dependent on the level of safety within the plant, and particularly on the capabilities of the plant design, its layout and the ability of the trained workforce to cope safely with the effects of the minor accidents, and thereby to prevent escalation into a major disaster.

At the stage when outline planning approval is sought

for a hazardous development, the detailed design and layout of the plant are generally not yet available. A prudent Planning Authority will, therefore, normally make it a condition that a hazard audit must be undertaken at a later stage, to confirm the safe operability of the plant. Although the detailed design will generally not be known, much detail about the outline design (or the design concept) and the outline layout ought to be available. Thus aspects such as the number of process modules, their separation and disposition, can be considered, as well as the inventories within various sections of the plant. Usually, the risk of an incident is greatest in the process modules, where many items of equipment are concentrated in a relatively small area and where conditions of internal temperature and pressure can often be extreme. In contrast, the largest amounts of hazardous materials usually occur in the storage areas. It is therefore, particularly important to ensure, at an early stage in the project, that the separation between the process modules and the storage areas is sufficient to minimize the risk of an incident on the process modules causing damage to the storage tanks, with this damage in turn leading to spillage of large quantities of hazardous materials.

No definitive criteria can be advanced for the risk of a failure on one part of the plant causing secondary damage to another part of the plant, since the acceptability of that risk depends crucially on the acceptability of the consequences of the secondary damage. Of course, industry codes of practice and national standards often contain guidance on layout and separation, but this is in the form of minimum general requirements. Technology is advancing at a speed such that the particular development proposed may not have been foreseen at the time the code or standard was drawn up.

It is self evident that the workforce on a hazardous plant should be exposed to the minimum level of risk compatible with the proper running of the plant. In the UK, the Advisory Committee on Major Hazards has made particular reference to this topic in their Second Report (Health and Safety Commission, 1979). The Committee has established categories of people among the workforce and categories of hazardous areas in relation to the risk, and has discussed in very general terms:

(i) The siting and construction standards of buildings within the plant boundaries;

(ii) The appropriate limitations of access to the higher risk areas (such as the process modules).

For example, many of the relatively large numbers of people customarily accommodated in the site office block (housing the administrative and technical personnel) have virtually no reason ever to visit the high risk areas; others may need to visit these areas occasionally or only for short periods. These members of the workforce should, therefore, be protected by locating the office block as far as is practicable away from the high risk areas. Such people do, however, benefit directly (in terms of their earnings) from the hazardous plant, and it is unrealistic to suppose that they can be insulated from the hazards of the business to such an extent that they run no greater risk than those employed in a conventional office remote from the hazardous plant.

Another category of people which the Advisory Committee on Major Hazards recognizes is the control room staff. The protection afforded to these staff and their control equipment must be such that they can shut the plant down following an accident, and thus minimize the risk of escalation of the damage. Again, no definitive risk criteria can be set down, since each situation must be assessed individually.

5. A QUANTITATIVE RISK ASSESSMENT METHOD

A formalized, sequential procedure has been developed in which the overall risk from a hazardous installation is analysed by a cumulative consideration of individual plant items, their location, possible modes of failure, and the consequences of such failure (Comer _et al_., 1980). This procedure was developed in response to interest from industry in optimizing plant layout, and the need of regulatory authorities to have a method which aids decision-making.

At the heart of the procedure is a computer programme which performs the main consequence calculations. The results of the analysis are presented as _Risk Contour Plots_, which show the frequency with which specified levels of damage would be expected to occur. These risk contour plots allow the appropriateness of the layout, inter-plant separation, and community safety zones to be assessed. The main stages of the analysis comprise:

- Preparation of a quantitative description of plant hardware;
- Identification of failure cases (to generate the "Failure List");

- Assessment of primary failure probability;
- Quantification of discharge rates for each failure case;
- Prediction of the dispersion of the discharged hazardous materials in the external environment;
- Evaluation of the impact of the dispersed material, and determination of hazard distances for particular damage levels relevant to the community impact and impact on neighbouring industrial facilities and their workforces;
- Assembly of probability distributions (generally two-dimensional, in the form of risk contour plots) of various intensities of damage in and around the plant or the entire complex;
- Assessment of the probability of domino effects, and modification of the primary failure probabilities if necessary;
- Calculation of the total impact on the local community and neighbouring industrial facilities (risk contour plots modified to take account of domino effects);
- Formulation of conclusions (if appropriate) about the suitability of the layout and chosen site for the hazardous installation.

These stages are considered in turn.

5.1 Quantitative plant description

The analysis requires outline engineering design data for all key plant items and their interconnections, as well as the location of the item. The facility to change the location co-ordinates of individual items, (such as storage tanks) and groups of items (such as process modules) is included in the computer model, and this allows the risk associated with various plant layouts to be assessed rapidly. The data requirements are:

(i) Location and layout of the site;

(ii) Process flow schemes (the relationship of one item of plant to another);

(iii) Equipment list (number of reaction vessels, tanks, pumps);

(iv) Inventories and normal conditions (how much, what is it, temperature, pressure?);

(v) Number and diameters of pipework connections;

(vi) Remote operated valves (or other factors which

allow the amount of spillage to be decreased).

Approximate data on all these topics should be available from a developer at the outline planning stage, although it may require a considerable amount of discussion between the hazard analyst and the developer's engineering team, since the data will inevitably not have been assembled into a consistent package at that stage. Such discussion often helps to crystallize a developer's ideas on the design concept, and can identify potential problem areas at an early stage in the project. For existing installations, the data will be available, but it will be necessary to check that the engineering drawings are up-to-date and are a true representation of the plant hardware.

5.2 Identification of failure cases

For each plant item, the possible modes of failure are identified. These are then assembled into a single list, called the "Failure List", for each plant. There are many ways of identifying the failure cases, and a few are discussed below.

5.2.1 Hazard and operability study

This is the most formalized method, but requires a very great amount of effort and is inappropriate until the detailed design has been undertaken, since a full detailed process description is necessary. The study involves the systematic questioning of every part of the process and utility plant, to discover how deviations from the design intention may occur. It is then decided whether the deviation is hazardous, and whether remedial action is necessary (such as a design modification or a change in the operating procedures). This method of study is described in a booklet produced by the UK Chemical Industries Association (1977), and involves the application of "guide words" (such as no, not, more, less, reverse) to the process intention (such as no flow, less temperature) in order to direct the thinking of the study team.

5.2.2 Technical audit

This involves experienced engineers in a study of the design, construction and operating procedures. A checklist approach may be used to identify problem areas where hazardous failures might occur.

5.2.3 Examination of the historical record

The types of hazardous or disastrous incidents which have occurred in the past are considered. This approach draws directly from real experience, and includes the effects of human error or omission, although the consequences are generally reported in terms of the resultant release of hazardous material from the plant hardware.

In order to be able to undertake risk assessments at the conceptual design stage, the basic approach used in the risk assessment computer programme is to draw upon general historical experience and thus to postulate three basic categories of failure.

(i) Sudden and complete rupture of a vessel or tank;
(ii) "Guillotine break" failures of connection pipework;
(iii) Internal explosion within reaction vessels (if relevant).

The computer programme models these failures for each item of plant equipment, using relatively simplistic calculations, since experience has shown that the uncertainties about the basic plant design render it unnecessary to use more sophisticated models. Care, however, is taken to ensure that each failure case is specified in such a way that it is compatible with the inherent limitations on the calculation procedures. In instances where failure cases are identified which are outside the limits of the calculation procedures, the consequence calculations are carried out separately using more detailed methods, and the results are included in the failure list as "specified cases".

Postulated failures on long pipelines, such as are common between different plants on a petrochemical complex, also need to be included in the failure list as "specified cases".

5.3 Primary failure frequences

There is a statistical probability that the plant failures postulated will occur in a given period of time. The risk assessment procedure requires an estimate of this probability, in order that a frequency of occurrence can be assigned to each postulated failure case.

As with the identification of the failure cases, there

are different ways of deriving the primary failure frequencies. Two possible ways are mathematical computation by fault tree analysis, or extraction of relevant data from the historical record.

5.3.1 Fault tree analysis

This involves synthesis of a logic tree, starting at the top with the failure case, and working downwards through all the events and circumstances which can contribute to the occurrence of the failure, until each contributory event is described in terms of "base events" whose frequencies are known or can be adequately estimated. The base event frequencies are then added or multiplied together according to the logic gates, until a value for the frequency of the failure case has been derived.

While fault tree anlaysis is invaluable for the analysis of complex control systems or sequences, it is too lengthy and costly a procedure to be applied to a whole plant. It also has the serious drawback that major errors can be caused by omission of a branch of the tree. Nevertheless, its selective application can be very useful in helping to identify the weakest links in a system, and highlighting areas for improvement. It should also be noted that human factors, such as operator performance, can be taken into account in fault trees (albeit approximately), since data are available on human performance and failure rates under different stress levels.

5.3.2 Historical data

If data from historical experience are to be used, it is necessary to ensure that the records are based on a statistically meaningful population of plant items and associated failures. Alternatively, if there is insufficient experience with a particular type of item, a somewhat arbitrary failure frequency can be selected as a minimum performance criterion at the conceptual design stage, and the assessment can then proceed. It is obviously essential that the design team later demonstrate that this performance criterion can be met.

The assessment method used by the author for conceptual design studies is based predominantly on data from recent historical experience, with the application of modifying factors to take account of atypical circumstances or extreme environments as required. If available, failure frequency data which are more specific to the particular plant items can be substituted, but the basic frequency values used are as follows:

Item	Failure Frequency (per million years)
Process pressure vessels	3
Pressurized storage tanks	1
Refrigerated storage tanks	1
Connection pipework	
(diameter ⩽ 25 mm)	30
" 40 mm	10
" 50 mm	7.5
" 80 mm	5
" 100 mm	4
" ⩾ 150 mm	3

For process pipelines, the failure frequency values used vary from 0.9 per metre per million years for small diameter pipelines exposed to mechanical damage, to 0.01 per metre per million years for large diameter lines which are protected from most damage (buried lines).

These failure frequency data are applicable to well designed and constructed facilities, which are competently operated and maintained. A developer may sometimes decide to incorporate special high integrity features at the detailed design stage, and this will result in an associated decrease in the risk. Conversely, it may be decided to use less stringent criteria for non-critical plant items, and to concentrate resources on the higher hazard items.

It is important to realize that there are other probability factors involved in the assessment of the final outcome of a release of hazardous material, not just the probability of the release occurring. Thus, for example, wind direction effects and possible ignition sources can both influence the probability of damage, injury and fatalities at any particular point.

5.4 Discharge of hazardous material

The failures postulated above lead to discharge of the contents of the vessel or pipework, followed by further spillage as material flows from neighbouring parts of the plant equipment. In terms of the subsequent hazard impact, it is important to know how much material escapes, how rapidly, and in what form. Sudden and complete rupture of a vessel or tank will lead to an instantaneous release, but the released material may behave in three typical ways:

(i) Disperse as a gas cloud, without any rain-out of liquid;

(ii) Form a mixture of gas and liquid, some of which may collect on the ground before vaporizing further;

(iii) Form a pool of liquid, spreading within the limits of any retaining walls (such as tank bunds) as it vaporizes.

The first case involves instantaneous release into the vapour phase, but a time-varying vaporization rate must be calculated for the other two cases, before the dispersion model can be used to evaluate the cloud mass and the distance travelled until dilution to a safe concentration.

For discharges from broken pipework connections on a vessel or tank, the peak discharge rate is calculated according to the nature of the leaking material and the location of the failure. For pipework connections below the level of the liquid on the vessel, the discharge rate is calculated assuming single-phase liquid flow. The total pressure head is taken into account. For failures in the vapour space of pressurized equipment containing liquefied gases, the two-phase sonic discharge rate is calculated. For vessels containing only gas under pressure, a conventional sonic flow equation is used. Ultimately, a steady state situation may be reached in which the rate of release of material is equal to the rate at which the material is rendered innocuous by dispersion through its lower flammable limit or toxic threshold concentration. If the supply of material runs out before this state is reached, the cloud will not attain the maximum size, and the hazard is decreased. It is, therefore, necessary to take into account the facilities (actual or intended) for isolating the section of plant in which the failure has occurred, thereby reducing the amount of material spilled. The reliability of such facilities and their modes of activation are obviously important.

5.5 Dispersion of the released material

The immediate consequence of release of material from a failed section of plant is that the material will start to disperse into the surroundings, either actively under the control of its momentum or density, or passively under the control of the turbulence in the surrounding atmosphere. Some releases will result in immediate ignition, and the material will burn typically as a jet flame which poses a local danger to the workforce, but in most cases presents little threat outside the plant unless it causes further plant damage, and escalation of the incident.

Three main types of release may be identified, each of which has different dispersion characteristics:

(i) A continuous release of material from a pressurized source resulting in a momentum jet (typically from connection and pipe failures);

(ii) An instantaneous release of material following catastrophic pressure vessel failure;

(iii) A continuous release of cold vapour from the vaporization of a spill of refrigerated liquefied gas.

In the first type, the dispersion is dominated by the momentum of the released material, and a jet mixing model is used. In the latter two cases, a model for the dispersion of dense vapour clouds is applied, after allowance for some initial mixing in the case of catastrophic pressure vessel failure. The density of the cloud is a particularly important factor in determining the hazard impact, as dense clouds can disperse considerable distances along the ground before the material becomes diluted to safe concentrations. The earliest dispersion models used in hazard assessments were borrowed directly from work on the dispersion of trace pollutants in the atmosphere, and thus did not take any account of the density of concentrated clouds. Such models, therefore, failed to represent the important physical processes in a realistic way, and gave poor correlations with actual incidents or tests.

In the case of catastrophic pressure vessel failure, there is substantial initial mixing due to the violence of the release as the material flashes to atmospheric pressure. Air entrained in this process causes the cloud to cool as any liquid evaporates, and so increases the density of the cloud relative to the ambient air. The density effects are often sufficient to cause the expanded puff from the initial release to slump and spread under gravity, forming a pancake shaped cloud. For releases from refrigerated spills, a shallow broad plume is produced, which is advected downwind as a continuous, elongated plume. An appropriate dispersion model taking these density effects into account has been described by Cox and Carpenter (1979).

5.6 Impact and hazard distances

For release of flammable materials, the most important parameters are the distances to the lower flammable limit (LFL) and the amount of material in the combustible cloud. These are calculated using the dispersion models mentioned above.

Certain releases of flammable materials will disperse harmlessly, but many reach a source of ignition and then burn, with the possibility of causing explosion blast damage as well as fire damage. In the absence of specific information about the ignition sources on the plant, a reasonable estimate of vapour cloud ignition probabilities can be based on the following generalized data:

Distance to Lower Flammable Limit	Ignition Probability
≤ 100 m	0.05
101 - 300 m	0.30
≥ 300 m	0.95

These data have been extracted from several risk assessment studies in which the author has been involved, and in which detailed account was taken of individual ignition sources. The data are considered to be applicable to vapour clouds formed within petrochemical complexes. In other situations, notably those involving spillages from ships in coastal waters and rivers or estuarine ports, it is often essential to consider the actual ignition sources which will be present. The hazard impact in such situations is dominated by a few large-scale potential spillage incidents, and those which do not ignite at source or on the ship, may yield a vapour cloud which travels a considerable distance before the next potential ignition source is encountered.

The extent of a burning vapour cloud is calculated as the distance to the LFL, increased to allow for the expansion of the cloud as it burns. Unless an explosion occurs, the main damage and fatalities will occur within the area of the burning cloud. In the case of very large vapour clouds containing hundreds of tonnes of combustible materials, radiant heat effects at points beyond the cloud edges can become significant, but explosion blast damage will be the over-riding effect which may constrain the choice of site for the hazardous plant.

The physical processes involved in the explosive burning of unconfined vapour clouds are not fully understood, and a theoretical model has not yet been produced which can adequately predict the damage patterns from an unconfined vapour cloud explosion (UVCE). Various empirical models have been proposed, often using comparisons with TNT, but recently a particularly useful model has been developed by Dutch State Mines in The Netherlands and has been incorporated into a Dutch national hazard analysis manual prepared by their national research laboratory (TNO). This model can be expressed mathematically as:

$$R(s) = C(s) \times (kE_e)^{\frac{1}{3}}$$

relating the radius of the maximum extent of a defined damage level to a constant for that damage level, multiplied by the cube root of a fraction of the theoretical combustion energy of the material in the vapour cloud. The model is applicable for clouds containing between 100 kg and 100 tonnes of combustible vapour, but it is sometimes expedient to extrapolate the model with caution to even larger clouds.

For releases of toxic materials, the distances to toxic threshold criteria can be calculated. The analysis can be more complicated than is necessary for flammable clouds, as account needs to be taken of the duration of the exposure. The nature of toxic hazards is such that a person may survive a short exposure at high concentration, yet succumb if exposed to a lower concentration but for a longer time. To predict impact at any point, an appropriate function of the cloud concentration can be integrated with respect to time.

5.7 Risk contour plots

From the preceding steps in the analysis, the computer programme generates and prints out the failure list, giving hazard distances corresponding to selected damage levels and the associated probabilities for each case on the failure list. The location of the plant item giving rise to the hazard is known, and from this information is calculated, for each point on and around the site, the frequency with which a given damage level is exceeded at that point. This calculation is repeated for all points on a regular grid, and a contour map of damage frequency values is constructed using a standard contour plotting programme. This risk contour map shows in a directly useable way the zone of influence of any particular hazardous facility.

The calculations undertaken in producing the risk contour maps from the hazard distances and probability data are not difficult in principle, but if there are large numbers of plant items the volume of work mounts to the point where the computer method becomes essential.

The risk contour plots for the selected damage levels are overlaid on a site plan to illustrate the frequencies of damage at various points around the plant. At this stage it is necessary to assess the probability of domino damage, where the consequences of one failure cause another. If the risk of failure of an item due to domino damage is significant in comparison with the primary failure frequency of that item, the analysis should be repeated using a failure frequency which is the sum of the primary frequency and the domino failure frequency.

By superimposing the modified risk contour plots on a map of the area, the total impact on the local community and neighbouring industrial facilities can be illustrated, thereby aiding the decision-making process.

6. CASE STUDIES

The following two examples are drawn from environmental impact assessments undertaken at the conceptual design stage when application was being made by the developer for outline planning permission.

6.1 Seabed-supported NGL plant in the Orkney Islands

In July 1979, the Edinburgh-based consortium of Ben Line Steamers Ltd., Liquid Gas Equipment Ltd. and Edinburgh Financial and General Holdings Ltd. applied to Orkney Islands Council for outline planning consent for a seabed-supported facility in an inlet on the island of Flotta The application was later withdrawn, pending evaluation of other sites.

The proposed facility consisted of two very large crude oil carrying ships (VLCCs) which were to be converted to provide platforms for the installation of process plant, storage tanks and associated equipment. The facility was to be built in a shipyard, delivered to Flotta and ballasted to rest upon prepared seabed foundations. With the aim of minimizing the disturbance to the local environment, the facility was to be completed with its own living accommodation, power generation plant, workshops, control centre, fire and safety systems and ship loading and unloading berths. The facility was intended to fractionate 2 million tonnes of natural gas liquids (NGL) per year, with a total storage capacity of 50,000 m^3. At the end of its anticipated useful life of about 10 years, the facility was to be deballasted and removed, and the site restored to its original condition.

A preliminary hazard assessment was undertaken using the computerized risk contour plotting techniques described earlier (Cremer and Warner Scotland, 1979). NGL and its separated components are flammable and potentially explosive when mixed with air.

6.1.1 Community risk

The results showed that there was a risk of less than 1 in 1 million years (10^{-6} per year) of an explosion overpressure causing serious structural damage to 5% of typical houses exposed. This damage would cause injuries, with the possibility of deaths. The risk is, however, well within the public acceptability criterion discussed earlier.

6.1.2 Risk to neighbouring industry

The Occidental Oil Terminal on Flotta is at a distance of about 2 kilometres from the proposed facility. The direct risk to the Occidental workforce was demonstrated to be absolutely minimal, and the risk of an explosion overpressure of sufficient severity to cause domino damage to the Occidental storage tanks was less even than 0.1 in 1 million years. This domino risk is insignificant in relation to the assumed primary failure frequency of such storage tanks.

6.1.2 Domino and layout implications

Figure 2 presents the risk contour plot for heavy plant damage, for the part of the NGL facility which was to have the storage tanks, the offloading berth for liquefied gas carriers up to 6000 m^3 capacity, and the accommodation.

Most of the risk comes from the localized but relatively

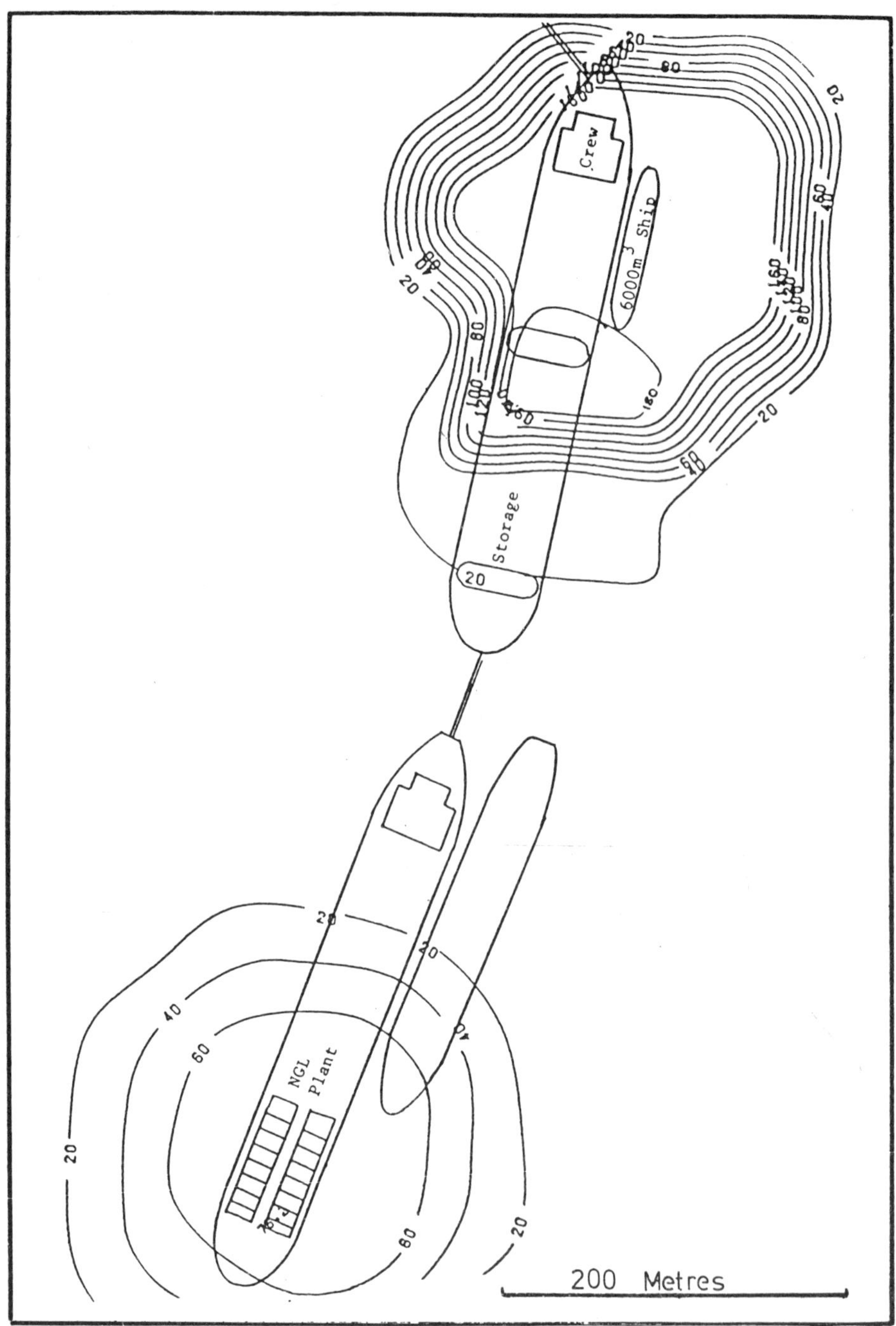

Figure 2: Heavy Plant Damage Risk (Per Million Years)

frequent incidents at the offloading berth, where the safe containment of the hazardous materials is inherently least reliable. The consequences of such incidents, however, seriously threaten the integrity of the storage tanks (with a domino risk of almost 200 in 1 million years, as opposed to a primary failure frequency of 3 times per million years for each of the pressurized storage tanks). The consequences also seriously threaten the safety of the plant personnel, whom it was proposed to accommodate in the aft superstructure of the VLCC.

It was, therefore, concluded that the layout and configuration of the ships and cargo loading facilities needed to be modified to decrease the risk. It was envisaged that the increased separation required between the loading and unloading berths and the other facilities, would involve construction of a more conventional jetty to service the facility. Provided that the layout could be modified such that domino interactions became negligible, it was concluded that the total hazard impact of the proposed development would be well within limits generally considered acceptable elsewhere, and the increase in risk faced by the local community and neighbouring industry would be negligible.

6.2 Gas reception terminal and SNG plant at St. Fergus

In May 1980, the British Gas Corporation (BGC) applied for outline planning permission for a coastal gas reception terminal at St. Fergus in north-east Scotland. This terminal is planned to receive gas from the UK Offshore Gas Gathering System at a peak rate of 85 million standard cubic metres of gas per day. This gas will contain about 40% weight of NGL, about 8 million tonnes per year.

The site chosen at St. Fergus is to the north of the existing BGC, Total Oil Marine Ltd. and Shell UK Ltd. terminals.

The function of the terminal will be to process the gas to meet the quality specifications for "natural gas" supplied to domestic and industrial consumers in the UK. To achieve this, it will be necessary to separate the NGL components of the gas delivered to shore (that is, ethane, propane, butane and gasoline) from the methane fraction which, together with some ethane to increase the calorific value, constitutes the "natural gas". Additionally, outline planning permission was sought for an adjacent plant to convert some of the NGL into substitute natural gas (SNG)

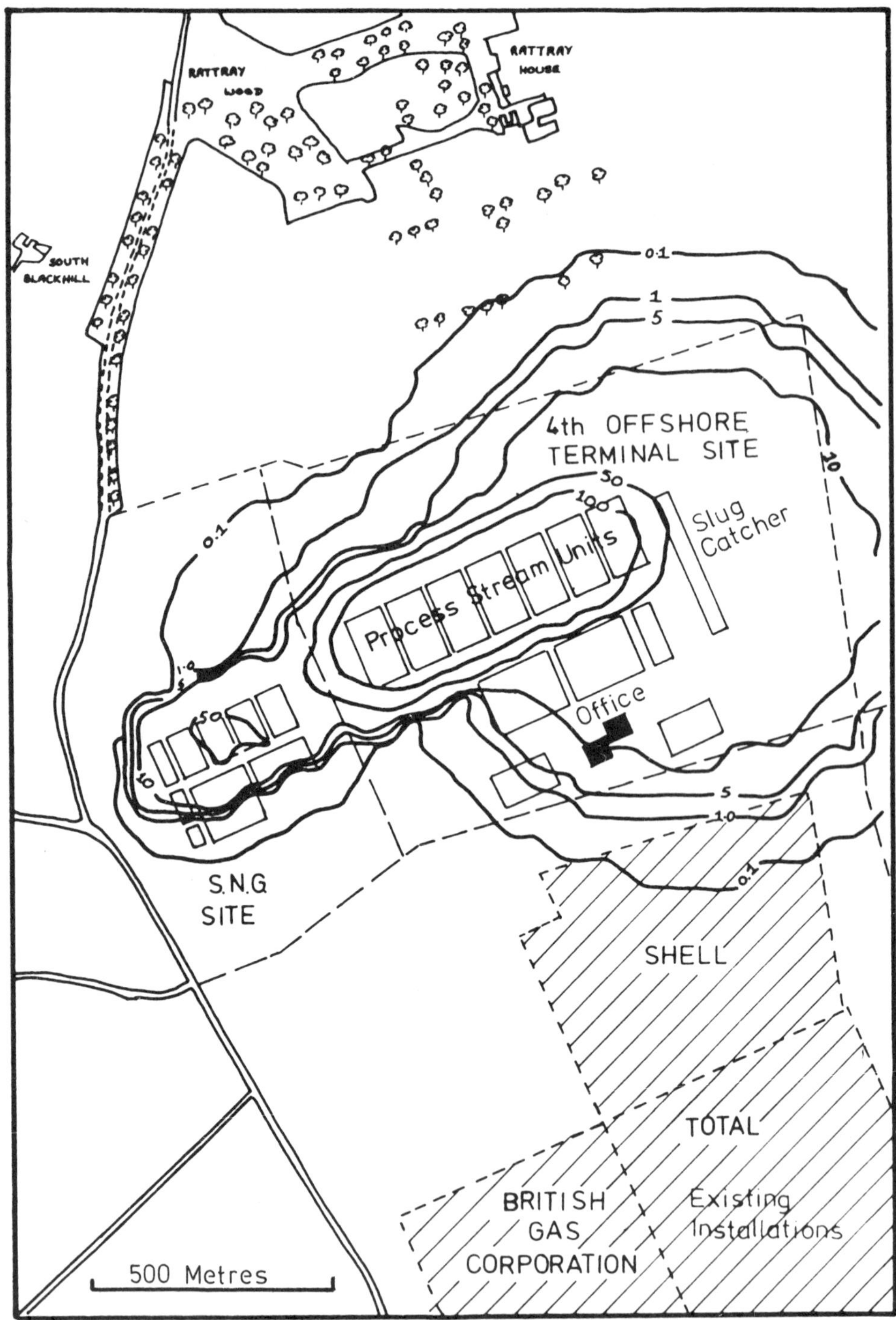

Figure 3: Heavy Plant Damage Risk (Per Million Years)

comprising methane together with a small amount of hydrogen. This plant would produce about 3.8 million standard cubic metres of SNG per day.

A conceptual hazard assessment was undertaken using the computerized risk contour plotting techniques described (Cremer and Warner Scotland, 1980). Approximately 900 postulated failure cases were taken into account. The impact from the proposed new terminal and SNG plant was first evaluated for each plant separately, and then in combination when it had been checked that the separation between the two plants was adequate to prevent inter-plant domino damage. The SNG plant was found to be of much lower potential hazard than the terminal. Risk contour plots for four damage levels are shown in the figures:

Figure 3 - Heavy Plant Damage Risk
Figure 4 - Repairable Building Damage Risk
Figure 5 - Burning Vapour Cloud Risk
Figure 6 - Risk of Window Breakage

6.2.1 Heavy plant damage risk

The higher contours are centred on the process modules, where there is the greatest risk of explosion. In contrast, the lower contours (which indicate potential damage at greater distances, but less frequently) are centred on the slugcatcher which acts as buffer storage for "slugs" of liquid which are swept through the incoming pipeline. The slugcatcher has a high inventory, and thus the potential to release large amounts of hydrocarbons which, if ignited, could cause damage beyond the plant boundaries.

The calculated risk to process facilities in the neighbouring Shell plant to the south is negligible, since the expected frequency of heavy plant damage is less than 1 in 10 million years (0.1 x 10^{-6} per year). It was, therefore, concluded that the separation of the high hazard facilities on the proposed new terminal and SNG plant from susceptible items on the neighbouring plants is sufficient to render negligible the possibility of inter-plant domino effects on process facilities.

An explosion overpressure which is sufficiently severe to cause heavy plant damage will also cause collapse of buildings, and typical mortality statistics indicate that about half of the people inside would be killed. The risk of this happening at the office building of the new gas reception terminal is shown in Figure 3 to be about 10 in 1

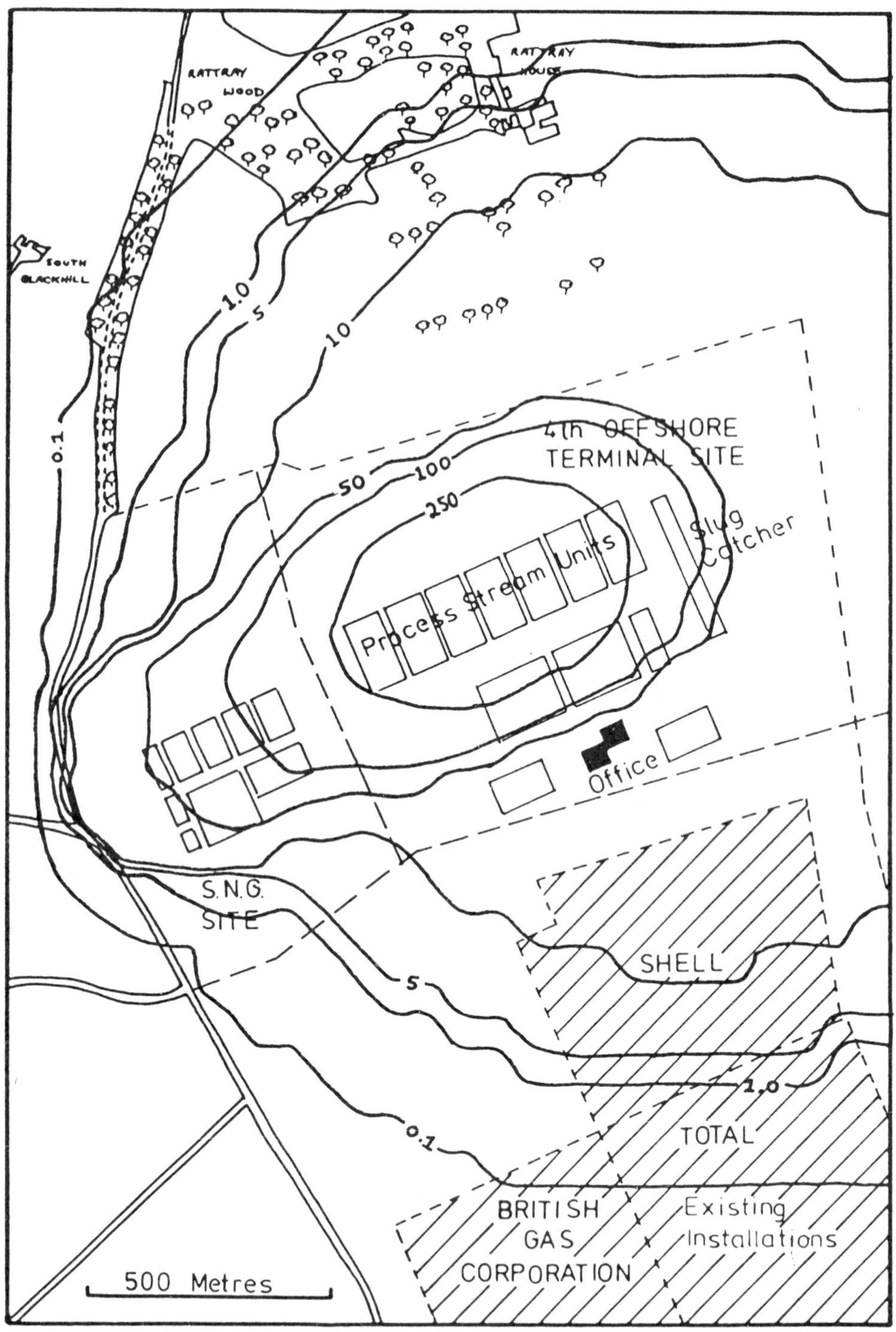

Figure 4: Repairable Building Damage Risk (Per Million Years)

million years. The associated risk of death to any occupant of the office building can be considered acceptable as it is low in comparison with the FAFR for related industries.

6.2.2 Repairable building damage risk

The risk contours shown in Figure 4 follow a similar pattern to those for heavy plant damage, since the same range of explosion incidents gives rise to both damage levels (the lesser damage occurring at greater distance). The contour values again indicate satisfactory spacing between the various plants.

The same overpressure level which will cause repairable building damage (approximately 0.1 bar overpressure) can also cause damage to atmospheric pressure storage tanks. The oil storage tanks on the south-east part of the Total terminal are at insignificantly low risk of such damage from the proposed new development (being outside even the 0.1 in 1 million years contour shown in Figure 4). Furthermore, the risk contours for such damage extend only very slightly outside the St. Fergus complex, to include only a very few isolated dwellings.

6.2.3 Burning vapour cloud risk

People caught in the open, inside a burning vapour cloud would be killed. In the case of very large burning clouds, the radiated heat beyond the cloud limits could also cause fatalities. The contours in Figure 5, however, indicate negligible risk to local inhabitants.

6.2.4 Risk of window breakage

The risk of this very low level of damage at the nearest inhabited building outside the terminal complex is less than 50 in 1 million years, and the risk at St. Fergus village to the south is less even than 1 in 1 million years (Figure 6).

On the basis of these contour plots, it was concluded that the siting and layout of the proposed new gas reception terminal and SNG plant is appropriate in terms of keeping the risk to the external community and neighbouring plants (and their workforces) very low.

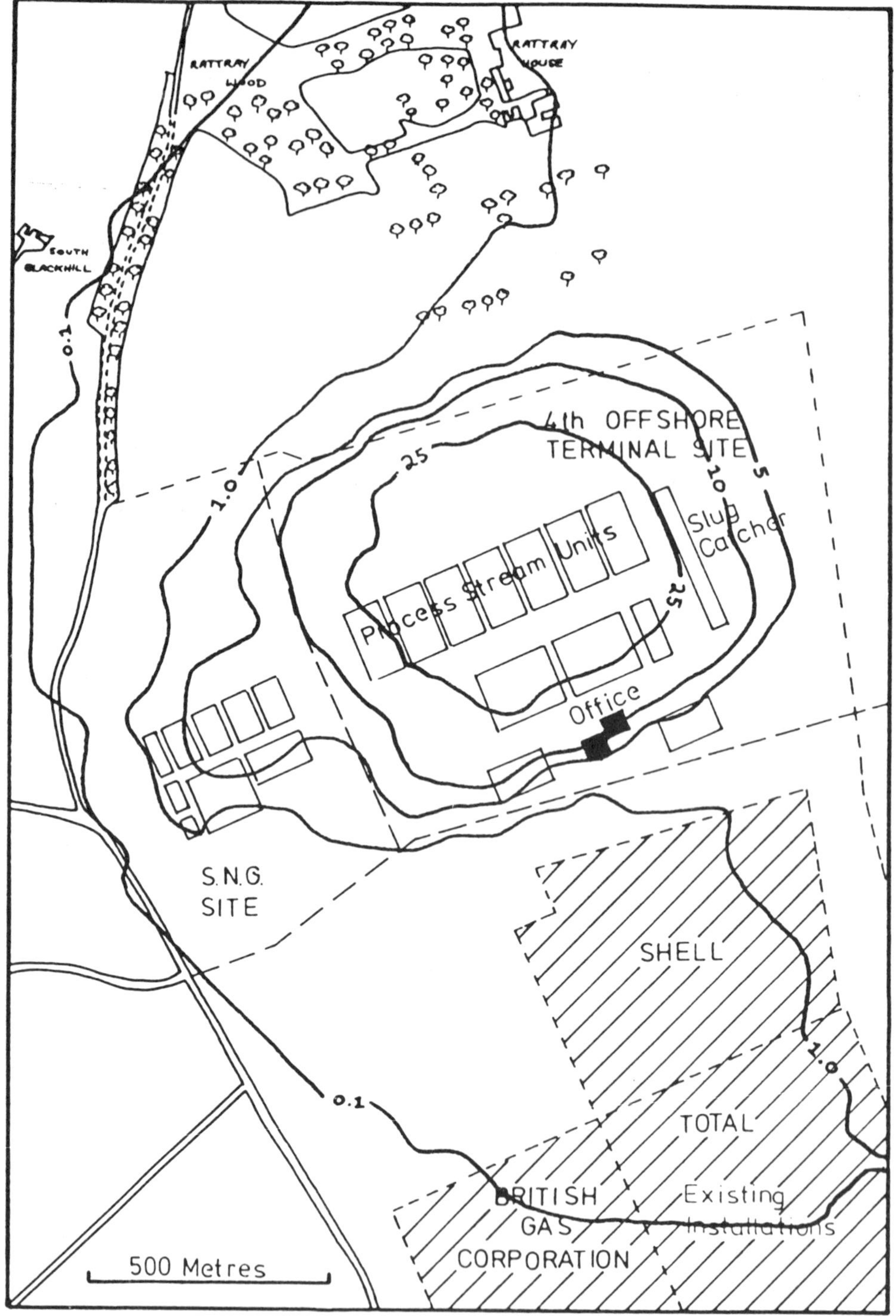

Figure 5: Burning Vapour Cloud Risk (Per Million Years)

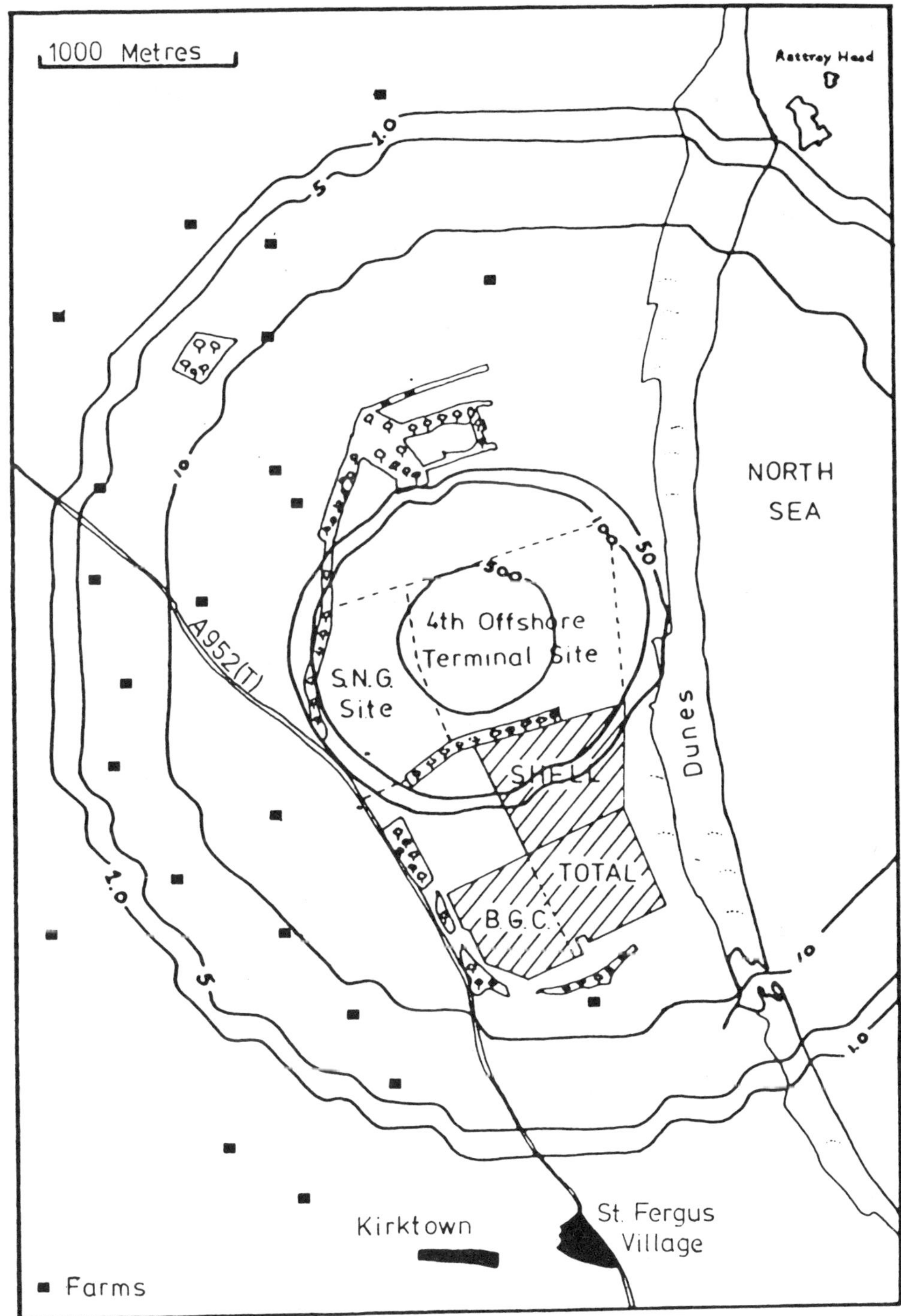

Figure 6: Risk of Window Breakage (Per Million Years)

ACKNOWLEDGEMENTS

The author is indebted to the Partners of Cremer and Warner for permission to present this paper. Acknowledgement is also made to W.J. Cairns and Partners and to British Gas Corporation, who commissioned the hazard impact assessments from which the above case studies are extracted.

REFERENCES

Chemical Industries Association: 1977, A Guide to Hazard and Operability Studies, Chemical Industries Association, London.

Comer, P.J., R.A. Cox, R. Sylvester-Evans: 1980, 'A Formalised Procedure for the Analysis of Risks Posed by Complex Petrochemical Developments', Bulletin of the Institute of Mathematics and its Applications 16, May/June, p. 121.

Cox, R.A. and R.J. Carpenter: 1979, 'Further Development of a Dense Vapour Cloud Dispersion Model for Hazard Analysis', Proceedings of Symposium "Schwere Gase", 3-4 September 1979, Battelle Institute, Frankfurt am Main.

Cremer and Warner: 1980, An Analysis of the Canvey Report Oyez Publishing Ltd., London, (ISBN 0 85120 465 1).

Cremer and Warner Scotland: 1979, 'Preliminary Hazard Assessment of Ship Mounted NGL Facility at Pan Hope, Flotta, Orkneys', in W.J. Cairns and Partners, 'Outline Environmental Appraisal' Edinburgh.

Cremer and Warner Scotland: 1980, Safety Assessment of a Proposed Gas Reception Terminal and Substitute Natural Gas Plant at St. Fergus, Scotland, a report commissioned by British Gas Corporation.

Health and Safety Commission: 1979, Second Report of the Advisory Committee on Major Hazards, HMSO, London (ISBN 0 11 883299 9).

Health and Safety Executive: 1978, Canvey: An Investigation of Potential Hazards from Operations in the Canvey Island/Thurrock Area, HMSO, London (ISBN o 11883200 X).

Slater, D.H.: 1979, 'Siting of Hazardous Plant', Health and Safety at Work, January 1979.

AFFILIATION

Dr Cameron Ramsay now of Technica, Consulting Scientists and Engineers, Aberdeen, was the Principal Scientist at Cremer and Warner Scotland at the time of writing.

PART C:

ASSESSMENT METHODS AND TECHNIQUES

Paul Tomlinson

THE USE OF METHODS IN SCREENING AND SCOPING

1. INTRODUCTION

To be effective an environmental impact assessment (EIA) procedure need only be applied to those actions which are considered to cause significant environmental consequences. It is, consequently, important to establish mechanisms for the selection of actions requiring EIA. Such a process of selection is termed "screening". The next stage in EIA is determining which issues should be examined in the EIA, this activity is often termed "scoping". In reality screening and scoping activities overlap, in that some methods not only provide a screening function, but also allow the identification of issues which require detailed examination. In addition, methods developed to assist the identification of potential impacts (eg impact matrices) may also be employed in impact assessment activities.

The object of this paper is to describe a number of possible approaches to screening and scoping activities and to discuss their strengths and weaknesses. It is not the objective of the paper to examine methods for the evaluation of alternatives or environmental impacts, nor to recommend any particular approach to screening and scoping.

2. SELECTION OF PROJECTS FOR EIA (SCREENING)

In some countries, as in the UK, no guidance exists to determine whether an EIA is required: a decision, in these cases, often being dependent upon the scale of the proposal, its environmental setting and the likely degree of public opposition. In other countries, various screening procedures exist and the differing requirements of each procedure has resulted in considerable disparities as to the number of EIAs undertaken in individual countries.

In this section, some general considerations relating to screening will be presented before attention is focused on a number of methods which may be used to determine when an EIA is required.

B. D. Clark et al. (eds.), Perspectives on Environmental Impact Assessment, 163–194.

FIGURE 1 Idealized Screening Procedure

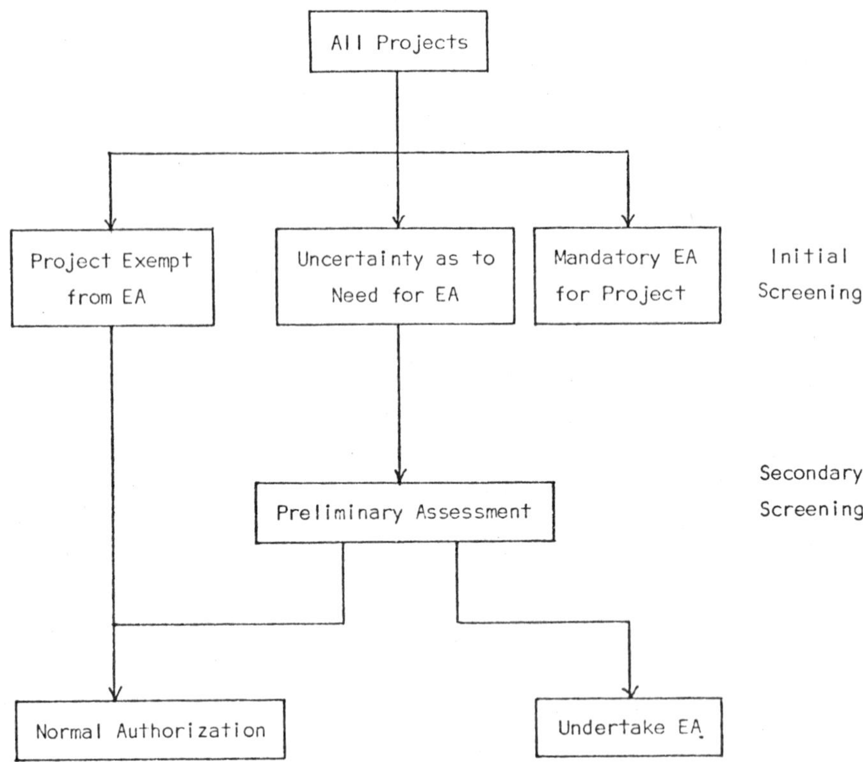

2.1. General considerations

During the development of screening procedures the following two objectives require consideration:
(i) clear identification of projects requiring an EIA;
(ii) quick and easy operation to avoid unnecessary delay.

Three project categories may be identified as a result of screening, namely:
(i) projects clearly requiring an EIA;
(ii) projects not requiring an EIA; and
(iii) projects for which the need for an EIA is unclear, that is, an intermediate category.

One solution to the uncertainty regarding the need for an EIA for "intermediate category" projects is to subject all such projects to an EIA even though it might be shown that no significant environmental effects were likely. Such an approach may prove costly in time and resources, as well as lead to criticism of the EIA procedures. An alternative is to employ a two stage screening process in which an initial screening can be used to quickly identify those projects which do and do not require an EIA. A secondary screening can then be used, subsequently, to determine whether EIAs are needed for the remaining projects (see figure 1).

2.2. Methods for screening

Five main methods are available to assist screening. They are:
(i) project thresholds;
(ii) sensitive area criteria;
(iii) positive and negative lists;
(iv) matrices;
(v) initial environmental evaluations.

Projects may be categorized by a number of parameters (see table 1), the values of which will be dependent upon the project. Particular values can then be selected to determine whether an EIA might be required. For example, projects with a capital cost greater than say £5 million could then be subject to an EIA. Project parameters are, by one means or another, incorporated into the five main screening methods.

2.2.1. Project thresholds

This method relies upon the establishment of thresholds for

Table 1: Key Project Parameters
(from: Battelle, 1978)

The Project	Project type Plant specification, equipment, layout, operation etc. Size of project - output volume - raw materials consumed - energy requirements - labour force - capital cost - physical size
Project Location	- by land use category industrial estate, residential, agricultural, etc. - by geographical criteria such as coast, mountain, etc. - by carrying capacity including air and water pollution dispersion characteristics.
Project Phasing	- site preparation - construction - operation - modification - induced development - decommissioning

Courtesy of Battelle and reproduced with permission from the Commission of the European Communities.

key features of the project or its environment. If a threshold were exceeded then an EIA would be required. Such thresholds can range from environmental factors such as the amount of agricultural land used for development, to project factors such as project size, cost or infrastructure demands. Table 2 illustrates those environmental consequences which may be related to project size.

Table 2 Environmental Impacts often Associated with Project Size

Probable Effects	- Visual intrusion - Direct impact on flora and fauna on the site and surroundings - Lost opportunity for alternative land uses
Possible Effects	- Potential pollution problems - Potential local employment impact - Potential local transport consequences - Influence on local land use patterns (induced development) - Changes in social values of local communities.

Project size may also be described in the following terms:

(i) surface area of the project;

(ii) area of land influenced by the project, eg by noise, zone of visual influence;

(iii) volume and height of the project.

To apply thresholds based simply on these project factors alone could lead to anomalies, since a 1000 m^2 project in one area may give rise to greater adverse effects than a 5000 m^2 project depending on its location. It is consequently of greater utility to create general guidance for such thresholds, for example:

(i) If the project land take is considered to be large in comparison to the scale or character of the surrounding environment then an EIA is required.

(ii) If the area affected is of more than local importance or if a large number of people outside the project area would be adversely affected then an EIA is required.

(iii) If the project is highly visible then an EIA is required.

Project cost or capital investment may be used as a threshold since it indicates that the project may encompass the following elements:

(i) have a large land area;
(ii) involve high value land (consequently often in densely populated or coastal areas);
(iii) have a large project size;
(iv) require advanced machinery or processing plant.

The application of financial criteria creates the situation in which projects may cost less than the financial threshold, yet still give rise to significant impacts. Alternatively, some projects exceeding the parameter may give rise to no significant impacts. It is, consequently, important to develop appropriate thresholds which achieve the following:-

(i) identify, at the lowest financial level, the greatest number of projects having significant impacts;
(ii) minimise the number of projects likely to be incorrectly identified as having significant impacts;
(iii) minimise the number of projects falling below the threshold which have significant environmental consequences.

Reliance upon one threshold may give rise to a number of incorrect decisions, it is, therefore, normal practice to link a series of different thresholds together. For example, in France a project cost of 60 million Francs, is linked to size thresholds, such as housing schemes over 3000 m^2 require an EIA. Other thresholds such as input-output parameters, for example raw material requirements or pollution generation, may be developed. Care will, however, be needed to prevent a series of thresholds becoming cumbersome and time consuming to use. Thresholds while, being simple, may need frequent revision particularly in the light of experience and inflation when financial criteria are applied.

2.2.2. Sensitive areas

Since the environmental consequences of a project is a function of both the project and the receiving environment, the sensitivity of that environment provides a means for determining whether an EIA is required. Environmentally sensitive areas (ESAs) may be determined in two distinct, but complementary ways. One approach is to determine the carrying capacity of the area in relation to the degree or intensity of interference or disturbance. This approach implies a pre-determined series of values which describe the resilience of the environment in relation to defined perturbations. For example, the concentration of pollutants which can be discharged without giving rise to adverse effects. A major problem with this approach is the relationship of the various generators of pollution to the defined assimilative capacity or limit which the environment can accept. Proposed developments giving rise to pollutant discharges below the carrying capacity of the environment need not require an EIA. Yet, when other polluting activities are proposed which cause the cumulative discharges to exceed the carrying capacity, only the latter development would require an EIA, even though its discharge may be of a lower amount than the former development. This approach requires considerable amounts of information concerning the environment and, given mans' limited understanding of many environments, criteria indicating a carrying capacity in relation to a particular pollutant or development activity is likely to be subject to error, and hence may give rise to controversy over its application or relevance.

An alternative approach to the identification of ESAs is to determine the importance of individual components of the area. In this approach, the characteristics of the environment, in terms of its objective and subjective values, rather than purely its ability to withstand perturbations, are given emphasis. Battelle (1978) produced, for the Commission of the European Communities, a classification scheme by which ESAs could be identified (see figure 2). In order to identify specific zones to be designated as an ESA, Battelle suggested that the following characteristics should be considered:

(i) Quality of the zone in terms of its own value, for example, recreational use, and in relation to other zones, for example by providing seasonal feeding grounds for wildlife from another locality.

(ii) The abundance of the particular zone. A zone may be valued due to its uniqueness or relative scarcity due to competing demands for its use, for example a forested zone may be valued due to multiple land uses of recreation, wildlife or silviculture.

(iii) Sensitivity to change in terms of the ability of the zone to withstand or spread the consequences of human activity, for example transfer of pollutants to another zone.

Figure 2: List of Potential Critical Areas (from Battelle, 1978)

- POTENTIAL CRITICAL AREAS
  - NON-PRODUCTIVE RESOURCE AREAS
    - NATURAL RESOURCE AREAS
      - WATER RELATED
        - COASTAL
        - INLAND
      - LAND RELATED
    - BUILT RESOURCE AREAS
  - PRODUCTIVE RESOURCE AREAS
    - NATURAL RESOURCE AREAS
      - WATER RELATED
        - COASTAL
        - INLAND
      - LAND RELATED
        - FOREST
        - AGRICULTURE
        - MINING QUARRIES
    - BUILT RESOURCE AREAS
  - HAZARDOUS AREAS

COASTAL	INLAND	LAND RELATED	BUILT RESOURCE AREAS	COASTAL	INLAND	FOREST	AGRICULTURE	MINING QUARRIES	BUILT RESOURCE AREAS	HAZARDOUS AREAS
COASTLINE	LAKES	SIGNIFICANT ECOLOGICAL COMMUNITIES	ARCHEOLOGICAL HISTORICAL SITES	OFF-SHORE	LAKES	FOREST	CROPS	MINERAL DEPOSITS	HOUSING	NOT DEVELOPED
BEACHES	RIVERS	WILDLIFE HABITATS	VILLAGES	CONTINENTAL PLATEAU	RIVERS	BUSHES	GRAZING	SAND + GRAVEL DEPOSITS	COMMERCIAL	
ESTUARIES	SHORELINES	SIGNIFICANT RECREATIONAL AESTHETIC OPPORTUNITIES	RUINS	COASTLINES	STREAMS	SAVANNA		QUARRIES	INDUSTRIAL	
MARSHES	MARSHES		INDIVIDUAL GROUP OF BUILDINGS	ESTUARIES	AQUIFER SHEETS				AGRICULTURAL	
	STREAMS									
	WATERFALLS									

Courtesy of Battelle, and reproduced with permission from the Commission of the European Communities.

Eagles (1981) also presented a series of criteria and steps to aid the identification of ESAs (see tables 3 and 4). Regardless of the approach adopted, projects located in ESAs could be required to undertake an EIA.

The use of ESAs in a screening exercise has the advantage of being simple and easy to use, however, if it were applied alone, it does have the disadvantage of ignoring project characteristics. This may give rise to EIAs being performed on activities having no significant environmental consequences. It is, therefore, useful to balance project criteria with sensitive area criteria, for example, a project threshold of x mg/m^3 heavy metal concentration in the effluent within a fishery resource area would cause an EIA to be required, while such a concentration may be acceptable in less sensitive areas.

2.2.3. Positive and negative lists

The positive list approach can be illustrated by the proposed directive of the Commission of the European Communities on EIA (Commission of the European Communities, 1980). In this document two lists of projects have been proposed. The first list (table 5) details those projects which would be subject to EIAs while the second list (table 6) contains those projects which shall be made subject to an EIA "where the Member States consider that their characteristics so require". Member States may, consequently, specify certain types of projects as being subject to an EIA, or may establish the criteria and/or thresholds necessary to determine those projects which are to be subject either to an EIA, another form of assessment or are to be exempt from any assessment.

Positive lists may be compiled by a review of existing developments, identifying those giving rise to significant environmental damage. Some indication of the environmental importance of categories of projects may also be gained by examining the existing licensing/consent requirements. Those project categories requiring many consents may be worthy of closer examination to determine whether they should be entered on a positive list. Similarly, projects which seldom give rise to adverse environmental consequences, for example agricultural buildings, can be identified and entered on a negative list, that is a list of projects not requiring an EIA. In the case of those projects in which classification is difficult, an intermediate status can be

Table 3 Criteria to Aid Identification of Environmentally Sensitive Areas

1. Distinctive and unusual land forms;
2. Importance of the ecological function of the areas to the maintenance of a natural system beyond its boundaries;
3. Unusual or high quality plant and/or animal communities;
4. Unusual habitat with a rarity value;
5. Unusual high diversity of biological communities due to a variety of geomorphological features etc.;
6. Provision of a habitat for rare or endangered species;
7. Large area providing a habitat for species that require such extensive areas;
8. Area location in combination with natural features providing a resource in scientific research or education terms;
9. Aesthetic value of the locality.

Table 4 Stages in the Identification of Environmentally Sensitive Areas

1. Assemble interdisciplinary study team;
2. Adopt standardized criteria;
3. Locate sources of information eg government studies, local experts;
4. Utilise sensitive area criteria to identify possible sensitive areas;
5. Field validation of sensitive areas;
6. Compile background files on important elements within the sensitive area eg endangered species, hydrographic regime etc.;
7. Final screening to identify sensitive areas;
8. Delineation of sensitive areas on topographic maps;
9. Detail reasons for criteria fulfillment of each delineated area;
10. Publish the results.

(Adapted from Eagles, 1981)

TABLE 5 Proposed List of Projects to be Subject to Mandatory EIA

1. Extractive industry
Extraction and briquetting of solid fuels
Extraction of bituminous shale
Extraction of ores containing fissionable and fertile material
Extraction and preparation of metalliferous ores

2. Energy industry
Coke ovens
Petroleum refining
Production and processing of fissionable materials
Generation of electricity from nuclear energy
Coal gasification plants
Disposal facilities for radioactive waste

3. Production and preliminary processing of metals
Iron and steel industry, excluding integrated coke ovens
Cold rolling of steel
Production and primary processing of non-ferrous metals and ferro-alloys

4. Manufacturing of non-metallic mineral products
Manufacture of cement
Manufacture of asbestos-cement products
Manufacture of blue asbestos

5. Chemical industry
Petrochemical complexes for the production of olefins, olefin derivatives, bulk monomers and polymers
Chemical complexes for the production of organic intermediates
Complexes for the production of basic inorganic chemicals

6. Metal manufacture
Foundries
Forging
Treatment and coating of metals
Manufacture of aeroplane and helicopter engines

7. Food industry
Slaughter-houses
Manufacture and refining of sugar
Manufacture of starch and starch products

8. Processing of rubber
Factories for the primary production of rubber
Manufacture of rubber tyres

9. Building and civil engineering
Construction of motorways
Intercity railways, including high speed tracks
Airports
Commercial harbours
Construction of waterways for inland navigation
Permanent motor and motorcycle racing tracks
Installation of surface pipelines for long distance transport

(FROM: Commission of the European Communities, 1980)

TABLE 6 Proposed List of Projects Which May be Subject to EIA

1. Agriculture
   Projects of land reform
   Projects for cultivating natural areas and abandoned land
   Water management projects for agriculture (drainage, irrigation)
   Intensive livestock rearing units
   Major changes in management plans for important forest areas

2. Extractive industry
   Extraction of petroleum
   Extraction and purifying of natural gas
   Other deep drillings
   Extraction of minerals other than metalliferous and energy-producing minerals

3. Energy industry
   Research plants for the production of fissionable and fertile material
   Production and distribution of electricity, gas, steam, and hot water (except the production of electricity from nuclear energy)
   Storage of natural gas

4. Production and preliminary processing of metals
   Manufacture of steel tubes
   Drawing and cold forging of steel

5. Manufacture of glass fibres, glass wool and silicate wool

6. Chemical industry
   Production and treatment of intermediate products and fine chemicals
   Productions of pesticides and pharmaceutical products, paint and varnishes, elastomers and peroxides
   Storage facilities for petroleum, petrochemical and chemical products

7. Metal manufacture
   Stamping, pressing
   Secondary transformation treatment and coating of metals
   Boilermaking, manufacture of reservoirs, tanks and other sheet-metal containers
   Manufacture and assemby of motor vehicles (including road tractors) and manufacture of motor vehicle engines
   Manufacture of other means of transport

8. Food industry
   Manufacture of vegetable and animal oils and fats
   Processing and conserving of meat
   Manufacture of dairy products
   Brewing and malting
   Fish-meal and fish-oil factories

9. Textile, leather, wood, paper industry
   Wool washing and degreasing factories
   Tanning and dressing factories
   Manufacture of veneer and plywood
   Manufacture of fibre board and of particle board
   Manufacture of pulp, papers and board
   Cellulose mills

10. Building and civil engineering
    Major projects for industrial estates
    Major urban projects
    Major tourist installations
    Construction of roads, harbours, airfields
    River draining and flood relief works

TABLE 6 continued

Hydroelectric and irrigation dams
Impounding reservoirs
Installations for the diposal of industrial and domestic waste
Storage of scrap iron

11. Modifications to development project included in Annex I

(FROM: Commission of the European Communities, 1980)

Figure 3: Level 1 Matrix
(from: FEARO, 1978)

LEVEL 1 MATRIX

ACTIVITIES IN VARIOUS STAGES OF PROJECT DEVELOPMENT (APPENDIX 3) / AREAS OF POTENTIAL ENVIRONMENTAL EFFECTS (APPENDIX 4)

No.	Activity	IDENTIFICATION OF ACTIVITIES	Physical/Chemical Effects: Water	Physical/Chemical Effects: Noise	Physical/Chemical Effects: Land	Physical/Chemical Effects: Atmosphere	Physical/Chemical Effects	Physical/Chemical Effects	Ecological Effects: Species and Populations	Ecological Effects: Habitats and Communities	Ecological Effects	Ecological Effects	Ecological Effects	Aesthetic Effects: Land	Aesthetic Effects: Atmosphere	Aesthetic Effects: Water	Aesthetic Effects: Flora and Fauna	Aesthetic Effects: Man-made Objects	Aesthetic Effects: Compositions	Aesthetic Effects	Social Effects	Social Effects	Social Effects
	SITE INVESTIGATION AND PREPARATION																						
1.1	Access roads																						
1.2	Site surveying																						
1.3	Soil testing																						
1.4	Hydrological testing																						
1.5	Environmental survey																						
1.6	Site clearing	X			X				X	X				X			X		X		X		
1.7	Burning																						
1.8	Excavation																						
1.9	Drainage alteration																						
1.10	Stream crossing																						
1.11	Equipment																						
1.12	Pest control																						
1.13	Utilities																						
1.14	Waste disposal and recovery																						
1.15	Product storage																						
	CONSTRUCTION																						
2.1	Access roads																						
2.2	Site clearing																						
2.3	Excavation	X	X		X				X	X				X		X	X		X		X		
2.4	Blasting and drilling																						
2.5	Demolition																						
2.6	Building relocation																						
2.7	Cut and fill																						
2.8	Tunnels underground structures																						
2.9	Erosion control	X			X				X	X				X			X		X				
2.10	Drainage alteration	X	X						X	X						X			X		X		
2.11	Stream crossing																						
2.12	Channel dredging and straightening																						
2.13	Channel revetments																						
2.14	Dams and impoundments																						
2.15	Piers seawalls																						
2.16	Offshore structures																						
2.17	Equipment	X		X					X						X		X		X		X		
2.18	Pest control																						
2.19	Utilities																						
2.20	Labour force	X																	X		X		
2.21	Waste disposal and recovery																						
2.22	Product storage																						
2.23	Abandonment																						
2.24	Reclamation																						
2.25	Reforestation																						
2.26	Fertilization																						
2.27	Ancillary transmission lines and pipelines																						
	OPERATION AND MAINTENANCE																						
3.1	Forest clearing																						
3.2	Excavation																						
3.3	Spoil and overburden																						
3.4	Blasting and drilling																						
3.5	Dredging																						
3.6	Equipment operation																						
3.7	Operational failures	X	X						X	X	X				X	X	X				X	X	
3.8	Energy requirements																						
3.9	Energy generation																						
3.10	Automobile, aircraft, vessel movement																						
3.11	Pedestrian movement																						
3.12	Utilities																						
3.13	Waste disposal and recovery	X	X						X	X	X				X	X	X		X		X	X	
3.14	Product storage																						
3.15	Spills and leaks																						
3.16	Explosions																						
3.17	De-icing snow removal and disposal																						
3.18	Pest control																						
3.19	Dust control																						
3.20	Abandonment																						
	FUTURE AND RELATED ACTIVITIES																						
4.1	Urbanization																						
4.2	Industrial development																						
4.3	Transportation																						
4.4	Energy requirements																						

created in which other supporting screening methods can be applied, such as thresholds or ESAs. Alternatively, all such projects for which uncertainty exists could be subjected to an EIA, although an exemption to an EIA could be granted by the appropriate authority on a case-by-case basis. In Ontario, all plans, programmes, projects of activities are subject to an environmental assessment unless specifically exempted by the Minister of the Environment.

Lists are one of the simplest approaches to screening. Some research is required to prepare such lists, but they offer an 'easy-to-use' system which is readily understood by all concerned. The preparation of lists, involves one main problem area; individual projects of the same general type of class may have considerable variations in size, plant, process and layout which may give rise to varying environmental consequences. In addition, gaining the acceptance of all parties on the entry of individual project types on particular lists (mandatory positive, intermediate and negative) may be difficult and time consuming.

2.2.4. Matrices

In Canada, the matrix has been promoted by the Federal Environmental Assessment and Review Office (FEARO) as a screening method to overcome the weaknesses of both the project and environment based screening methods, while avoiding the need for extensive studies to determine the need for an EIA. Two levels of matrices are used, the Level 1 Matrix (see figure 3) provides a broad screening, while the Level 2 Matrix focuses upon specific environmental impacts (see figure 4). During the Level 1 Matrix stage, project activities are identified which are likely to occur during the four principal development phases, namely:

(i) site investigation and preparation;
(ii) construction;
(iii) operation and maintenance;
(iv) future and related activities.

Four main areas of potential environmental consequences are identified namely; physical/chemical, ecological, aesthetic and social aspects. Each of the interactions (potential impacts) between the "activities" and "effects" are then marked in the appropriate cell. For example, excavation during construction activities causes an interaction with habitats and communities (see figure 3).

Potential impacts identified in the Level 1 Matrix are

Figure 4: Section of a Level 2 Matrix (from: FEARO, 1978)

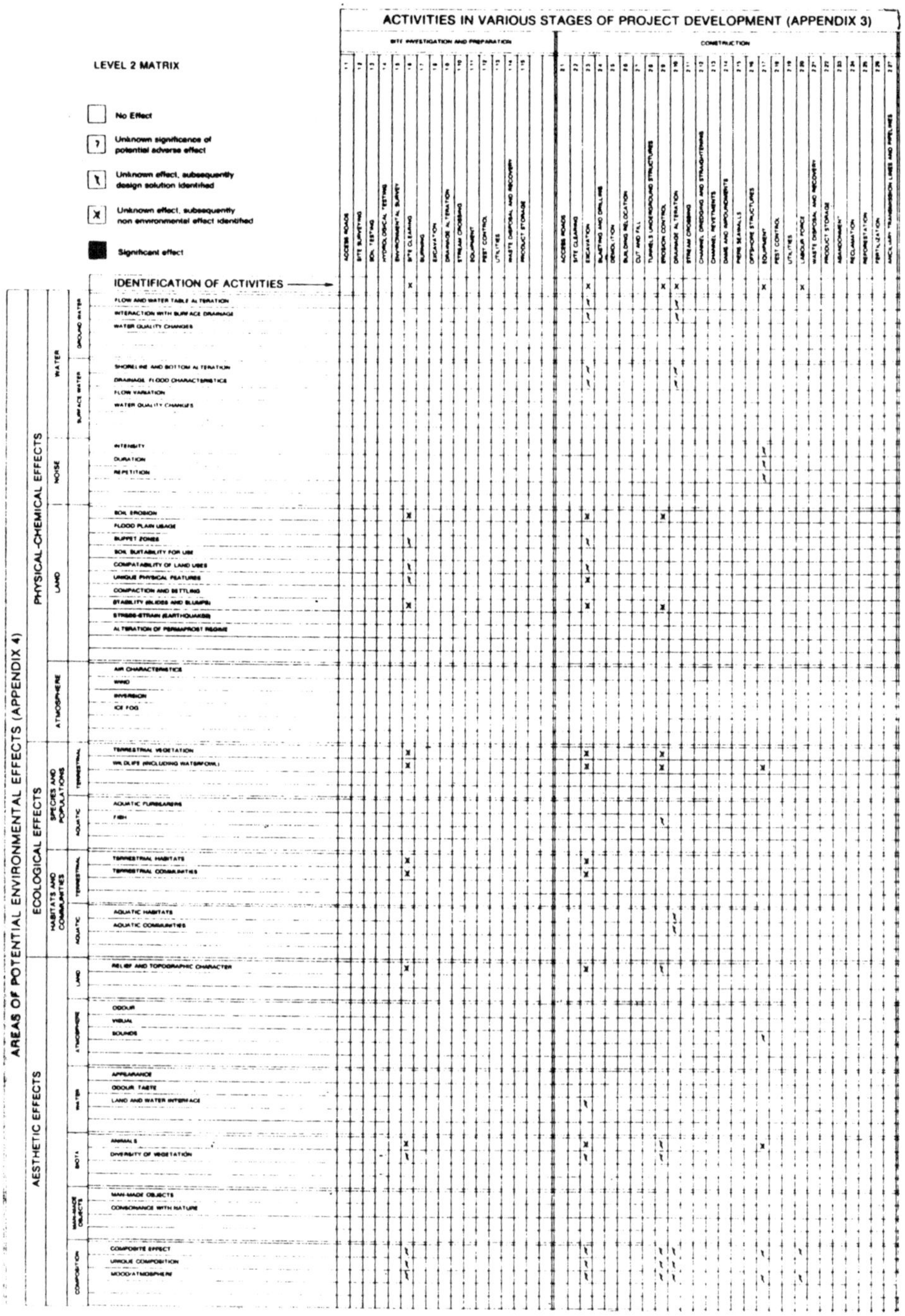

LEVEL 2 MATRIX

- [ ] No Effect
- [?] Unknown significance of potential adverse effect
- [ɿ] Unknown effect, subsequently design solution identified
- [x] Unknown effect, subsequently non environmental effect identified
- [■] Significant effect

ACTIVITIES IN VARIOUS STAGES OF PROJECT DEVELOPMENT (APPENDIX 3)

AREAS OF POTENTIAL ENVIRONMENTAL EFFECTS (APPENDIX 4)

SITE INVESTIGATION AND PREPARATION

Areas of potential environmental effects		1.1 ACCESS ROADS	1.2 SITE SURVEYING	1.3 SOIL TESTING	1.4 HYDROLOGICAL TESTING	1.5 ENVIRONMENTAL SURVEY	1.6 SITE CLEARING	1.7 BURNING	1.8 EXCAVATION	1.9 DRAINAGE ALTERATION	1.10 STREAM CROSSING	1.11 EQUIPMENT	1.12 PEST CONTROL	1.13 UTILITIES	1.14 WASTE DISPOSAL AND RECOVERY	1.15 PRODUCT STORAGE
	IDENTIFICATION OF ACTIVITIES →						x									
PHYSICAL-CHEMICAL EFFECTS – WATER – GROUND WATER	FLOW AND WATER TABLE ALTERATION															
	INTERACTION WITH SURFACE DRAINAGE															
	WATER QUALITY CHANGES															
PHYSICAL-CHEMICAL EFFECTS – WATER – SURFACE WATER	SHORELINE AND BOTTOM ALTERATION															
	DRAINAGE, FLOOD CHARACTERISTICS															
	FLOW VARIATION															
	WATER QUALITY CHANGES															
PHYSICAL-CHEMICAL EFFECTS – NOISE	INTENSITY															
	DURATION															
	REPETITION															
PHYSICAL-CHEMICAL EFFECTS – LAND	SOIL EROSION						x									
	FLOOD PLAIN USAGE															
	BUFFER ZONES						ɿ									
	SOIL SUITABILITY FOR USE															
	COMPATABILITY OF LAND USES						ɿ									
	UNIQUE PHYSICAL FEATURES						ɿ									
	COMPACTION AND SETTLING															
	STABILITY (SLIDES AND SLUMPS)						x									
	STRESS-STRAIN (EARTHQUAKES)															
	ALTERATION OF PERMAFROST REGIME															
PHYSICAL-CHEMICAL EFFECTS – ATMOSPHERE	AIR CHARACTERISTICS															
	WIND															
	INVERSION															
	ICE FOG															
ECOLOGICAL EFFECTS – SPECIES AND POPULATIONS – TERRESTRIAL	TERRESTRIAL VEGETATION						x									
	WILDLIFE (INCLUDING WATERFOWL)						x									
ECOLOGICAL EFFECTS – SPECIES AND POPULATIONS – AQUATIC	AQUATIC FURBEARERS															
	FISH															
ECOLOGICAL EFFECTS – HABITATS AND COMMUNITIES – TERRESTRIAL	TERRESTRIAL HABITATS						x									
	TERRESTRIAL COMMUNITIES						x									
ECOLOGICAL EFFECTS – HABITATS AND COMMUNITIES – AQUATIC	AQUATIC HABITATS															
	AQUATIC COMMUNITIES															
AESTHETIC EFFECTS – LAND	RELIEF AND TOPOGRAPHIC CHARACTER						x									
AESTHETIC EFFECTS – ATMOSPHERE	ODOUR															
	VISUAL															
	SOUNDS															
AESTHETIC EFFECTS – WATER	APPEARANCE															
	ODOUR, TASTE															
	LAND AND WATER INTERFACE															
AESTHETIC EFFECTS – BIOTA	ANIMALS						x									
	DIVERSITY OF VEGETATION						ɿ									
AESTHETIC EFFECTS – MAN-MADE OBJECTS	MAN-MADE OBJECTS															
	CONSONANCE WITH NATURE															
AESTHETIC EFFECTS – COMPOSITION	COMPOSITE EFFECT						ɿ									
	UNIQUE COMPOSITION						ɿ									
	MOOD/ATMOSPHERE						ɿ									

CONSTRUCTION

Areas of potential environmental effects		2.1 ACCESS ROADS	2.2 SITE CLEARING	2.3 EXCAVATION	2.4 BLASTING AND DRILLING	2.5 DEMOLITION	2.6 BUILDING RELOCATION	2.7 CUT AND FILL	2.8 TUNNELS UNDERGROUND STRUCTURES	2.9 EROSION CONTROL	2.10 DRAINAGE ALTERATION	2.11 STREAM CROSSING	2.12 CHANNEL DREDGING AND STRAIGHTENING	2.13 CHANNEL REVETMENTS	2.14 DAMS AND IMPOUNDMENTS	2.15 PIERS SEAWALLS	2.16 OFFSHORE STRUCTURES	2.17 EQUIPMENT	2.18 PEST CONTROL	2.19 UTILITIES	2.20 LABOUR FORCE	2.21 WASTE DISPOSAL AND RECOVERY	2.22 PRODUCT STORAGE	2.23 ABANDONMENT	2.24 RECLAMATION	2.25 REFORESTATION	2.26 FERTILIZATION	2.27 ANCILLARY TRANSMISSION LINES AND PIPELINES
	IDENTIFICATION OF ACTIVITIES →			x						x	x							x			x							
PHYSICAL-CHEMICAL EFFECTS – WATER – GROUND WATER	FLOW AND WATER TABLE ALTERATION			ɿ							ɿ																	
	INTERACTION WITH SURFACE DRAINAGE			ɿ							ɿ																	
	WATER QUALITY CHANGES																											
PHYSICAL-CHEMICAL EFFECTS – WATER – SURFACE WATER	SHORELINE AND BOTTOM ALTERATION			ɿ							ɿ																	
	DRAINAGE, FLOOD CHARACTERISTICS			ɿ							ɿ																	
	FLOW VARIATION																											
	WATER QUALITY CHANGES																											
PHYSICAL-CHEMICAL EFFECTS – NOISE	INTENSITY																	ɿ										
	DURATION																	ɿ										
	REPETITION																	ɿ										
PHYSICAL-CHEMICAL EFFECTS – LAND	SOIL EROSION			x						x																		
	FLOOD PLAIN USAGE																											
	BUFFER ZONES			ɿ																								
	SOIL SUITABILITY FOR USE																											
	COMPATABILITY OF LAND USES			ɿ																								
	UNIQUE PHYSICAL FEATURES			x																								
	COMPACTION AND SETTLING																											
	STABILITY (SLIDES AND SLUMPS)			x						x																		
	STRESS-STRAIN (EARTHQUAKES)																											
	ALTERATION OF PERMAFROST REGIME																											
PHYSICAL-CHEMICAL EFFECTS – ATMOSPHERE	AIR CHARACTERISTICS																											
	WIND																											
	INVERSION																											
	ICE FOG																											
ECOLOGICAL EFFECTS – SPECIES AND POPULATIONS – TERRESTRIAL	TERRESTRIAL VEGETATION			x						x																		
	WILDLIFE (INCLUDING WATERFOWL)			x						x								x										
ECOLOGICAL EFFECTS – SPECIES AND POPULATIONS – AQUATIC	AQUATIC FURBEARERS																											
	FISH									ɿ																		
ECOLOGICAL EFFECTS – HABITATS AND COMMUNITIES – TERRESTRIAL	TERRESTRIAL HABITATS			x																								
	TERRESTRIAL COMMUNITIES			x																								
ECOLOGICAL EFFECTS – HABITATS AND COMMUNITIES – AQUATIC	AQUATIC HABITATS										ɿ																	
	AQUATIC COMMUNITIES										ɿ																	
AESTHETIC EFFECTS – LAND	RELIEF AND TOPOGRAPHIC CHARACTER			x						ɿ																		
AESTHETIC EFFECTS – ATMOSPHERE	ODOUR																											
	VISUAL																											
	SOUNDS																	ɿ										
AESTHETIC EFFECTS – WATER	APPEARANCE																											
	ODOUR, TASTE																											
	LAND AND WATER INTERFACE			ɿ																								
AESTHETIC EFFECTS – BIOTA	ANIMALS			x						ɿ								x										
	DIVERSITY OF VEGETATION			ɿ						ɿ																		
AESTHETIC EFFECTS – MAN-MADE OBJECTS	MAN-MADE OBJECTS																											
	CONSONANCE WITH NATURE																											
AESTHETIC EFFECTS – COMPOSITION	COMPOSITE EFFECT			ɿ						ɿ	ɿ							ɿ			ɿ							
	UNIQUE COMPOSITION			ɿ						ɿ	ɿ																	
	MOOD/ATMOSPHERE			ɿ						ɿ	ɿ							ɿ			ɿ							

Reproduced by permission of The Minister of Supply and Services Canada.

then further defined and subdivided in Level 2 (see figure 4). This allows the identification of those activities which will have no effect, those for which an environmental design solution or mitigating measure can be identified, those having unknown and potential adverse effects and those having significant effects. For example, the effects of excavation upon terrestrial habitats may be unknown, so a questionmark is entered in that cell. On completion of this stage within the Level 2 Matrix and if significant effects are indicated, then a panel of experts is convened to review the project and coordinate an EIA. Generally, further information is needed to reduce the number of "unknown and potential adverse effect" screening decisions made. To assist this process, background information presenting a series of 'relevant factors' is supplied for both project activities and environmental effects. Table 7 presents an illustrative list of "relevant factors".

Table 7 <u>List of Relevant Factors for Excavation</u>

1. Extent and depth of excavation.
2. Character of underlying soil (sensitive soils, permafrost etc).
3. Proximity to noise sensitive areas.
4. Disruption of traffic patterns, services, etc.
5. Requirements for water table modification.
6. Surface water drainage.
7. Topography.
8. Quantity and type of spoil.
9. Nature of soil and susceptibility to water and wind erosion.
10. Adjacent land use.
11. Aesthetics.

(Adapted from FEARO, 1978).

In addition to the list of "relevant factors", criteria to assist in the determination of adverse effects and indicate the type of information required were developed (see table 8).

On the basis of these further investigations the cells marked to indicate "unknown and potential adverse effect" are modified to record either the identification of design solutions or the realization that significant effects will occur. If data were not available to undertake such

Table 8 Criteria for Making Screening Decisions

Several general criteria can be used when making a decision as to the environmental effect of an activity. These criteria are not mutually exclusive but are interrelated.

Magnitude: defined as the probable severity of each potential impact. Will the impact be irreversible? If reversible, what will be the rate of recovery or adaptability of an impact area? Will the activity preclude the use of the impact area for other purposes?

Prevalence: defined as the extent to which the impact may eventually extend as in the cumulative effects of a number of stream crossings. Each one taken separately might represent a localized impact of small importance and magnitude, but a number of such crossings could result in a widespread effect. Coupled with the determination of cumulative effects is the remoteness of an effect from the source activity. The deterioration of fish production resulting from access roads could affect sport fishing in an area many miles away and for months or years after project completion.

Duration and Frequency: the significance of duration and frequency can be explained as follows. Will the activity be long-term or short-term? If the activity is intermittent, will it allow for recovery during inactive periods?

Risks: defined as the probability of serious environmental effects. The accuracy of assessing risk is dependent upon the knowledge and understanding of the activities and the potential impact areas.

Importance: defined as the value that is attached to a specific area in its present state. For example, a local community may value a short stretch of beach for bathing or a small marsh for hunting. Alternatively, the impact area may be of a regional, provincial or even national importance.

Mitigation: Are solutions to problems available?

(Adapted from FEARO, 1978)

re-classification, then an Initial Environmental Evaluation would be required. For example, to resolve the unknown effect of excavation upon terrestrial habitat, an initial environmental evaluation would be undertaken to examine in more detail the potential effects. If at any stage, however, a decision had been made to refer the project to a Panel, then an initial environmental evaluation would not be required, as the Panel would examine all relevant issues.

2.2.5. Initial Environmental Evaluation (IEE)

The United Nations Environment Programme (1980) has developed a set of guidelines for the assessment of industrial activities, in which guidance is given on how to undertake an IEE. Figure 5 indicates the various steps of a general assessment procedure within which an IEE is incorporated. The initial impact identification activities commence in step 4, in which the relationships between the project and the environment are examined for possible interactions. An IEE is then conducted in step 5. This involves the application of tests in order to determine those elements (for example, sensitive areas) and sub-elements (for example, coastal zones) of the environment which may be subject to important impacts (see table 9). In this questionnaire checklist, the likely effects of projects are considered by answering a series of questions related to impact types. Answers to these questions provide a simplified view of the consequences of development and consequently allow a decision to be made as to whether an EIA is required.

In the US a decision to prepare an EIA is made on the basis of whether the proposed "major action significantly affecting the quality of the human environment". This is determined by the Federal agencies which have specific criteria for identification of those typical categories of action which:

(i) normally require an environmental impact statement (EIS);

(ii) normally do not require either an EIS or an environmental assessment (IEE);

(iii) normally require IEEs but not necessarily EISs.

If an action were not to need an EIS, then the agency is required to involve environmental agencies, project proponents and the public, to the extent practicable, in preparing the environmental assessment which is the US term for an IEE. An environmental assessment will show whether an EIA

Figure 5: The UNEP Guide to Assessing Proposed Developments

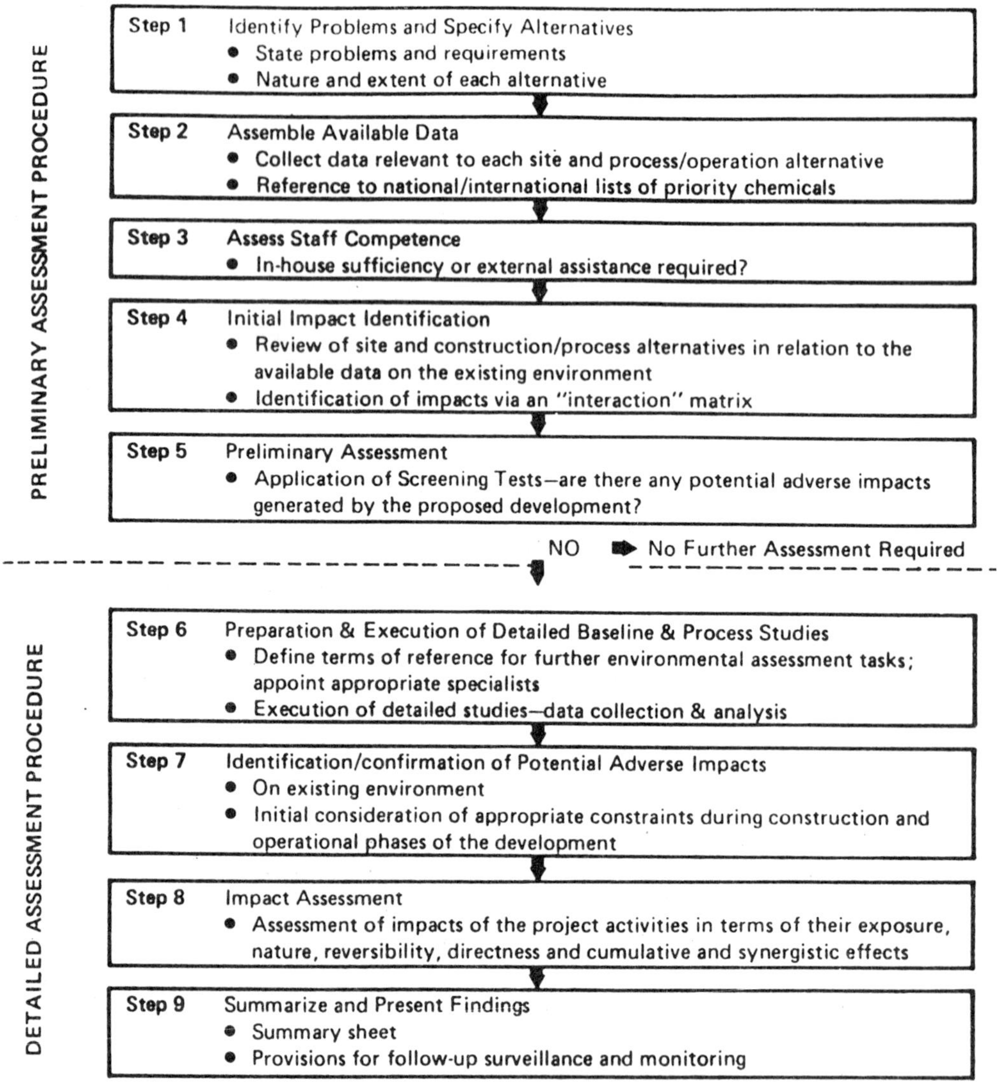

PRELIMINARY ASSESSMENT PROCEDURE

Step 1 Identify Problems and Specify Alternatives
- State problems and requirements
- Nature and extent of each alternative

Step 2 Assemble Available Data
- Collect data relevant to each site and process/operation alternative
- Reference to national/international lists of priority chemicals

Step 3 Assess Staff Competence
- In-house sufficiency or external assistance required?

Step 4 Initial Impact Identification
- Review of site and construction/process alternatives in relation to the available data on the existing environment
- Identification of impacts via an "interaction" matrix

Step 5 Preliminary Assessment
- Application of Screening Tests—are there any potential adverse impacts generated by the proposed development?

NO ➡ No Further Assessment Required

DETAILED ASSESSMENT PROCEDURE

Step 6 Preparation & Execution of Detailed Baseline & Process Studies
- Define terms of reference for further environmental assessment tasks; appoint appropriate specialists
- Execution of detailed studies—data collection & analysis

Step 7 Identification/confirmation of Potential Adverse Impacts
- On existing environment
- Initial consideration of appropriate constraints during construction and operational phases of the development

Step 8 Impact Assessment
- Assessment of impacts of the project activities in terms of their exposure, nature, reversibility, directness and cumulative and synergistic effects

Step 9 Summarize and Present Findings
- Summary sheet
- Provisions for follow-up surveillance and monitoring

TABLE 9 Screening Table for Water Impacts

Subelement	Potential Impact(s)	Required Information	Sources of Information
Hydrological balance	Will the project alter the hydrological balance?	Extent of project; source of water - ground or other. Importance of groundwater in maintaining area rivers, streams, lakes, ponds, wells, flora and fauna.	Developer Hydrologist/ hydrogeologist
Groundwater regime	Will the project affect the groundwater regime, e.g. in terms of quality, quantity, depth/gradient of water table and direction of flow? Will alterations to water table depth alter structural qualities of soil? Will dewatering methods be necessary to undertake excavations?	Extent of project; source of water supply; waste disposal practices; proposed surface cover. Ground conditions - permeability, percolation, water table, location of recharge area, slope proximity to streams or other waterbodies.	Developer Geological maps/ survey; local well-drillers; soils engineer.
Drainage/channel pattern	Will the project impede the natural drainage pattern and/or induce alteration of channel form?	Existence, nature and pattern of drainage; soil characteristics.	Site visit; geological maps.
Sedimentation	Will the project induce a major sediment influx into area water bodies?	Location of construction and cleaning activities. Erosion potential of site soils. Direction of runoff flow, % slope on site. Erosion and sediment control plan for site.	Developer Soils surveyor Site visit; topographical map Developer
Flooding	Will there be risk to life and materials due to flooding?	Extent of project; 100-year flood plain.	Developer Geological survey
Water quality	Does potable water supply meet established standards - WHO, etc? Will receiving waters meet established standards? Will waters be adequately accommodated and treated? Will groundwater suffer contamination by surface seepage, intrusion of saline or polluted water?	Whether existing water quality meets standards for intended usage; capacity of waste treatment plant/sewerage system to accommodate project wastes. Water disposal plan; source of water. Location of groundwater recharge.	Health/waste disposal authorities Developer Hydrologist/ hydrogeologist
Surface waters	Will the project impair existing surface waters through filling, dredging, water extraction or discharge; waste discharge or other detrimental practices? Will recreation or aesthetic values be endangered? Will the project affect dry weather flow characteristics?	Location of project; location of construction and clearing activities. Source of water supply and site of waste disposal; dams/obstructions; flow characteristics over an extended period. Ecological characteristics; recreation uses.	Developer Civil engineer/ hydrologist Aquatic biologist - area survey

(FROM: UNEP © 1980)

Table 10 Main Considerations in the IEE Guidelines for Mineral Projects

1.	Overview Summary	- describe the project, probable major environmental impacts, mitigating measures on significance of unmitigated environmental impacts and information deficencies.
2.	Project Rationale	- the need, alternatives within the project and associated projects.
3.	Project Proposal	- development concept, geological description of mineral, mining method, processing, waste disposal, supporting services etc.
4.	Description of Existing Environment	- baseline information of the environment, climate, terrain, hydrology, biological aspects, human considerations, existing uses.
5.	Environmental Impacts	- identification and comparison of expected environmental impacts, mitigating measures and beneficial effects.
6.	Major Impacts and Mitigating Measures	- analysis of major impacts, mitigating measures and plans for surveillance and monitoring.
7.	Residual Impacts	- unmitigated impacts should be discussed.

(Adapted from EAP, 1976)

is required or not. If no EIS is to be prepared, then a formal document based on the environmental assessment is required to establish an administrative record. This document, called a Finding of No Significant Impact (FONSI), is then available for public review for 30 days before the agency makes a final decision. FONSI's shall include a brief discussion of the needs for the proposal, of alternatives, of the environmental impacts of both the action and its alternatives, and a list of agencies and persons consulted. Environmental assessments, as a result, not only determine the need for an EIS, but also give rise to improvements in those projects which do not eventually require an EIS.

As indicated in the previous section, IEEs are used in the Canadian Environment Assessment and Review Process. The purpose of these IEEs, is to determine whether unknown potentially adverse environmental effects are significant or whether mitigating measures can be adopted to prevent adverse effects occurring. The Canadians have developed a series of guidelines to assist the preparation of IEEs for a number of project types, eg mining developments (see table 10). IEEs require deeper analysis than other screening methods and consequently more time and resources. This means that IEEs ought only to be applied to those projects in which considerable uncertainty exists as to the need for an EIA. The advantages of the IEE approach is that allows for some public input and may also result in improvements in project design for those not subject to an EIA.

2.3. Overview on Screening Methods

As will have become apparent, each of the methods has different strengths and weaknesses, consequently the combined use of such methods can overcome individual weaknesses.

In Malaysia, the EIA screening activities utilise positive and negative lists, sensitive area criteria and a screening questionnaire. An examination is first made of the positive list in order to identify those projects requiring an EIA. Then the negative list is examined. Projects on this list are further screened to ensure that they are not situated in an environmentally sensitive area. If projects were located in a sensitive area then an EIA would be required. Those projects not on either list are then subject to the questionnaire. This questionnaire comprises six main areas of inquiry, namely:

(i) siting of project;
(ii) resource demand;
(iii) waste production;
(iv) labour requirements;
(v) infrastructure needs;
(vi) regulations, guidelines and codes of practice.

Within each subject area the answers to a detailed series of questions then indicates the need for an EIA.

Associated with the different screening methods are varying degrees of simplicity and ease of operation. The application of a rigorous screening method to all projects could lead to delays to the system, consequently a gradual increase in complexity, in accordance with the difficulty in determining whether an EIA is needed for an individual project, may prove a beneficial approach.

3. SCOPING

Scoping is the term given to the process of developing and selecting alternatives to a proposed action and identifying the issues to be considered in an EIA. Essentially it is a procedure designed to establish the terms of reference for an EIA. Its aims are:

(i) to identify concerns and issues regarding consideration;
(ii) to facilitate an efficient EIS preparation process;
(iii) to enable those responsible for EIA to properly brief the study team on the alternatives and impacts to be considered at different depths of analysis;
(iv) to provide an opportunity for public involvement;
(v) to save time.

Scoping is not a discrete exercise as it may continue well into the planning and design phase depending on whether new issues arise for consideration. Equally, some of the activities described in the screening section of this paper may serve the same function as scoping. For example, an IEE or matrix approach results in the identification of those concerns and issues requiring examination.

Before discussing specific aspects of scoping, a number of procedural aspects require consideration, these include:

(i) Responsibility for scoping?
(ii) How is the scoping of an EIA defined in terms of the identification and selection of alternatives and the determination of significant issues?

(iii) How are interested parties and the public to be involved in scoping?

3.1. Responsibility for scoping

Responsibility for scoping may rest with the authorizing agency, in the case where guidance is given to the project proponent on what issues should be examined - the terms of reference (TOR) for the EIA. In other cases scoping may be implemented by the proponent. In situations where there exists limited exchange of information and concerns between the proponent, relevant agencies and the public, there is the possibility that significant issues may be overlooked. This may then give rise to the rejection of the EIS or delay to the project while omitted issues are investigated.

3.2 Identification and selection of alternatives

The consideration of alternatives to the proposed project is one of the key aspects of EIA, in that it provides a means by which project assumptions, goals and needs can be examined. Consideration of alternatives provides for the examination of different mechanisms to achieve a stated objective and assists the decision makers in the choice of an alternative which has the least adverse and greatest beneficial environmental social and economic consequences. While this concept may be readily accepted, problems invariably occur when consideration is given to the following aspects:

(i) What is a reasonable range of alternatives to be considered either in terms of number or variety?

(ii) How should alternatives be identified?

(iii) What level of examination should be applied to each alternative?

3.2.1. Range of alternatives

Ensuring that a range of alternatives are examined is important, since failure to do so may result in the EIS being challenged on the basis of viable alternatives being omitted. This issue is often addressed by requiring that only "practical" rather than all "feasible" alternatives are considered. The U.S. Fish and Wildlife Service (1980) provide the following advice:

> 'All reasonable alternatives should be at least initially considered. However, only practical

Figure 6: Overlay Mapping
(from: Leonard and Partners, 1977)

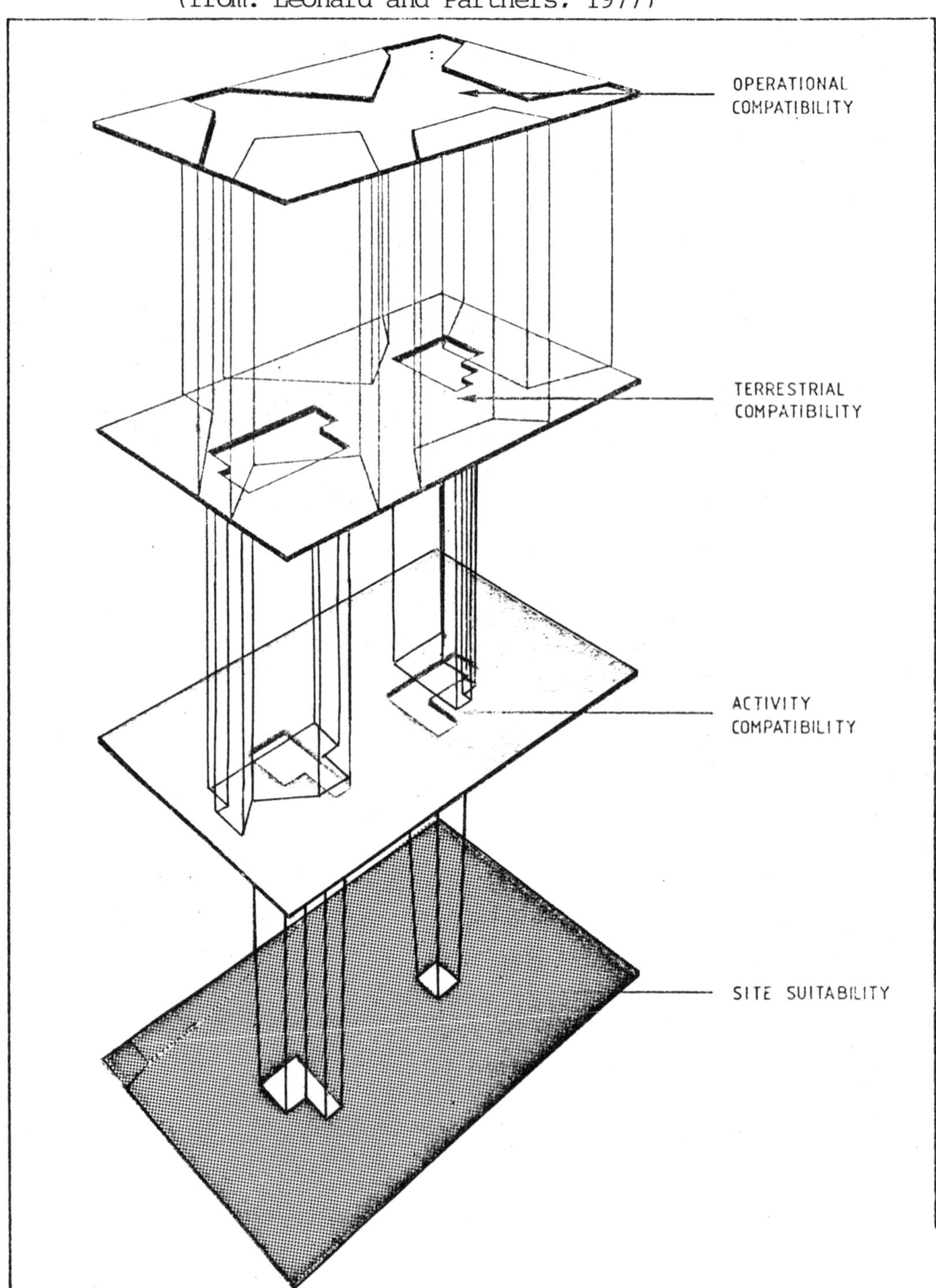

Courtesy of Leonard and Partners.

> alternatives, including the no action and proposed action need be discussed in detail in the EIA.'

The Californian Environmental Quality Act (1972) requires that:

> 'Alternatives should be capable of accomplishing the objectives in a successful manner, within a reasonable period of time, taking into account economic, social, environmental and technological factors'.

A report by Environmental Resources Ltd (ERL, 1981) recommended to the Dutch Ministry of Health and Environmental Protection that the following categories of alternatives should be given some consideration.

(i) demand alternatives;
(ii) activity alternatives;
(iii) location alternatives;
(iv) process alternatives;
(v) scheduling alternatives;
(vi) input alternatives eg raw materials;
(vii) residual control alternatives eg mitigating measures.

Such alternatives may then be assessed in relation to their cost, feasibility and conflict with prior or future policies or decisions.

3.2.2. Identification of alternatives

Few methods exist to assist the identification of alternatives, those which do, principally relate to the identification of alternative sites or routes. The overlay method, originally developed for a planning study of the town of Billerica, Massachusetts (Manning, 1913), has become established as a method for transportation alternative route selection, and has also been used to identify alternative sites. At the simplest level the key restraining factors of a development, such as, engineering factors, the existence of areas of ecological or landscape value are mapped on a transparent overlay sheet. Alternative sites or routes are then identified by placing one map on top of another. Clear areas then represent potential alternatives (see figure 6).

At this simple level, difficulties are encountered in dealing with more than about a dozen overlays. In addition, it is difficult to accommodate the concept of varying degrees of restraint, although to some degree, colour or tone intensity may be used. It is with the application of computerized systems that the greatest opportunities lie in relation to the identification of alternative sites or routes. The computer, using a common data set, representing the local environment under investigation can allow different restraining factors to be weighted in accordance with their importance, degrees of restraint can be easily accommodated and a final aggregate map can be produced as a numerical or shade intensity output. The computer is able not only to determine an environmental index for each alternative route, but also identify new alternative routes, which result in least environmental impacts.

Two main problems arise with this approach. First, access to data and a computer, although increasing availability of micro-computers will reduce the latter constraint. The second problem relates to the quality of the data in terms of boundary definition and heterogenity within the grid units used to co-ordinate the data. For example, boundary definition in biological communities is generally difficult since gradients rather than distinct boundaries exist. Equally, a grid unit may be classified as one type of vegetation, yet contain other important types at lower densities.

3.3. Determination of significant issues

Two important considerations arise when considering the determination of significant issues, namely:

(i) the range of potential significant issues; and,

(ii) the identification of significant issues.

The legislative framework in which EIA operates may constrain the range of issues which may or may not be included in an EIA, for example social and economic issues may, in some countries be excluded, while in others certain aspects, such as impacts on health, may be mandatory considerations.

Scoping addresses itself to the identification of significant issues by the careful consideration of existing information relevant to the assessment, as well as the organized involvement of other agencies and consultations with the public. Beanlands and Duinker (1983) have identified two categories of scoping. Social scoping is considered to provide the social value concerns of the public and these can be gathered by recognised methods of communication such as public meetings and structured questionnaires. Social scoping can be considered as the establishment of the terms of reference in which impacts should be considered.

Ecological scoping, on the other hand established the terms under which the impacts can be effectively studied, or need to be studied. Beanlands and Duinker suggest that the ecological scope of an EIA can be approached by focusing on four questions:

(i) Is there reason to believe that the valued ecosystem components will be effected either directly or indirectly by the project?
(ii) Is it realistic to attempt to study the effects on the valued ecosystem components directly?
(iii) How can the effects on valued ecosystem components be studied indirectly?
(iv) Is it necessary or helpful to use indicators of impact?

Through the application of these two scoping activities significant issued may be determined. It is important, however, to remember that the term impact attaches a value to remember that the term impact attaches a value to change either positive or negative and thus related to social scoping. Change, on the other hand has no intrinsic value and it is the role of social scoping to determine the issues of social importance, while it is the role of ecological scoping to determine which changes may be predicted or measured.

As suggested, there is no method for identifying significant issues, but rather an assemblage of interactions and discussions between the public, various agencies and the project proponent.

3.4. Public involvement in scoping

Each country evolves its own approach to scoping depending upon the abilities and willingness of all parties to become involved in the exercise. Many papers have been written on the wider subject of public involvement in EIA and the mechanisms which may be adopted for example Bishop (1975), Grima (1977) and Wood (1978). Given the extensive literature concerning public involvement which exists, this paper will only describe one approach, that of the Canadian Federal Environmental Assessment Panels. The Panels, which comprise four to six experts, are established to examine the environmental and related implications of a particular project. The Panel is responsible for issuing guidelines for preparing an EIS and reviewing the completed EIS. The Panel is responsible for the scoping activity in that it identifies the issues which must be considered in the EIS. To aid this task, Panels will often request public views by written comment, by workshops or public meetings, before completing the preparation of the guidelines (FEARO, 1980). These public meetings are structured to allow:

(i) the Panel to receive the views and questions of the local people;

(ii) the project proponents have an opportunity to reply to questions and comments;

(iii) the agencies having an interest in the project can present their views concerning the project;

(iv) contentious issues can be discussed in order that the Panel may take account of all views on the subject.

Great care is required in the mechanisms adopted for seeking public involvement and it is important that the process is reinforced by feedback to the public of the results of their expressions of concern. In the Canadian context this is achieved by publishing the EIS guidelines, the EIS is published for comment and if necessary a statement of any deficiencies in the EIS is also made public.

3.5. Overview on scoping

Both screening and scoping activities are important stages in EIAs, although their exact boundaries are often blurred. A variety of approaches are available each with various strengths and weaknesses. It is, therefore, important that appropriate methods, perhaps in combination, relevant to the needs of individual countries are adopted.

While approaches to screening are reasonably well documented, scoping does not have any specific methods, and hence only a few documents focusing upon this subject exist. Those which do tend to report the advantages which can be gained from scoping. In an efficient scoping procedure, the unnecessary expenditure of time and financial resources on irrelevant issues are minimized, essentially streamlining the EIA. With the aid of involvement of the public, acceptable terms of reference for the EIA can be developed thereby reducing the likelihood of a major controversy once an EIS has been prepared. The scoping activities should also assist in the coordination of action from the various agencies involved in the assessment of a proposed project. This is, however, the theory, it is equally possible that early public involvement and inter-agency politics could cause delays to the EIA rather than minimize them. It is impossible to suggest which view may be correct, since little evidence exists for either at present. As scoping relies to a great extent upon the exchange of information and concerns between the interested parties, including the public, an appropriate organization is essential if difficulties are to be avoided.

REFERENCES

Battelle: 1978, The Selection of Projects for Environmental Impact Statements., Commission of the European Communities, Environment and Consumer Protection Service, Brussels.

Beanlands, G.E. and P.N. Duinker: 1983, An Ecological Framework for Environmental Impact Assessment in Canada, Institute for Resource and Environmental Studies, Dalhousie University, Halifax, Canada.

Bishop, A.B.: 1975, 'Public Participation in Environmental Impact Assessment', in M. Blisset (ed.), Environmental Impact Assessment, Lyndon B. Johnson School of Public Affairs, University of Texas at Austin, pp. 219-236.

Commission of the European Communities: 1980, Draft Directive Concerning the Assessment of the Environmental Effect of Certain Public and Private Projects, 7973/80: /COM(80313 Final), Commission of the European Communities, Brussels.

Eagle, P.F.J.: 1981, 'Environmentally Sensitive Area Planning in Ontario, Canada', APA Journal July 1981, 313-323.

EAP: 1976, Guidelines for Preparing Initial Environmental Evaluations, Environmental Assessment Panel, Ottawa.

ERL: 1981, Environmental Impact Assessment Studies on Methodologies, Scoping and Guidelines, Final Report Volume 2 Scoping and Guidelines, Ministry of Health and Environmental Protection, Leidschendam, The Netherlands.

FEARO: 1978, Guide for Environmental Screening, Federal Activities Branch, Environmental Protection Service and Federal Environmental Assessment Review Office, Ottawa.

FEARO: 1980, Environmental Assessment Panels - What they are - what they do, Federal Environmental Assessment Review Office, Ottawa.

Grima, A.P.: 1977, 'The Role of Public Participation in the Environmental Impact Process' in M. Plewes and J.B.R. Whitney (eds.), Environmental Impact Assessment in Canada Processes and Approaches. Environmental Studies, University of Toronto, Toronto, pp. 61-76.

Leonard and Partners: 1977, Belvoir Prospect, Surface Works Report Volume 2 Plans, Leonard and Partners, Croydon.
Manning, W.: 1913, 'The Billercia Town Plan', Landscape Architecture, 3, 108-118.
UNEP: 1980, Guidelines for Assessing Industrial Environmental Impact and Environmental Criteria for the Siting of Industry, Industry and Environment Guidelines Series, Volume 1, United Nations Environment Programme, Paris.
U.S. Fish and Wildlife Service: 1980, NEPA Planning and Documentation FWS Handbook, Office of Environmental Co-ordination, U.S. Fish and Wildlife Service, Department of Interior, Washington DC.
Wood, W.M.: 1978, 'Public Involvement Techniques Utilized in Highway Transportation Planning', in S. Bendix and H.R. Graham (eds.), Environmental Assessment: Approaching Maturity, Ann Arbor Science, Ann Arbor, Michigan, pp. 205-213.

AFFILIATION

Paul Tomlinson is an Environmental Scientist in the Centre for Environmental Management and Planning, Department of Geography, University of Aberdeen, Scotland.

Ronald Bisset

METHODS FOR ASSESSING DIRECT IMPACTS

1. INTRODUCTION

Procedures for Environmental Impact Assessment (EIA) are, in themselves, not sufficient to ensure that Environmental Impact Statements (EISs) are prepared in such a way that ensures proper and adequate consideration of all likely environmental impacts. Nor can they ensure that EISs present complex data in a concise and informative manner to both decision-makers and the public. EISs which fail to attain a satisfactory standard may lead to public disquiet about the utility of EIA. The difficulties involved in preparing comprehensive and adequate EISs have received considerable attention in the literature. As a result a variety of structured approaches to the implementation of EIA and the preparation of EISs has accumulated (see Canter, 1977 and Clark et al., 1980). Such approaches have gained the titles "method" or "methodology", (often used interchangeably) although the former will be used in this paper.

Methods which have been developed for the identification of direct impacts will be discussed in this paper. Direct impacts can be defined as those environmental changes which exhibit an initial, first-order relationship with an impact-causing factor. For example, damage to crops or vegetation as a result of an air pollution episode. Identification of direct impacts only provides a partial picture of the variety of impacts which can result from a major project. Many components of environmental systems are linked to each other in a variety of complex inter-relationships. A direct change in one component (a direct impact) may, therefore, result in repercussions elsewhere in the system, as indicated in figure 1. Methods which have been designed explicitly to deal with such secondary impacts, often termed indirect impacts, are discussed in the paper by Wathern in this volume.

B. D. Clark et al. (eds.), Perspectives on Environmental Impact Assessment, 195–212.

Figure 1: Diagram illustrating environmental linkages

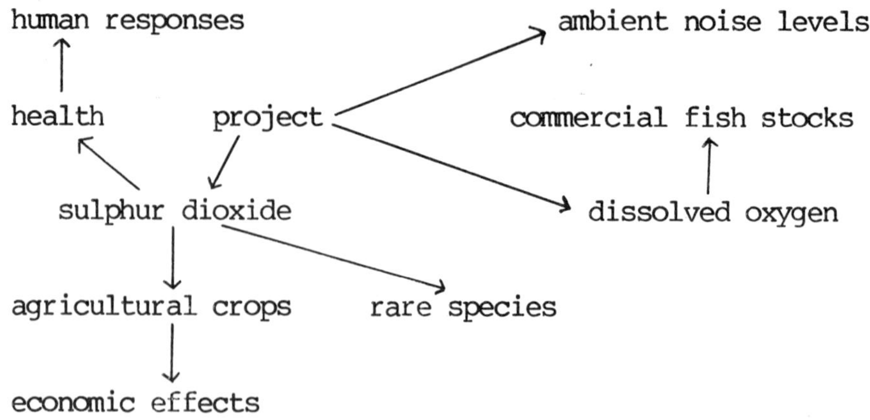

This paper will examine the conceptual differences between two types of EIA method, checklists and matrices, and will consider their benefits and disadvantages as shown by past use.

2. ACTIVITIES INVOLVED IN THE IMPLEMENTATION OF EIA

Before examining checklists and matrices, it is useful to consider those EIA activities for which methods can offer assistance and guidance. It is generally agreed that there are five basic activities involved in EIA. These are:

(i) Impact identification - the need to determine those impacts requiring investigation. Initially this may seem an easy task, but in practice limited knowledge of the effects of different developments on a variety of environments makes this a more complex task.

(ii) Impact measurement - there is little agreement on what measurement means in EIA literature. Usually, it refers to quantitative estimates of environmental parameters, for example, noise. However, other aspects such as the extent of an impact in time and space, and the likelihood of the impact occurring can also be measured.

(iii) Impact interpretation - the need to determine the importance of an impact. Sometimes the relative importance of impacts in comparison with others of a different nature is considered under impact interpretation.

(iv) Impact communication - presentation of information to assist decision-makers and interested lay people reach a conclusion on the merits of a proposed project.

(v) Impact monitoring and mitigation - the identification of impacts to be monitored. Monitoring impacts provides information on the nature of impacts essential for determining measures to mitigate unwanted effects and increase beneficial effects.

As will become apparent, not all methods are suitable for all of these activities.

3. CHECKLISTS

The term "checklist" covers a variety of methods having widely varying characteristics and degrees of complexity. Three types will be discussed below. They are; simple checklists, scaling - weighting checklists and questionnaire check - lists.

3.1. Simple checklists

This type of checklist consists of a list of environmental, economic and social factors which may be affected by specific types of projects or actions (for example, dredging). Some of these checklists provide lists of various development actions such as "ground clearance" or "channelization", which may give rise to impacts. This simple type of checklist is only able to aid the identification of impacts and ensure that impacts are not overlooked. Table 1. presents a checklist developed by the U.S. Department of Housing and Urban Development (1975). While such a checklist acts as a framework for identifying impacts, additional guidelines are needed to perform some of the other EIA tasks.

A comprehensive checklist accompanied by specific guidance on the measurement of impacts was developed by Schaenman (1976). It contains 47 factors which should be considered in an EIA. For each factor information is provided on data requirements, sources of information, and predictive techniques which can be used for assessing impacts (see table 2). A significant feature of this method is the emphasis placed upon relating all impacts to the people likely to be affected. The guidance provided for each factor allows this checklist to be used for impact identification, measurement and presentation of results. It cannot, however, be used to develop and utilize means of determining the relative importance of different impacts.

Table 1: Typical Simple Checklist
(From Interim Guide for Environmental Assessment 1975)

PHYSICAL

1. Geology
   - 1.1 Unique Features
   - 1.2 Mineral Resources
   - 1.3 Slope Stability/Rockfall
   - 1.4 Depth to Impermeable Layers
   - 1.5 Subsidence
   - 1.6 Consolidation
   - 1.7 Weathering/Chemical Release
   - 1.8 Tectonic Activity/Vulcanism
2. Soils
   - 2.1 Slope Stability
   - 2.2 Foundation Support
   - 2.3 Shrink-Swell
   - 2.4 Frost Susceptibility
   - 2.5 Liquefaction
   - 2.6 Erodibility
   - 2.7 Permeability
3. Special Land Features
   - 3.1 Sanitary Landfill
   - 3.2 Wetlands
   - 3.3 Coastal Zones/Shorelines
   - 3.4 Mine Dumps/Spoil Areas
   - 3.5 Prime Agricultural Land
4. Water
   - 4.1 Hydrologic Balance
   - 4.2 Ground Water
   - 4.3 Ground Water Flow Direction
   - 4.4 Depth to Water Table
   - 4.5 Drainage/Channel Form
   - 4.6 Sedimentation
   - 4.7 Impoundment Leakage and Slope Failure
   - 4.8 Flooding
   - 4.9 Water Quality
5. Biota
   - 5.1 Plant and Animal Species
   - 5.2 Vegetative Community
   - 5.3 Diversity
   - 5.4 Productivity
   - 5.5 Nutrient Cycling
6. Climate and Air
   - 6.1 Macro-Climate Hazards
   - 6.2 Forest and Range Fires
   - 6.3 Heat Balance
   - 6.4 Wind Alteration
   - 6.5 Humidity and Precipitation
   - 6.6 Generation and Dispersion of Contaminants
   - 6.7 Shadow Effects
7. Energy
   - 7.1 Energy Requirements
   - 7.2 Conservation Measures
   - 7.3 Environmental Significance

SOCIAL

8. Services
   - 8.1 Education Facilities
   - 8.2 Employment
   - 8.3 Commercial Facilities
   - 8.4 Health Care/Social Services
   - 8.5 Liquid Waste Disposal
   - 8.6 Solid Waste Disposal
   - 8.7 Water Supply
   - 8.8 Storm Water Drainage
   - 8.9 Police
   - 8.10 Fire
   - 8.11 Recreation
   - 8.12 Transportation
   - 8.13 Cultural Facilities
9. Safety
   - 9.1 Structures
   - 9.2 Materials
   - 9.3 Site Hazards
   - 9.4 Circulation Conflicts
   - 9.5 Road Safety and Design
   - 9.6 Ionizing Radiation
10. Physiological Well-Being
    - 10.1 Noise
    - 10.2 Vibration
    - 10.3 Odor
    - 10.4 Light
    - 10.5 Temperature
    - 10.6 Disease
11. Sense of Community
    - 11.1 Community and Organization
    - 11.2 Homogeneity and Diversity
    - 11.3 Community Stability and Physical Characteristics
12. Psychological Well-Being
    - 12.1 Physical Threat
    - 12.2 Crowding
    - 12.3 Nuisance
13. Visual Quality
    - 13.1 Visual Content
    - 13.2 Area and Structure Coherence
    - 13.3 Apparent Access
14. Historic and Cultural Resources
    - 14.1 Historic Structures
    - 14.2 Archaeological Sites and Structures

Courtesy of the U.S. Department of Housing and Urban Development.

ENVIRONMENTAL FACTORS

DATA REQUIRED	INFORMATION SOURCES/PREDICTIVE TECHNIQUES
Air Quality	
Health	
Change in air pollution concentrations by frequency of occurrence and number of people at risk.	Current ambient concentrations, current and and expected emissions, dispersion models, population maps.
Nuisance	
Change in occurrence of visual (smoke, haze) or olfactory (odor) air quality nuisances, and number of people affected.	Baseline citizen survey, expected industrial processes, traffic volumes.
Water Quality	
Changes in permissible or tolerable water uses and number of people affected - for each relevant body of water.	Current and expected effluents, current ambient concentrations, water quality model.
Change in noise levels and frequency of occurrence, and number of people bothered.	Changes in nearby traffic or other noise sources, and in noise barriers; noise propagation model or nomographs relating noise levels to traffic, barriers, etc.; baseline citizen survey of current satisfaction with noise levels.

Table 2: Section of Schaenman Checklist. (Modified after Schaenman, 1976)

Table 3: Checklist used in EES. (From Dee et al., 1973)

ECOLOGY
Terrestrial species and populations
Browsers and grazers (14)
Crops (14)
Natural vegetation (14)
Pest species (14)
Upland game birds (14)

Aquatic species and populations
Commercial fisheries (14)
Natural vegetation (14)
Pest species (14)
Sport fish (14)
Waterfowl (14)

Terrestrial habitats and communities
Food web index (12)
Land use (12)
Rare and endangered species
Species diversity (14)

Aquatic habitats and communities
Food web index (12)
Rare and endangered species (12)
River characteristics (12)
Species diversity (14)

Ecosystems
Descriptive only

AESTHETICS
Land
Geologic surface material (6)
Relief and topographic character (16)
Width and alignment (10)

Air
Odour and visual (3)
Sounds (2)

Water
Appearance of water (10)
Land and water interface (16)
Odour and floating material (6)
Water surface area (10)
Wooded and geologic shoreline (10)

Biota
Animals—domestic (5)
Animals—wild (5)
Diversity of vegetation types (9)
Variety within vegetation types (5)

Man-made objects
Man-made objects (10)

Composition
Composite effects (15)
Unique composition (15)

PHYSICAL/CHEMICAL
Water quality
Basin hydrologic loss (20)
Biochemical oxygen demand (25)
Dissolved oxygen (31)
Faecal coliforms (18)
Inorganic carbon (22)
Inorganic nitrogen (25)
Inorganic phosphate (28)
Pesticides (16)
pH (18)
Streamflow variation (28)
Temperature (28)
Total dissolved solids (25)
Toxic substances (14)
Turbidity (20)

Air quality
Carbon monoxide (5)
Hydrocarbons (5)
Nitrogen oxides (10)
Particulate matter (12)
Photochemical oxidants (5)
Sulphur oxides (10)
Other (5)

Land pollution
Land use (14)
Soil erosion (14)

Noise pollution
Noise (4)

HUMAN INTEREST/SOCIAL
Education/scientific
Archeological (13)
Ecological (13)
Geological (11)
Hydrological (11)

Historical
Architecture and styles (11)
Events (11)
Persons (11)
Religions and cultures (11)
" Western Frontier " (11)

Cultures
Indians (14)
Other ethnic groups (7)
Religious groups (7)

Mood/atmosphere
Awe inspiration (11)
Isolation/solitude (11)
Mystery (4)
" Oneness " with nature (11)

Life patterns
Employment opportunities (13)
Housing (13)
Social interactions (11)

3.2. Scaling-weighting checklists

A complex checklist which incorporates scaling and weighting of impacts, has been developed to overcome this problem (Dee et al., 1973). This scaling-weighting checklist is called the Environmental Evaluation System (EES) and was developed for water resource projects, although it can be used for other projects. EES consists of a checklist of 78 parameters mostly covering the physical environment, although some social and economic factors are also included (see table 3). Essentially, this method is concerned with the determination of environmental quality of an area without a project, and then predicting a net positive or negative change should the project proceed. This involves predicting the future state of the environment without the project in the same time-scale as that used to predict the project-affected environment. Environmental quality is measured by the use of "value functions" for each of the 78 parameters thereby providing a numerical index of "overall quality". To achieve this index it is assumed that the "quality" state of each parameter can be expressed on an arbitrary scale of 0-1, where 1 represents "high quality" and 0 represents "low quality". The value function for dissolved oxygen (see figure 2) indicates that a level of 10 mg/l equals a "quality" score of 1, while 2 mg/l equals about 0.1. The determination of these "quality scores" for a variety of environmental components is termed "scaling".

To assess the impacts of a project it is necessary to determine the relative importance of those parameters. This is achieved by using experts to assign numerical weights to each parameter (see table 3). A thousand units are divided between the parameters according to their importance as determined by the experts. To obtain the numerical index for the "overall quality", the weighting is simply multiplied by the "quality" score for each parameter. The summation of each would then indicate the future "overall quality" state of that area. It is then necessary to predict the likely future state of each parameter assuming the project took place. This may be achieved by mathematical modelling or the use of expert opinion. For example, if the dissolved oxygen concentration were to fall from 6 mg/l to 4 mg/l then the calculation would be as follows:

Figure 2: Value functions for dissolved oxygen. (From Dee et al., 1973).

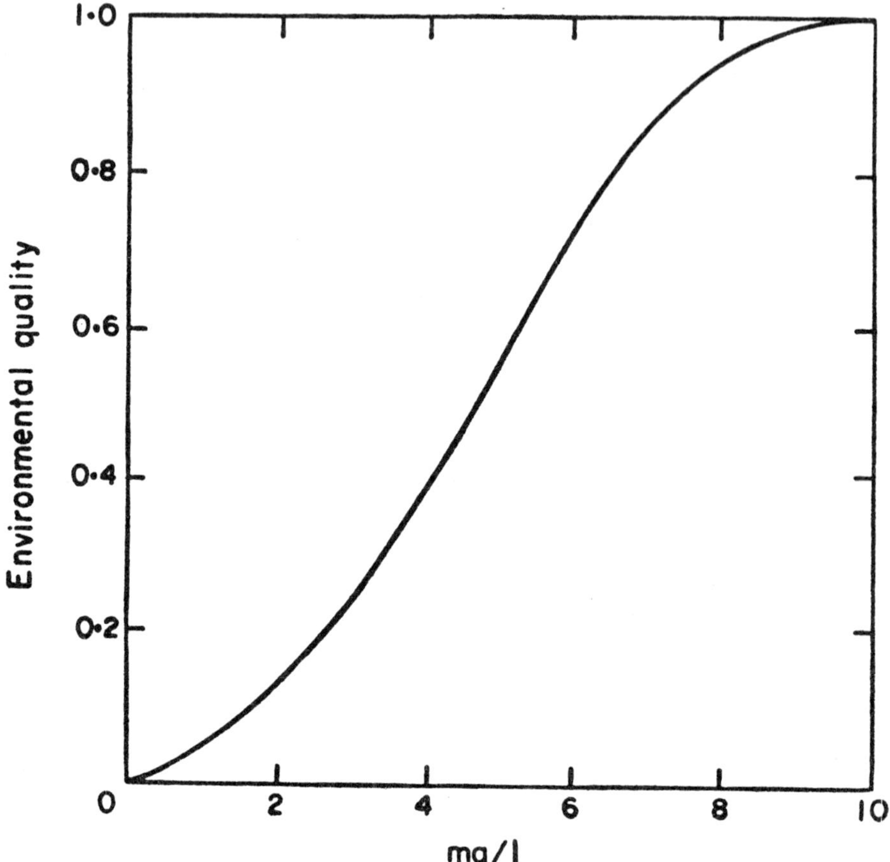

Reproduced with permission from the Journal of Environmental Management, Academic Press Inc. © .

Predicted state without project	$=0.72 \times 31$
	$=22.32$
Predicted state with project	$=0.32 \times 31$
	$=11.78$

A reduction of 2 mg/l would therefore result in nearly a 50% fall in environmental quality. When the total scores for the "with" and "without" project situations are compared, the overall impact of the project can be measured. From such an analysis it would then be possible to select the least damaging alternative. Another benefit of this system is that it has a system of "red flags" by which serious impacts are highlighted.

This complex checklist can undertake all five EIA activities and is one reason why so much effort has been placed on developing such methods. While the method provides a straight-forward numerical means of comparing the benefits and disadvantages of projects, a political price must be paid. The use of value judgements derived from a small group of experts may not be representative of the views of those likely to be affected by a proposal.

EES is complex to operate and any EIS prepared using this or a similar method, for example, the Water Resources Assessment Methodology developed by the U.S. Army Corps of Engineers (Solomon *et al*., 1977) and the Sondheim method (Sondheim, 1978), would not be comprehended easily by non-experts. Not only would the technical information on impacts have to be understood, but also the principles of the method would need to be comprehended. This seems to be an unfair burden for the public. Finally, the complexity of this approach results in high manpower, time and resource requirements especially when determining the value functions.

3.3. Questionnaire checklists

Another type of checklist is the "questionnaire" which presents a series of questions relating to the impacts of a project. Table 4 presents two sections (there are 31 in total) from a recent questionnaire checklist developed by the U.S. Agency for International Development (1981). This method was designed for the assessment of rural development projects in developing countries. The questions are listed under generic categories such as "terrestrial ecosystems" and "disease vectors". Those assessing projects must attempt to answer all the questions. Three answers are possible depending upon how much is known about a particular impact.

Table 4: Section of U.S. AID Questionnaire Checklist. (Developed by the U.S. AID, 1981).

Disease Vectors

a) Are there known disease problems in the project area transmitted through vector species such as mosquitos, flies, snails, etc.?	Yes___	No___	Unk___
b) Are these vector species associated with:			
Aquatic habitats?	Yes___	No___	Unk___
Forest habitats?	Yes___	No___	Unk___
Agricultural Lands?	Yes___	No___	Unk___
Degraded habitats?	Yes___	No___	Unk___
Human settlements?	Yes___	No___	Unk___
c) Will the project:			
Increase vector habitat?	Yes___	No___	Unk___
Decrease vector habitat?	Yes___	No___	Unk___
Provide opportunity for vector control?	Yes___	No___	Unk___
d) Will the project work force be a possible source of introduction of disease vectors not currently found in the project area?	Yes___	No___	Unk___
e) Will increased access to and commerce with the project area be a possible source of disease vectors not presently occurring in the project area?	Yes___	No___	Unk___
f) Will the project provide opportunities for vector control through improved standards of living?	Yes___	No___	Unk___

ESTIMATED IMPACT ON DISEASE VECTORS.......ND..HA..MA..LA..O..LB..MB..HB

Public Health

a) Are vector-borne diseases an important part of the local public health situation?	Yes___	No___	Unk___
b) Are there clinics or other disease control programs in operation or planned for the area?	Yes___	No___	Unk___
c) Will the project decision result in an increase in disease vector density or distribution?	Yes___	No___	Unk___
d) Will the project decision result in workers or other persons entering the area with contagious or vector-borne diseases?	Yes___	No___	Unk___
e) Will the project decision result in clearing operations that could expose workers to disease vectors?	Yes___	No___	Unk___

Courtesy of the U.S. Agency for International Development.

If insufficient information was available for a definite response, then "unknown" (unk in the table) would be recorded. If an answer could be provided then it would be possible to use the scale for designating the expected impact. For example, an impact may be "high adverse" or "medium benefit". The assessment of a project therefore consists of a systematic "question and answer" procedure with the addition of quantitative and qualitative information where appropriate.

Once the initial question in a particular category has been dealt with subsequent questions relate to other aspects of that impact category. For example, the final question in the category "terrestrial ecosystems" forces consideration of indirect impacts. In addition, the consequences of possible mitigating measures are considered.

A conceptually similar checklist is contained in an assessment manual produced for use by U.K. local authorities, central government and developers (Clark et al., 1981) (see table 5).

Checklists of environmental components are limited as they tend to be either checklists of development actions or environmental components. This was perceived to be a weakness by many involved in EIA. Consequently, these two types of lists have been brought together to form a two-dimensional matrix to aid impact identification.

4. MATRICES

The most famous of these methods is probably the matrix developed by Leopold and his colleagues (Leopold et al., 1971). Figure 3 shows a section of this matrix which consists of two checklists. A list of development actions is displayed horizontally, while a list of environmental components is displayed vertically. The inclusion of these two checklists in a matrix aids impact identification, as items in one list can be related systematically to all items in the other list in order to identify likely impacts. Once completed, that is when all likely interactions between a development activity and environmental components have been identified, the matrix is a useful and easily-understood visual summary. This matrix can be used to measure and interpret impacts as well as identify them.

This is achieved by incorporating guidelines to characterise impacts in terms of magnitude and importance on a common 1-10 scale, where 1 is least magnitude or importance and 10 the greatest. The magnitude of an impact is taken to

Table 5: Section of Checklist Developed by Clark *et al.*, 1981.

EXISTING LEVELS OF ENVIRONMENTAL POLLUTION

Air pollution

a. Will the installation significantly alter the levels of atmospheric pollutants in the local authority area?
b. Will the release of atmospheric pollutants be a hazard to human health, crops, livestock, wildlife and stonework?
c. Will inversion lead to a local build up of high levels of pollutants from the new installation?
d. Will there be significant synergistic effects with existing pollutants in the atmosphere?
e. Will the distribution of wind direction cause significant fumigation of areas sensitive to atmospheric pollution?
f. Will the installation produce significant quantities of particulate matter which will be a nuisance to the local community?
g. Will the installation produce offensive odours?

Water pollution

a. Will effluents, treated or untreated, have a significant effect on the flora and fauna of the river, canal, lake, estuary or coastal waters?
b. Will effluents find their way into surface water by means of underground waters?
c. Are there stretches downstream where effluents are likely to change the flora and fauna?
d. Will there be significant synergistic effects with existing pollutants in the receiving waters and/or between constituents of the effluents?
e. Will there be significant potentiation effects with existing constituents of the effluents or receiving waters?
f. Will the discharges lead to the build up of locally high levels of pollutants?
g. Will variations in water flow (e.g. seasonal) cause a significant increase in the concentration of pollutants?
h. Will salinity gradients and/or current movements in estuaries lead to locally high build up of pollutants and cause problems of dispersion?
j. Will fishing (commercial and recreational) be affected by discharges?
k. Will other water-based activities such as water-skiing, canoeing, sailing etc. be affected by discharges?
l. Will there be any odour likely to cause offence?
m. What dependent communities or species of animals and birds are likely to be affected by a change in the aquatic flora and fauna?
n. Are there any sensitive plant communities dependent on the receiving waters for their supply which are likely to be adversely affected by discharges (including fines) from the development?
p. Do any horticultural or agricultural enterprises use receiving waters for irrigation?

Courtesy of the Controller of Her Majesty's Stationery Office.

INSTRUCTIONS

1- Identify all actions (located across the top of the matrix) that are part of the proposed project.

2- Under each of the proposed actions, place a slash at the intersection with each item on the side of the matrix if an impact is possible.

3- Having completed the matrix, in the upper left-hand corner of each box with a slash, place a number from 1 to 10 which indicates the MAGNITUDE of the possible impact; 10 represents the greatest magnitude of impact and 1, the least, (no zeroes). Before each number place + if the impact would be beneficial. In the lower right-hand corner of the box place a number from 1 to 10 which indicates the IMPORTANCE of the possible impact (e. g. regional vs. local); 10 represents the greatest importance and 1, the least (no zeroes).

4- The text which accompanies the matrix should be a discussion of the significant impacts, those columns and rows with large numbers of boxes marked and individual boxes with the larger numbers.

SAMPLE MATRIX

	a	b	c	d	e
a		2/1			8/5
b		7/2	8/8	3/1	9/7

		PROPOSED ACTIONS	A. MODIFICATION OF REGIME: a. Exotic flora or fauna introduction	b. Biological controls	c. Modification of habitat	d. Alteration of ground cover	e. Alteration of ground water hydrology	f. Alteration of drainage	g. River control and flow modification	h. Canalization	i. Irrigation	j. Weather modification	k. Burning	l. Surface or paving	m. Noise and vibration
A. PHYSICAL AND CHEMICAL CHARACTERISTICS	1. EARTH	a. Mineral resources													
		b. Construction material													
		c. Soils													
		d. Land form													
		e. Force fields and background radiation													
		f. Unique physical features													
	2. WATER	a. Surface													
		b. Ocean													
		c. Underground													
		d. Quality													
		e. Temperature													
		f. Recharge													
		g. Snow, ice, and permafrost													
	3. ATMOSPHERE	a. Quality (gases, particulates)													
		b. Climate (micro, macro)													
		c. Temperature													
	4. PROCESSES	a. Floods													
		b. Erosion													
		c. Deposition (sedimentation, precipitation)													
		d. Solution													
		e. Sorption (ion exchange, complexing)													
		f. Compaction and settling													
		g. Stability (slides, slumps)													
		h. Stress-strain (earthquake)													
		i. Air movements													

Figure 3: Section of the Leopold Matrix.
Courtesy of the U.S. Geological Survey.

be an expression of its scale, for example, the geographical area of the impact. The score for importance may, however, be different. Importance refers to the significance of the impact. If a visual impact were to occur in an area of poor landscape quality then a score of only 2 or 3 may be awarded instead of 8 or 9 in an area of high quality.

Assigning scores for magnitude and importance to all identified impacts depends on the subjective views of those assessing a proposal. Leopold _et al_., (1971) do not provide detailed guidance on how these scores can be assigned on a standardised basis. No criteria to assist the assessor decide between particular scores are provided. Also, since the scores are ordinal, they cannot be manipulated arithmetically.

Leopold _et al_., recommend that scores for magnitude and importance should be included in the appropriate cell of the matrix identified as representing a likely impact. The cell should be bisected by a diagonal line with the score for magnitude being placed in the top left-hand corner. The score for importance is placed in the bottom right-hand corner. These scores can be accompanied by a plus sign (+) to indicate a beneficial effect. A completed matrix would consist of a number of cells containing two numerical scores and, in some cases, a plus sign. Although the Leopold matrix contains 8800 cells, Leopold _et al_., estimate that, for most projects ,not more than 25-50 cells will be identified as representing impacts. Thus, this matrix can indicate in a structured, self-contained format, a considerable amount of information regarding impacts. However, the subjective nature of the information must be kept in mind. In addition, this matrix is unable to deal explicitly with the time aspects of impacts, although Leopold _et al_., recognise the importance of this factor. Despite these drawbacks, Leopold-type matrices have been, arguably, the most commonly used EIA method. Their continued popularity attests to the utility of matrices in EIA.

The usefulness of a Leopold-type matrix has been demonstrated in the assessment of exploratory drilling sites in Antarctica (Parker and Howard, 1977). Matrices were used to assess the impacts of drilling at alternative sites, and to identify the location where drilling could take place with the minimum environmental disturbance. It was considered that using two scales of 1-10 was unnecessarily complicated and a single scale of 0-5 was used. This scale measured both the time and importance aspects of the identified impacts. No impact was represented by 0, 1 represented an impact commencing on the first day of an activity, whereas 5 represented a severe long term impact (figure 4). Magnitude was not considered, as it was believed that it would not vary greatly between the sites. For each impact, scores from the 0-5 scale are used to characterise the behavior of impacts over time. For example, in the case of the impact of drainage alteration on soils, immediate effects are not considered serious (score 1). Nevertheless, the effects increase in importance with time and the last score (4) represents an important long-term impact. In contrast, the effects of the same development action on micro-organisms decline in severity over time (from a score of 3 to 1).

Twenty matrices were prepared to identify the "best" site for drilling. Once a decision had been made, monitoring was implemented to ascertain the impacts arising from drilling operations. Information on the severity of impacts and other changes over time was collected to improve future predictive exercises using the matrices. Those who used this matrix state that their innovative adaption of the Leopold matrix was beneficial in their assessment work. Despite this demonstration of the usefulness of Leopold-type matrices,they exhibit serious, theoretical and practical limitations.

These matrices focus upon direct impacts between two items - the impact-causing factor and the initial target environmental component affected by the impact-causing factor. Those assessing a project have to structure their thought by mentally linking a target component and a development activity in order to ascertain whether an impact is likely. It is then necessary to consider the next component in a similar manner. Thus, impacts are identified by a series of discrete, two-way linkages between activities and components of the environment, thereby indicating the likely direct impacts.

Individual components are, however, linked to each other via a complex web of interactions, therefore direct impacts can result in further changes which may require some time before they become apparent. To enable the project assessors to identify indirect, second-order impacts, methods which are specially constituted to deal with the "networks" of impacts arising from an initial direct impact are required.

Checklists and Leopold-type matrices are useful for identifying and characterising direct impacts, but present a fragmented view of environmental systems.

Figure 4: Section of Matrix Developed by Parker and Howard. (Adapted from Parker and Howard, 1977).

	SOILS	SURFACE, INCLUDING STREAMS	STABILITY (SLIDES & SLUMPS)	MICROORGANISMS	BENTHIC ORGANISMS
DRILLING FLUID DISCHARGE	32110	33211	11000	44432	-----
EMPLACEMENT OF TAILINGS AND OVERBURDEN	33333	11110	11000	33211	-----
WELL DRILLING AND FLUID REMOVAL	11000	33211	-----	11111	-----
ALTERATION OF DRAINAGE	12344	11111	11110	33211	-----
MODIFICATION OF SURFACE OR PAVING	11111	11111	11111	21111	-----
MODIFICATION OF HABITAT	11111	11111	12321	44432	-----

5. CONCLUSIONS

Checklists and matrices have the same strengths and weaknesses. They are useful for the identification of direct first-order impacts, which ensures that once identified they are not then overlooked in the analysis. It has been shown that checklists, with the aid of additional guidance, can be used to undertake most of the tasks required in EIA. This is achieved at the expense of simplicity and, one suspects, public comprehension. Matrices do not achieve, generally the same complexity as checklists, and do relate project action to environmental features in an easily-understood format. Evidence from EISs shows that simple checklists and matrices are often used. The more complex scaling-weighting checklists (such as EES) are used, but much less frequently. This would seem to indicate the relative merits of the two methods, at least in terms of their actual use. (This assertion is not based on quantified evidence, but is the impression gained by the author as a result of his involvement in EIA).

Both methods have similar failings in that they cannot easily take account of indirect impacts. The division of the environment into discrete components which are examined in turn for project-induced change, inevitably compartmentalizes the environment. This creates a false impression of the nature of environmental systems portraying simplicity rather than the existing complexity. However, before identifying indirect impacts, it is necessary to identify all direct impacts; consequently, these methods fulfil a useful role. Perhaps the most useful of the two is the interaction matrix.

REFERENCES

Canter, L.W.: 1979, Water Resources Assessment: Methodology and Technology Sourcebook, Ann Arbor Science, Ann Arbor, Michigan.

Clark, B.D., Bisset, R. and Wathern, P.: 1980, Environmental Impact Assessment: A Bibliography With Abstracts, Mansell/Bowker, London.

Clark, B.D., Chapman, K., Bisset, R., Wathern, P. and Barrett, M.: 1981, A Manual For The Assessment of Major Developments, HMSO, London.

Dee, N., Baker, J.K., Drobny, N.L., Duke, K.M., Whitman, I. and Fahringer, D.C.: 1973, 'An Environmental Evaluation System For Water Resources Planning', Water Resources Research 9 (3), pp. 523-535.

Leopold, L.B., Clarke, F.E., Hanshaw, B.B. and Balsley, J.R.: 1971, A Procedure For Evaluating Environmental Impact, U.S. Geological Survey Circular 645, U.S. Geological Survey, Washington D.C..

Parker, B.C. and Howard, R.V.: 1977, ´The First Environmental Monitoring and Assessment in Antarctica: The Dry Valley Drilling Project´, Biological Conservation, 12 (2), pp. 163-177.

Schaenman, P.S.: 1976, Using An Impact Measurement System to Evaluate Land Development, The Urban Institute, Washington D.C..

Soloman, R.C., Colbert, B.K., Hansen, W.J., Richardson, S.E., Canter, L.W. and Vlachos, E.C.: 1977, Water Resources Assessment Methodology (WRAM): Impact Assessment and Alternative Evaluation´, Technical Report Number Y-77-1, U.S. Army Corps pf Engineers, Vicksburg, Mississippi.

Sondheim, M.W.: 1978, ´A Comprehensive Methodology for Assessing Environmental Impact´, Journal of Environmental Management, 6 (1), pp. 27-42.

U.S. Agency for International Development.: 1980, Environmental Design Considerations For Rural Development Projects, U>

Projects, U.S. Agency for International Development, W Washington D.C..

U.S. Department of Housing and Urban Development.: 1975, Interim Guide For Environmental Assessment, U.S. Departmen of Housing and Urban Development, Washington D.C..

AFFILIATION

Ronald Bisset is Senior Environmental Scientist in the Centre for Environmental Management and Planning (formerly PADC), University of Aberdeen, Scotland.

Peter Wathern

METHODS FOR ASSESSING INDIRECT IMPACTS

1. INTRODUCTION

Many environmental impacts can be readily determined as they result directly from the effects of particular development actions upon individual environmental components. Such environmental components, however, are not isolated, but aggregated into hierarchical groupings with varying intra- and inter- group dependencies. Development may cause a change in one particular environmental component which induces a sequence of change involving a number of inter-related components, with impacts occurring in components apparently remote from the direct effects of the development. Impacts which act through a number of intermediary components of the physical or biological environment, are called indirect impacts. Indirect impacts often occur unexpectedly upon implementation of a project because the environmental assessment has been undertaken with imperfect knowledge of both the functioning of particular systems, and their response to perturbations. Methods have been developed to help predict such indirect impacts by exploring the effects of inter-relationships within environmental systems.

Many direct and indirect impacts are experienced at varying intensity over a wide geographical area, indeed, this spatial variation may be their dominant feature. The use of some form of mapping technique seems appropriate for their appraisal and subsequently in presenting information on their extent and intensity. Similiarly, it is often better to express the distribution of costs and benefits spatially. Although there has been some methodological development on spatial aspects of EIA, it has long been a significant problem in planning and there is a considerable amount of research experience for those involved in environmental assessment to draw upon. Spatial aspects can only be analysed and displayed effectively on a base map, and the use of overlays has extended the number of components that can be represented.

B. D. Clark et al. (eds.), Perspectives on Environmental Impact Assessment, 213–231.

	Currents	Wind	Water temperature	Light	Intertidal vegetation	Upland vegetation	Bacteria	Insects	Larvae	Shell fish	Crabs	Other crustaceans	Pelagic fish	Bottom fish	Waterbirds	Birds of prey	Song birds	Marsh & shore birds	Upland game birds	Aquatic mammals	Upland mammals
Currents		1																			
Wind						1															
Water temperature	1	1		1																	
Light																					
Intertidal vegetation	1		1	1																	
Upland vegetation				1			1	1									1	1	1		1
Bacteria			1		1	1		1	1	1	1	1	1	1	1	1	1	1	1	1	1
Insects	1	1	1		1	1	1														
Larvae	1		1		1		1			1	1	1	1	1							
Shell fish	1		1		1		1														
Crabs			1		1				1	1		1	1	1							
Other crustaceans			1		1		1	1	1	1	1		1	1							
Pelagic fish			1					1	1												
Bottom fish			1		1			1	1	1	1	1	1								
Waterbirds					1			1	1	1	1	1	1	1							
Birds of prey						1		1			1	1	1	1	1		1	1	1		1
Song birds						1		1		1		1	1	1							
Marsh & shore birds						1		1		1	1	1	1	1							
Upland game birds						1		1													
Aquatic mammals										1	1	1	1	1							
Upland mammals						1				1	1	1					1		1		

Note: A(1) in any cell indicates that the row component is dependent on the column component.

Figure 1: Component Interaction Matrix (From Environment Canada, 1974)

2. IDENTIFYING INDIRECT IMPACTS

Indirect impacts can be predicted only when all direct dependencies within a system are known. A dependency exists when one environmental component acts as a resource, such as food or shelter, for another component, or when it acts as a vector for an impact-causing agent.

A direct or first order dependency can be denoted as; $A \rightarrow B$, signifying that B is dependent upon A. This is a unidirectional relationship, the reverse being denoted by $B \rightarrow A$. If C were dependent upon B ($B \rightarrow C$) there would be an indirect second order dependency between A and C ($A \rightarrow C$), as changes in A would induce indirect changes in C. This chain of dependency is extended through many intermediaries to give higher order dependencies in some systems. It is possible to deduce indirect relationships by following individual chains of dependency, but in complex systems this rapidly becomes difficult and subject to error.

2.1. Interaction matrix

Ross (1974) has argued that although the Leopold matrix (Leopold *et al*., 1971) is extremely useful in revealing the incidence and magnitude of probable disruptions, the ability to detect only primary impacts severely restricts its utility. Environment Canada (1974) proposed an alternative matrix method which enables indirect impacts to be identified using matrix algebra. All components of the system are arranged along both the horizontal and vertical axes of the matrix (see figure 1). Cells of the matrix are scored if the horizontal component is dependent upon the vertical component. Thus, for example, song birds are a food source for birds of prey, but there is no reverse dependency so that cell remains blank. If the component interaction matrix is then squared using matrix algebra, cells representing dependencies through two-link chains are identified. Multiplying the squared and the original matrices together produces a cubed matrix in which third order (three link) dependencies are identified. This process can be repeated indefinitely. In the Environment Canada study, matrix multiplication was terminated when fifth order interactions were detected, because at this stage it was considered that all significant linkages had been detected. Environment Canada (1974) also produced a secondary matrix in which the numbers in the cells indicated the links in the shortest chain of dependency between pairs of components.

CURRENTS	WIND	WATER TEMPERATURE	LIGHT	INTERTIDAL VEGETATION	UPLAND VEGETATION	BACTERIA	INSECTS	LARVAE	SHELLFISH	CRABS	OTHER CRUSTACEANS	PELAGIC FISH	BOTTOM FISH	WATER BIRDS	BIRDS OF PREY	SONG BIRDS	MARSH & SHORE BIRDS	UPLAND & GAME BIRDS	AQUATIC MAMMALS	UPLAND MAMMALS	
4	1	4	3	4	2	3	3	4	4	4	4	4	4	4	4	3	3	3	4	3	CURRENTS
3	3	3	2	3	1	2	2	3	3	3	3	3	3	3	3	2	2	2	3	2	WIND
1	1	4	1	4	2	3	3	4	4	4	4	4	4	4	4	3	3	3	4	3	WATER TEMPERATURE
0	0	0	0	0	0	0	0	0	0	0	0	0	0	0	0	0	0	0	0	0	LIGHT
1	2	1	1	5	3	4	4	5	5	5	5	5	5	5	5	4	4	4	5	4	INTERTIDAL VEGETATION
2	2	2	1	2	2	1	1	2	2	2	2	2	2	2	2	1	1	1	2	1	UPLAND VEGETATION
2	2	1	2	1	1	2	1	1	1	1	1	1	1	1	1	1	2	1	1	1	BACTERIA
1	1	1	2	1	1	1	2	2	2	2	2	2	2	2	2	2	2	2	2	2	INSECTS
1	2	1	2	1	2	1	2	2	1	1	1	1	1	2	2	2	3	2	2	2	LARVAE
1	2	1	2	1	2	1	2	2	2	2	2	2	1	2	2	2	3	2	2	2	SHELL FISH
2	2	1	2	1	3	2	2	1	1	2	1	1	1	3	3	3	4	3	3	3	CRABS
2	2	1	2	1	2	1	1	1	1	1	2	1	2	2	2	2	3	2	2	2	OTHER CRUSTACEANS
2	2	1	2	2	2	2	1	1	2	2	2	2	2	3	3	3	3	3	3	3	PELAGIC FISH
2	2	1	2	1	2	2	1	1	1	1	1	1	2	3	3	3	3	3	3	3	BOTTOM FISH
2	2	2	2	1	2	2	1	1	1	1	1	1	1	3	3	3	3	3	3	3	WATER BIRDS
2	2	2	2	2	1	2	1	2	2	1	1	1	1	1	3	1	1	1	3	1	BIRDS OF PREY
2	2	2	2	2	1	2	1	2	1	2	1	2	1	3	3	2	2	2	3	2	SONG BIRDS
2	2	2	2	2	1	2	1	2	1	1	1	2	1	3	3	2	2	2	3	2	MARSH & SHORE BIRDS
2	2	2	2	2	1	2	1	3	3	3	3	3	3	3	3	2	2	2	3	2	UPLAND & GAME BIRDS
2	3	2	3	2	3	2	2	2	1	1	1	1	1	3	3	3	4	3	3	3	AQUATIC MAMMALS
2	3	2	2	2	1	2	2	2	1	1	1	2	2	3	3	1	2	1	3	2	UPLAND MAMMALS

Figure 2: Minimum Link Matrix
(From Environment Canada, 1974)

This is referred to as a minimum link matrix (figure 2). Using this matrix it is possible to identify indirect impacts, for example, crabs are dependent on larvae as a food source, while larvae depend on bacteria which in turn depend on upland vegetation. The crab thus has a first order relationship to the larvae, a second order relationship to the bacteria and a third order relationship to the vegetation. If a development caused a change in an environental component, such as vegetation, then changes would also be expected for all those components having an indirect dependency, for example the crabs.

The matrices were used to assess the level of disruption to environmental components resulting from the development of a lumber transhipment facility at Nanaimo, British Columbia, Canada. The degree to which each direct and indirect interaction would be affected was assessed in a subjective manner, and alternative proposals were ranked according to their impact. Ross (1976) carried out an independent re-appraisal of these data based on a non-metric multi-dimensional scaling. Combining the estimated relative contribution of each dependency to total biomass production and the relative degree of dependence between components gave quantitative importance weights and significance of disruption for each dependency. Alternatives were re-ranked using these data which confirmed the original subjective ranking.

The use of a binary system to indicate dependency can be criticized because it assumes that all direct dependencies are of similar magnitude. Within an ecosystem the magnitude of individual linkages differs with varying importance for the functioning of the ecosystem. This problem can be accommodated if quantitative data on the degree of dependence is used in the analysis. Lennon (1979) has used this approach in considering the effects of an impoundment scheme on an upland grassland ecosystem in mid-Wales. One of the major effects of the development would be to remove low-lying land from agricultural production. Use of this land is essential to maintain the current level of grazing on the surrounding unimproved grasslands. Relaxation of grazing would have profound effects upon the composition of the grassland. A component interaction matrix based on quantitative dependency data was used to investigate the importance of sheep in upland ecosystems. Perkins (1978) produced energy flow data within a comparable upland ecosystem, which seem relevant to this problem (figure 3). These data show not only the direct dependencies, but also the magnitude of each in terms of

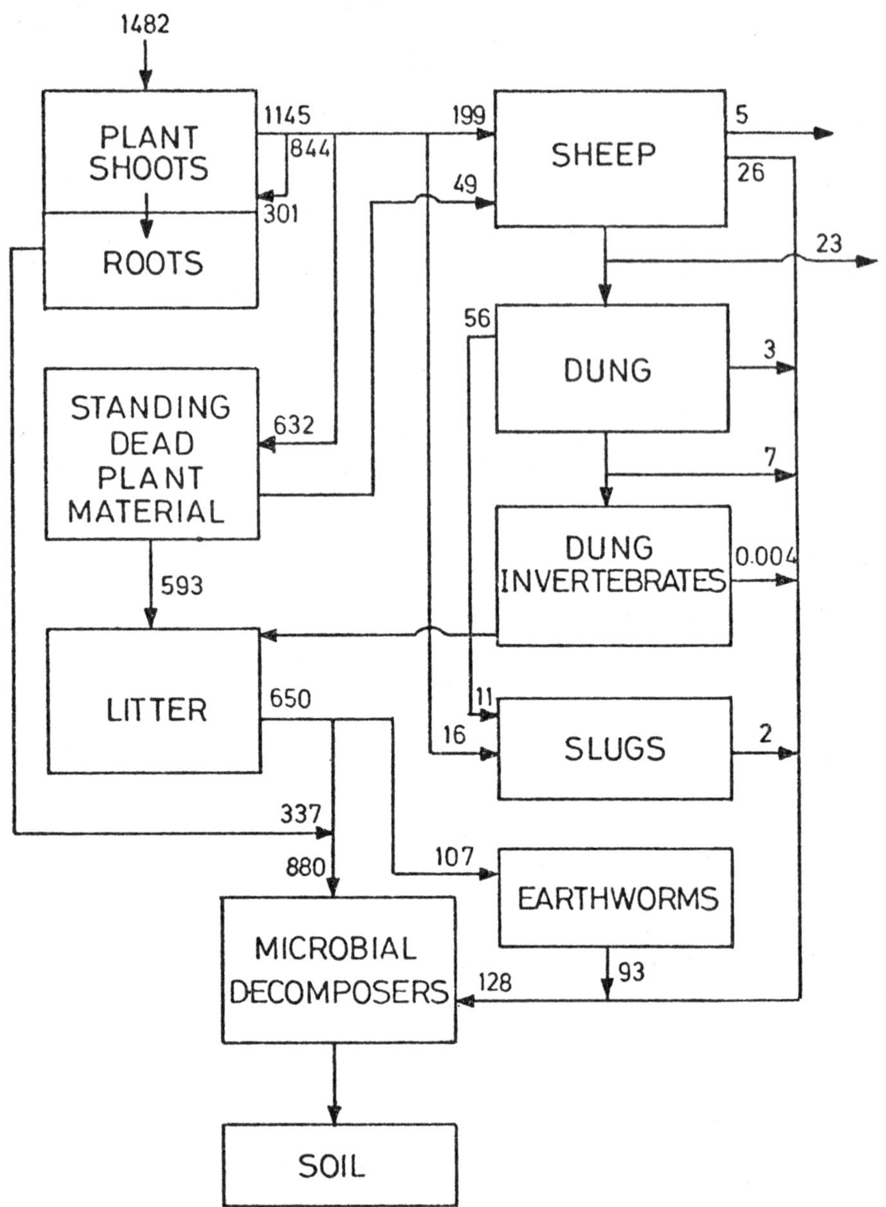

Figure 3: Energy Flow for an Upland Grassland Ecosystem (Adapted from Perkins, 1978)

energy flow between components. The component interaction matrix constructed from these data had numerical values representing the relative dependence of components inserted in appropriate cells (figure 4). These values indicated the energy entering a component from a particular source as a proportion of the total energy flow into that component. The matrix was multiplied until a fifth power was obtained. This level of termination was an arbitrary decision, but can be justified because at this stage energy flows were small, representing less than 3% of the total energy flow into individual components. Each matrix produced by successive multiplications was summed to give the aggregate dependency between components incorporating all interactions up to and including fifth order dependencies (figure 5). Values in excess of 1.0 were obtained due to the existence of loops within the system which permit energy, in effect, to circulate repeatedly through a component.

There are practical dificulties in applying quantitative parameters to component interaction matrices. For example, data on energy or nutrient flow can only be gathered slowly as a result of detailed studies by a specialist team. The length of time necessary to compile such data exceeds the time scale of most development proposals. Lennon (1979) was able to use quantitative parameters only because the Perkins data, collected as part of an unrelated study, already existed. Thus use of this method will be dependent upon the availability of appropriate data.

2.2. Flow diagrams and network analysis

Flow diagrams and network analysis, by following the ramifications of linkages within systems, represent alternative methods for identifying indirect impacts. Two approaches can be adopted. First, the structure and functioning of the system under consideration can be described in detail and points where development is likely to impinge upon the system can be identified. Alternatively, the characteristics of the development can be used to trace potential environmental change, producing an expanding sequence of activity-change-impact linkages.

Gilliland and Risser (1977) give a full account of the use of a systems diagram in the assessment of a missile testing range in New Mexico. The complete diagram contains 61 links between components and whenever possible links are expressed in kilocalories of energy flow. When this is not

	Live plants	Roots	Standing Dead	Litter	Soil	Sheep	Sheep Dung	Dung Invertebrates	Slugs	Earthworms	Leachates
Live Plants											.169
Roots	1.0										
Standing Dead	1.0										
Litter			.597				.086		0.17		
Soil		.334		.645		.026	.010		.002	.092	
Sheep	.802		.198								
Sheep Dung						1.0					
Dung Invertebrates							1.0				
Slugs	.812						.188				
Earthworms				1.0							
Leachates	1.0										

Figure 4: Component Interaction Matrix for an Upland Grassland Ecosystem. (From Perkins, 1978)

	Live Plants	Roots	Standing Dead	Litter	Soil	Sheep	Sheep Dung	Dung Invertebrates	Slugs	Earthworms	Leachates
Live Plants	.198										.203
Roots	.198										.298
Standing Dead	.198										.298
Litter	.815		.615			.089	.089		.017		.135
Soil	.926	.334	.460	.737		.102	.076		.015	.092	.157
Sheep	1.19		.198								.198
Sheep Dung	1.17		.198			1.00					.192
Dung Invertebrates	1.14		.198			1.00	1.00				.169
Slugs	1.09		.037			.188	.188				.055
Earthworms	.802		.615	1.00		.089	.089		.017		.115
Leachates	1.20										.198

Figure 5: Upland Grassland Ecosystem Matrix summed to fifth order dependencies (From Perkins, 1978)

feasible, links are expressed in terms of appropriate measures of energy, such as decibels or curies for noise and radioactivity respectively. This basic data was collected from published sources. Each impact can be measured in terms of its effects on the amount of energy flows beween various components. The combined effects of all impacts affecting those links which can be expressed in kilocalories can be added together. The result can be given as a percentage of known gross primary production of the system. This enables a comprehensive, althought not complete, picture of impacts to be outlined. The magnitude of most impacts can be compared directly with the undisturbed system. It was found that in this case environmental impacts would result in the loss of 1% of the energy flow through the system. This was not considered to be of great importance. Figure 6 is a summary of the systems diagram.

The study by Gilliland and Risser was limited to ecological impacts; however, there have been studies showing how systems diagrams can link environmental and economic systems by translating all types of energy flow in the natural and man-made (economic) environment, into a common unit, generally coal equivalents. These can be converted into money terms by use of known or calculated energy/money ratios for different national or regional economies. Not all parameters can be reduced to energy flow values, even though they may be expressed quantitatively. Hydrological data, for example, are expressed in terms of the volume of water transferred between components. This inability to use energy data throughout reduces the value of the approach, which was formulated to provide a universal unit for appraisal.

Sorensen (1971) formulated a hybrid method, combining a matrix and a network, for the assessment of environmental impact of fifty-five alternative development types in the coastal zone. Sorensen argued that these development types involve certain activities, causal factors, which are responsible for inducing change in the coastal environment. Some causal factors may be common to a number of different development types. They were identified from a pre-defined stepped matrix. The implications of a particular causal factor can be explored through a cause-condition-effect chain (figure 7). A causal factor may induce an initial environmental response or condition in, for the sake of convenience, six categories, namely; water, climate, geophysical, biota, access and aesthetics. Initial condition changes may lead to direct effects. Alternatively, such

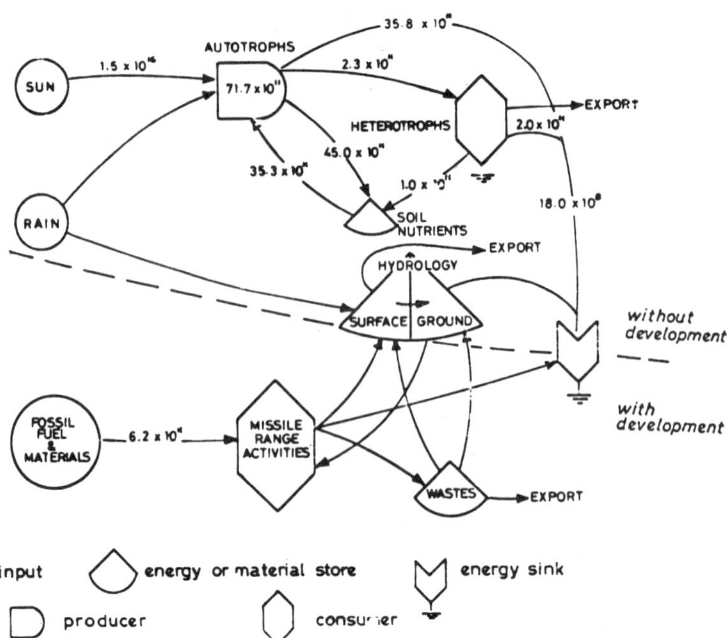

Figure 6: Simplified Systems Diagram (From Gilliland and Risser, 1977)

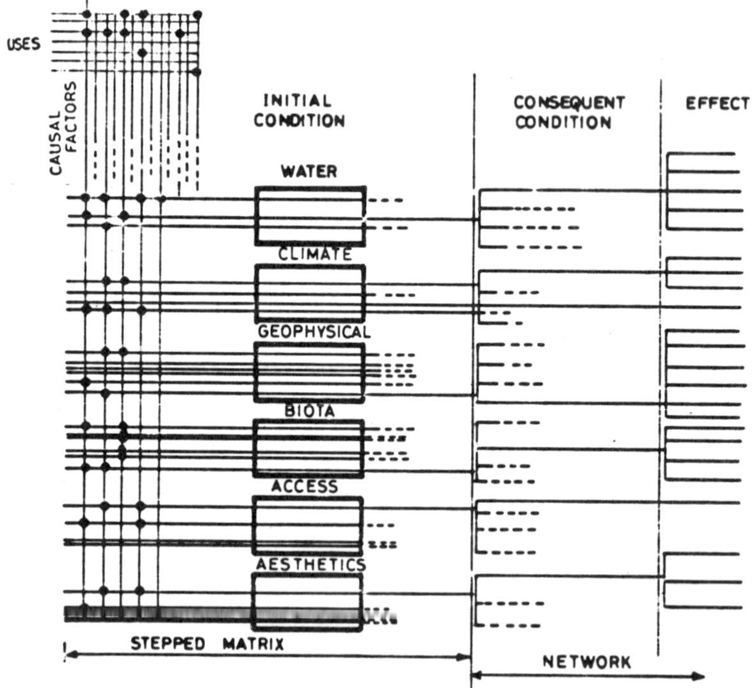

Figure 7: The Matrix - Network Approach. (Adapted from Sorenson, 1971)

changes may invoke a condition change leading to an indirect effect.

The Sorensen network represents an early attempt to provide information in a forward planning context, on the potential impacts of different development types as opposed to specific proposals. The data were computerized so that they were readily available and could be accessed in a variety of ways. Data could be provided with reference to specific uses, individual causal factors and cause-condition-effect sequences either independently or in combination. Alternatively, all causal factors affecting particular conditions, such as estuarine salinity, could be identified.

The Sorensen type of network appears to have two major uses in EIA. First, it could be used in an initial sieving exercise to identify those aspects of a particular development which should be subject to detailed scrutiny in subsequent analysis. Secondly, the information is presented in a format which can be readily comprehended. Thus, it would not only provide a suitable approach for building up a data base, but also a useful visual display of results for decision-makers.

Greater flexibility has been incorporated within the IMPACT computer based system formulated by the U.S. Forestry Service (Thor et al., 1978). It has been developed to assist forest planners in the preparation of EISs for proposals affecting forest land. IMPACT comprises a structured information base on a suite of interactive programs which enables the operator to access the available information from remote terminals. The data base consists of a keyword index of activities associated with particular development proposals. In addition, there is, for each activity, a generalized network containing a description of its possible environmental ramifications. The operator can use the computer to identify activities and follow individual networks. As the system is interactive and depends upon feedback from the operator, it provides a process of directed investigation. The operator uses information on the proposals to answer specific questions programmed into the computer. When the answers are not known, this may indicate that more detailed study is necessary. Alternatively, the operator may override particular lines of investigation which are not relevant to the proposal under consideration.

3. ASSESSING SPATIAL ASPECTS

Lapping (1975) argued that overlays should not be regarded primarily as an impact assessment method, but this belies the use and potential of the approach in appraising environmental impacts. Overlay techniques were developed originally for a planning study of the town of Billerica, Massachusetts (Manning, 1913). Subsequently, their use has been extended and the method has long been regarded as important in a wide variety of planning activities (Steinitz, Parker and Jordan, 1976). In the United States mandatory consideration of the environmental consequences of certain development proposals predates the National Environmental Policy Act. For example, minimizing environmental impact is an important consideration in route selection under the National Transportation Act (Matuszeski, 1975). Consequently, extension of the use of the overlay method to EIA was manifest initially in transportation studies, for example Alexander and Mannheim (1962) and McHarg (1968). Both manual and computer-based overlay methods have been used in EIAs. For the manual approach, a transparent overlay sheet is prepared for each parameter under consideration. On these sheets, the geographical extent and degree of environmental impact is represented by some colour or shade intensity coding. Aggregate impact is obtained by overlaying each of the transparancies on the base map. The aggregate impact on different areas is shown by the relative intensity of shading. Overlap can be combined in a variety of ways to show either total impact or impact on selected aspects, for example ecological or social impacts.

Although conceptually a simple method, there are practical difficulties in the manual application of overlays. An important constraint is the limited number of overlays that can be considered at any instant. Interpretation of more than a dozen overlays is often difficult and the results confused. Steinitz, Parker and Jordan (1976) claim that the aggregate map must be recoded and redrawn before it is of much value for analysing impacts, greatly increasing the inefficiency of the approach. The inflexibility and inefficiency created by the small number of parameters that can be included in any overlay analysis using a manual approach, is likely to prove an important constraint when dealing with large and complex development proposals. Alexander and Mannheim (1962) attempted to overcome this problem by a hierarchical grouping of parameters, in which several were aggregated to provide a single parameter. For example, soil properties such as nutrient status, porosity and stoniness may be aggregated to provide a parameter indicative of soil quality. The use of this aggregation process allowed the creation of a number of composite overlays, each comprising several parameters. Steinitz, Parker and Jordan (1976) adopted a different approach, using monochromatic overlays. A monochromatic overlay was produced for each parameter and then aggregated by a photographic method thereby greatly increasing the flexibility and efficiency of the method.

Many of the constraints of overlays are removed by adopting a computerized approach. This method has been used for appraising the impact of several developments, for example, Krauskopf and Bunde (1972), Sharpe and Williams (1973), Dudnik and Schachtel (1974), Newkirk and Throughton (1974) and Dooley and Newkirk (1976). The approach differs from manual overlays in that a computer file is produced which contains the digitized data for each parameter. These data can be accessed independently and analyzed in any combination for a particular proposal, the original data matrix remaining unmodified. Effectively, an unlimited number of parameters can be considered. A data file is prepared by subdividing the area under consideration into a number of grid squares and recording data from each square. Krauskopf and Bunde (1972) recorded 130 parameters related to the natural and cultural characteristics of the area. These were recorded on the basis of percentage cover or number of occurrences within each grid square. An example of a computerized overlay is shown in figure 8. Dudnik and Schachtel (1974) in contrast, analysed only a subset of grid squares and used computer programs to generate a complete data file for the whole area by interpolation. Data for each parameter was normalized, that is reduced to a common scale, for subsequent analysis.

The utility of the computer based approach lies in the ability to assess a range of alternatives rapidly from a common data set. This may involve ascribing weights to different parameters (the method varying in different cases). In contrast, McHarg's manual overlay method (McHarg, 1968) assumes that all parameters are equally important. Krauskopf and Bunde (1972) subjectively assess weights, whilst Dudnik and Schachtel (1974) use correlations between parameters and step-wise regression to identify the most important parameters. The ability of the computer to process data rapidly can be used to assess parameters in a variety of combinations. This allows various facets of the projected development impact to be appraised independently and with

Figure 8: Computerized Overlay Factor Maps and Composite Map (d).
(From Munn, 1979)

different weighting schemes. Aggregate impact can be represented as a numerical or shade intensity map in the computer print-out.

The use of computerized overlays imposes constraints on the format of data. As these techniques involve quantitative analysis of potential impacts, there are difficulties in accommodating qualitative data. In addition, grid boundaries rarely coincide with natural boundaries with the result that adopting a grid to record data introduces discontinuous spatial variation of parameters which may vary continuously in space. This may be an additional source of error. With electronic digitizers, however, map data may be transcribed directly to a computer data file without the imposition of a grid format. Although this eliminates potential error, it increases the difficulties of programming subsequent analysis.

One important potential use of computerized overlays in EIA appears to be in the assessment of linear developments, in particular alternative route locations (Krauskopf and Bunde, 1972, Dooley and Newkirk, 1976). Krauskopf and Bunde have gone beyond this in programming the computer to locate the route with the least cost in terms not only of construction costs, but also in terms of intangible social and environmental costs. Initially, there was little agreement between computer generated and design engineer preferred routes, because of the propensity of the engineer to avoid areas of particular sensitivity at the expense of greater aggregate impact elsewhere. These assumptions were incorporated into the program which then gave a good correlation with engineer designed routes.

Serious criticisms of the accuracy of the overlay approach have been made by MacDougall (1975). There are two major sources of error. The first is boundary definition. The primary data sources for producing overlays in developed countries generally are existing maps supplemented by aerial photographs, remote sensing imagery and fieldwork. Yet in many instances the boundaries between certain types of, for example, soil or vegetation, are drawn arbitrarily, because these is no distinct line between them, only a gradation of characteristics. When a large scale overlay transparancy is produced from a smaller scale map, such as a soil map, these inaccuracies may be substantial. The second source of error is tract heterogeneity. A tract on a vegetation map, classified as a particular vegetation type, may consist of different types of vegetation below the level of definition of the map. Even when tracts are defined with a purity of 95%

rarely achieved for many parameters, overlaying ten parameters may lead to an error of 40% in aggregate composition. An additional disadvantage of the manual approach is the inaccuracy in delimiting boundaries that may result from the transfer of data from maps and aerial photographs on to transparencies and drawing composite boundaries. With the wide adoption of modern cartographic techniques and high resolution remote sensing imagery, these errors will decrease.

4. CONCLUSION

Methodological developments have resolved some of the problems associated with identifying the indirect effects of proposals and demonstrating spatial aspects of impacts. Networks and systems diagrams have been advocated as a means of identifying higher order impacts because of their ability to show indirect relationships between environmental components. However, there seems to be little validity in this claim. Detailed knowledge of the direct and indirect linkages within a system and an understanding of the implications of these linkages are important prerequisites for the construction of a systems diagram or network. The methods are valuable for demonstrating the complexity of systems and for allowing linkages to be traced. In contrast, multiplication of a component interaction matrix does reveal indirect linkages in a system where only direct linkages are known.

These methods assume that a statement of indirect relationships indicates the likely effects of development upon a system. This assumption is valid only when the system is not able to adapt. Holling (1978) has argued that the complex structure of ecosystems often conceals their functions when perturbed. No component is directly or indirectly dependent upon all other components of the system. Ecosystems are often composed of a number of sub-systems which show considerable internal integration, but only a few clearly defined linkages to other sub-systems. Modification of one sub-system may lead to compensatory changes within other sub-systems which maintain the overall integrity of the ecosystem. Such homeostatic changes affect the response of ecosystems to environmental change induced by development and make certain ecosystems resilient (Simon, 1962). Therefore the tacit assumption that a statement of the existing inter-relationships infers the likely impacts on an ecosystem, may not be justifiable, even when the inter-relationships are expressed in a quantified manner. Projected indirect impacts may not occur because of the redirection of energy flow and the establishment of new equalibria.

Clark et al., (1976) have argued that EIA is likely to be effective only when it is fully integrated into the project formulation and decision-making process. Initially, overlay techniques were used for showing aggregate impacts on a geographical basis. Such techniques however, have a design capability which has the potential for integrating EIA into project formulation. This potential has been realized to a certain extent with linear developments such as transmission lines and roads, where there is a degree of route - and therefore design - flexibility. A computer can be programmed to select preferred routes with minimum aggregate impact or least impact on selected social or environmental parameters. The capacity of computers to rapidly assess a range of alternatives is an important capability which should be exploited in EIA regardless of any consideration of the method employed.

REFERENCES

Clark, B.D., K. Chapman, R. Bisset and P. Wathern: 1978, Environmental Impact Assessment in the USA: A Critical Review, Department of Environment Research Report No. 26, HMSO, London.

Dee, N., J.K. Baker, N.L. Drobny, K.M. Duke, I. Whitman and D.C. Fahringer: 1973, 'An Environmental Evaluation System for Water Resource Planning', Water Resources Research, 9 (3), 523-535.

Dooley, J.E. and R.W. Newkirk: 1976, Corridor Selection Methods to Minimize the Impact of an Electricity Transmission Line, James F. MacLaren Ltd, Toronto.

Dudnik, E.E. and W. Schachtel: 1974, 'Assessing and Evaluating the Desirability and Suitability of the Natural Environment for Human Activity or Development: A Methodology and Application', in D.H. Larson (ed.), Man-Environment Interactions, Evaluation and Application Part III, Dowden, Hutchinson and Ross, Stroudsberg, Pennsylvania, pp. 155-175.

Environment Canada: 1974, An Environmental Assessment of Nanaimo Port Alternatives, Environment Canada, Ottawa.

Gilliland, M.W. and P.G. Risser: 1977, 'The Use of Systems Diagrams for Environmental Impact Assesment: Procedures and An Application', Ecological Modelling, 3 (3), 188-209.

Holling, C.S.: 1978, Adaptive Environmental Assessment and Management, Wiley, Chichester.

Krauskopf, T.M. and D.C. Bunde: 1972, 'Evaluation of Environmental Impact Through a Computer Modelling Process', in R. Ditton and T. Goodale (eds.), *Environmental Impact Analysis: Philosophy and Methods*, University of Wisconsin Sea Grant Program, Madison, Winsconsin, pp. 107-125.

Lapping, M.B.: 1975, 'Environmental Impact Assessment Methodologies: A Critique', *Environmental Affairs*, 4 (1), 123-134.

Lennon, M.: 1979, *A Comparison of the Methods of Environmental Environmental Impact Analysis Using the Proposed Craig Goch Reservoir as an Example*, Unpublished B.Sc. thesis, University College of Wales, Aberystwyth.

Leopold, L.B., F.E. Clarke, B.B. Hanshaw and J.R. Balsley: 1971, *A Procedure for Evaluating Environmental Impact*, *U.S. Geological Survey Circular* No. 645, U.S. Geological Survey, Washington D.C..

Manning, W.: 1913, 'The Billercia Town Plan', *Landscape Architecture*, 3, 108-118.

Matuszeski, W.: 1975, 'The Scope of NEPA and its Relation to other Federal Legislation', in M. Blisset (ed.), *Environmental Impact Assessment*, Lyndon B. Johnson School of Public Affairs, University of Texas, Austin, pp. 91-99.

MacDougal, E.B.: 1975, 'The Accuracy of Map Overlays', *Landscape Planning*, 2 (1), 23-30.

McHarg, I.: 1968, *A Comprehensive Highway Route Selection Method*, Highway Research Record No. 246, Highway Research Board, Washington D.C..

Munn, R.E.: 1979, *Environmental Impact Assessment*, SCOPE 5, second edition, Wiley, Chitchester.

Newkirk, R.T. and M.J. Throughton: 1974, *Computer Based Environmental Assessment Applied to Environmental Assessment*, University of Western Ontario, London, Ontario.

Perkins, D.F.: 1978, 'Snowdonia Grasslands', in D.W. Heal and D.F. Perkins (eds.), *Production Ecology of British Moors and Montane Grasslands*, Springer-Verlag, Berlin, pp. 289-295.

Ross, J.H.: 1974, *Quantitative Aids to Environmental Impact Assessment*, Occasional Paper No. 3, Environment Canada Lands Directorate, Ottawa.

Simon, H.A.: 1962, 'The Architecture of Complexity', *Proceedings of the American Philosophical Society*, 106, 467-482.

Sorensen, J.C.: 1971, *A Framework for the Identification and Control of Resourse Degradation and Conflict in the Multiple Use of the Zone*, Masters thesis, Department of Landscape Architecture, University of California, Berkeley.

Steinitz, C., P. Parker and L. Jordan: 1976, 'Handdrawn Overlays: Their History and Prospective Uses', Landscape Architecture, 66, 444-455.
Sharpe, C. and D.L. Williams: 1972, 'The Making of an Environmental Fit', Landscape Architecture, 62, 210-215.
Thor, E.C., G.H. Elsner, M.R. Travis and K.M. O'Loughlin: 1978, 'Forest Environmental Impact Analysis - A New Approach', Journal of Forestry, 76, 723-725.

AFFILIATION

Dr. Peter Wathern is lecturer in Botany at the University College of Wales, Aberystwyth, and was formerly a research fellow with PADC at Aberdeen University.

Geoffrey A. Norton

SYSTEMS ANALYSIS AND ENVIRONMENTAL IMPACT ASSESSMENT

1. INTRODUCTION

The purpose of this paper is to present the systems analysis approach to environmental impact assessment (EIA), and other ecological assessment methods which we are attempting to develop at the Environmental Management Unit, Imperial College, London University. Before embarking on a detailed discussion of this subject, there are four points which ought to be made. Although some of these may seem elementary, they are worth repeating to provide the context to the paper. The systems approach to EIA, adopted by the Environmental Management Unit, will then be outlined and some case studies will be described to illustrate the approach.

1.1. Normative and positive ecology

The subject of ecology can be considered as consisting of two separate branches. The normative branch is concerned with moral aspects of ecology, and involves value judgements in deciding "what ought to be". A distinguishing feature of normative ecology is that its statements cannot be tested by the usual experimental techniques, but simply rely on a sense of what is good or bad. In contrast, the positive branch of ecology is concerned with such scientific questions as "what is or will be", and, at least in theory, its statements can be tested. This paper is concerned with positive ecology, although clearly normative aspects will have to be considered.

1.2. Defining the EIA problem

A crucial feature of an EIA is to determine the precise dimensions of the relevant problem. Although this may often be confused by institutional and other factors, it is suggested that, wherever possible, decision analysis should be used as the conceptual basis for this "scoping"

B. D. Clark et al. (eds.), Perspectives on Environmental Impact Assessment, 233–240.

process. Typically, this will consist of three phases:

(i) Structuring the problem. To determine the boundaries of EIA and the criteria on which the assessment is made, this initial phase consists of identifying the decision-makers and the groups affected. In developing a dialogue with these sectors, the purpose is to determine the range of choice available, and to obtain precise definitions of objectives and thus objective criteria.

(ii) Assessment. Within the framework set by structuring the problem, above, and using the objective criteria, this stage is concerned with the assessment of outcomes associated with the range of relevant planning, design and management options.

(iii) Presentation. At this stage, the assessment findings are presented in a form of value to the decision-making process.

This paper concentrates largely on the second phase of analysis, assessment.

1.3. Planning and EIA

Those decisions that have an impact on ecological processes and the environment fall into three categories: planning, design and management decisions. To assess the impact of planning decisions, a variety of integrated resource analysis techniques, consisting largely of land capability and suitability techniques, can be employed. While these techniques should be regarded as an integral part of EIA and will undoubtedly play an increasingly important role in the future. This paper will concentrate on EIA and the impact of design and management decisions.

1.4. An EIA screening procedure

The range of projects for which an EIA has to be carried out is likely to vary considerably, particularly with regard to the complexity of the problem, the institutional context and the time and expertise available for carrying out the study. On the assumption that the demand for EIAs will always exceed the capacity to carry them out, this raises two questions. How should resources be allocated between different projects? and secondly, which is the best technique for a particular problem? It is suggested that a screening technique consisting of the components shown in Table 1 could be of value in addressing both questions. How the assessment of the nature and scope of the decision problem might be achieved has already been indicated in the structuring phase of decision analysis. To assess the degree of potential impact, two factors have to be considered: the nature of the impacting activity and the nature of the ecosystem affected. This raises the issue of assessing ecological impact.

2. ECOLOGICAL IMPACT ASSESSMENT

The two components involved in assessing ecological impact are the nature of the impacting activity and the nature of the ecosystem affected. The features of the impacting activity that are of interest are the type of impact it can have (for example, heavy metal pollution, soil erosion, and habitat reduction) and the likelihood of these impacts occurring (Table 1). Two features, a normative and positive component, are associated with the nature of the ecosystem affected. The normative aspect is concerned with the importance of the ecosystem in social terms, such as conservational or recreational value, while the positive component is concerned with the ecological response of the ecosystem to impacting activities. This positive ecological assessment deals largely with the type of impact and the ecological response that results (Table 1).

Thus, the problem of ecological assessment consists of assessing the expected changes in the physical environment, the biotic component, ecosystem morphology and size, and the effect of these changes on those features of ecosystems that are valued by society, namely environmental quality, productivity, species composition, ecosystem behaviour and landscape characteristics. Traditionally, this problem of ecological assessment has been tackled by empirical investigations, often involving surveys, together with an appeal to the judgement of experts, based on "principles" of applied ecology. A particularly pressing need is for such ecological "principles" as exist to be identified and made explicit and workable; this point will be discussed later.

One way of organizing expert judgements is to adopt a flexible matrix approach. Alternative or complementary techniques, can also be used, all however, falling under the general heading of systems analysis.

Table 1. A Screening Process for EIA

WHAT IS THE DECISION PROBLEM?
1. Range of choice:- associated with site, design, construction, method and post-construction operation.
2. Choice of associated protection projects:- to alleviate expected impacts.
WHAT IS THE DEGREE OF POTENTIAL IMPACT BOTH DIRECT AND INDUCED?
1. Nature of impacting activities - type of impact:- Change in the physical environment e.g. introduction of toxic materials, change in water table.
Change in biotic components, e.g. planting, harvesting, pest control.
Change in spatial size and form of ecosystem, e.g. reduction of habitat, erection of barriers.
2. Nature of impacting activities - likelihood of impact:- Certainty or probability distribution.
3. Nature of ecosystem(s) affected:- (a) Importance of ecosystem e.g. critical resources such as nature reserves or national parks.
(b) Ecological consequences.

Table 2. Descriptive Analysis of the Savanna Ecosystem

STRUCTURE
1. Physical
2. Vegetation
3. Fauna
DYNAMICS
1. Changes over time - such as seasonal, regular, long term changes and pertubations due to fire, flood, drought, insects etc.
2. Principal pathways and flows in terms of material flows and control factors

3. THE SYSTEMS ANALYSIS APPROACH

The work of the Environmental Management Unit at Silwood Park, Imperial College has been concerned with the development and use of three stages of systems analysis, namely the descriptive stage, the qualitative modelling stage and the quantitative modelling stage.

3.1. The descriptive stage

In drawing on details of the problem and on available information about the site, simple descriptive techniques are employed to provide a systematic means of describing the interaction and dynamics of the ecosystem, and how it may respond to disturbance. The purpose of this "first shot" at an explicit overview is to serve as an initial screening technique as well as to identify key questions for further analysis. From our experience, this type of technique is extremely valuable in workshop sessions, its dynamic nature providing a real stimulus for discussion and allowing an improved description of the problem to be developed.

3.2. The qualitative modelling stage

At this stage, simple synoptic models of ecology and decision analysis are used to "tease-out" the key questions identified in the previous stage. In using these models to obtain a feel for the dynamics of the system, an assessment can be obtained of how various impacts might affect it, and the key parameters and variables involved can be identified. Again, the techniques involved at this stage of analysis can be employed for screening or for identifying research priorities.

3.3. The quantitative modelling stage

Here we are concerned with detailed simulation models that address far more specific questions of how impact may occur and how it might change with (say) different climatic conditions and changes in design. Since this level of analysis is far more demanding of data, time and expertise, it is only likely to be justified for major projects, and then only when relevant, for instance, for assessing watershed pollution.

Having set out the approach, it is now appropriate to consider two case studies that illustrate how it can be applied in practice.

4. THE SAVANNA ECOSYSTEM

The savanna ecosystem in southern Africa is subject to a variety of impacts, including stocking with cattle, and cutting of trees and shrubs. The intention of the South African Savanna Ecosystem Study is to obtain a sufficient understanding of ecological processes for the assessment of the impact of these activities. As well as the conventional empirical and experimental techniques of ecology, systems analysis techniques are also being employed. This example is used to illustrate how descriptive and qualitative systems analysis techniques have been used.

The overall form of the descriptive analysis, which is described in detail in Walker et al., (1978), is shown in Table 2. The description of the ecosystem is divided into two parts, structure and dynamics. The three components of structure - the physical environment (soils, hydrology and so on), vegetation, and fauna - are portrayed by conventional techniques, such as mapping and vertical profiles. To describe the dynamics of the system, in terms of changes over time and principal pathways and flows, less conventional techniques, such as time profiles and interaction matrices, are employed. As shown by Walker et al., (1978), this simple but explicit descriptive model of the savanna ecosystem has allowed a series of key questions requiring research investigation to be identified. It is also suggested here that such a model can be used to "think through" an impact assessment.

More recently, Walker and colleagues at the University of British Columbia have used qualitative modelling techniques to investigate the dynamics of the system. Using a simplified model with five state variables (two soil water zones, woody vegetation, grass, and herbivores) the authors build up an overall picture of the dynamics of the system by synthesising the constituent ecological processes. For instance, from the information that is available on the effect of grass cover on the infiltration rate and grass cover is derived. Similarly, a graphic relationship between the rate of change of grass simply as a function of grass cover is obtained. By combining these two functions, the more complex relationships between rate of change of grass and grass cover, including the effect on infiltration rate, can be derived.

In a similar fashion the relationships associated with woody vegetation can be obtained and combined with the grass relationships to obtain an overall picture of the dynamics of the wood-grass-subsystem, as driven by water competition, and in the absence of herbivores. At this stage of analysis, two stable points are apparent: complete woody vegetation cover and grass cover with scattered trees. The final step in the analysis is to introduce the relationships associated with grazing and browsing herbivores.

In this graphic form, or more practically in a computerized form, this model can provide a means of assessing the qualitative nature of the impact of such activities as clearing woody vegetation, and increasing stocking rate.

5. QUANTITATIVE MODELLING CASE STUDY

The role of quantitative simulation models in impact assessment has been recognized for some time now, particularly for assessing the impact of point sources of pollution on air and water quality. As non-point source pollution, particularly the effect of agricultural practice on water quality, becomes increasingly important (Loehr _et al_., 1979), quantitative simulation models are likely to be of equal value in this field as well. As an example, simulation techniques are being used in a project carried out in the Environmental Management Unit to investigate the effect of changes in land use and agricultural practice, combined with various climatic scenarios and water management options, on the nitrogen levels in potable water extracted from a river in the United Kingdom.

A second, and much simpler, type of quantitative model that can be of value in impact assessment, is the use of simple, empirically-derived, relationships. For instance, the quantitative expression of the relationship between species number and area of habitat can be used to assess the impact of habitat reduction on species diversity and, by the use of incidence functions for different species, the likely impact on species composition can also be derived. Such a study has been carried out at Imperial College, looking at the effect of woodland area on bird species.

6. CONCLUSION

It should be emphasised that the systems analysis techniques described above, are not intended to replace those that already exist but simply to add to the range of techniques available for carrying out EIA. For the initial stages of an EIA, the use of descriptive techniques, especially in the context of workshop sessions is strongly recommended. This descriptive stage can help in making decisions on the likely ecological response to impacting activities, as well as in identifying areas for further investigation and deciding on appropriate analysis techniques.

Where ecological dynamics play an important role in potential impact, qualitative techniques offer considerable potential. This is particularly true where a whole series of EIAs are likely to be carried out in the future for a particular ecosystem or generic group of ecosystems. Quantitative techniques, particularly simulation modelling, will continue to have an important role to play in EIA, especially as more attention is focused on non-point sources of pollution. In the future, we might also expect to see much simpler, empirical-quantitative, models being used to answer specific ecological assessment problems.

REFERENCES

Loehr, R.C., D.A. Haith, M.F. Walter and C.S. Martin (eds.): 1979, Best Management Practices for Agriculture and Silviculture, Ann Arbor Science, Ann Arbor, Michigan.

Walker, B.H., G.A. Norton, G.R. Conway, H.N. Comins and M. Birley: 1978, 'A Procedure for Multidisciplinary Ecosystem Research: With Reference to the South African Savanna Ecosystem Project', Journal of Applied Ecology 15, 481-502.

AFFILIATION

Dr Geoffrey A. Norton is a senior lecturer at Imperial College, University of London and a member of the Environmental Management Unit.

Ros Atkins

A COMPARATIVE ANALYSIS OF THE UTILITY OF EIA METHODS

1. INTRODUCTION

This paper presents the findings of a short research project into EIA methods. There are two components. The first summarizes a theoretical comparision of six EIA methods; and the second describes an attempt to evaluate the utility of these methods based on a case study.

2. CRITERIA FOR A THEORETICAL COMPARISON OF EIA METHODS

A study of EIA methods proposed in the U.S., and an examination of review literature, show a need to identify criteria which will produce a complete and effective EIA method, Beer (1977), Catlow and Thirlwall (1976), Clark *et al.*, (1978), Munn (1975), Ortolano and Hill (1972), Skutsch and Flowerdew (1976), Warner and Bromley (1974) and Winkle *et al.*, (1976), have addressed this issue. Disagreement surrounds the selection of criteria, due to problems inherent in predicting environmental impacts. In addition, the significance of impacts differs between different social groups, so that public involvement in the evaluation process is important.

A review of environmental literature identified the following criteria which various authors claim are necessary for a complete EIA.

(i) flexibility to accommodate differing objectives of decision-makers;
(ii) screening to identify projects likely to cause significant impacts;
(iii) comparison of alternatives;
(iv) identification of environmental factors likely to be affected in specified geographic areas;
(v) identification of interactions between impacts, including induced impacts, synergistic, potentiating and dampening interactions;
(vi) identification of both positive and negative effects;

B. D. Clark et al. (eds.), Perspectives on Environmental Impact Assessment, 241–252.

(vii) prediction of the degree and timescale of the reversal of the impacts and the commitment of natural resources;
(viii) evaluation of impacts in terms of magnitude and significance;
(ix) identification of hazard and risks;
(x) recognition of the uncertainties inherent in the information base and the predictive abilities;
(xi) consideration of the resiliance of systems;
(xii) identification of monitoring needs during implementation and operational phases of the project;
(xiii) consideration of the validity of the results;
(xiv) separate assessments for different timescales such as the construction, operation and post-operation periods;
(xv) efficient communication of these findings and use of time, money, data and personnel.

The original aim of EIA, in the U.S., was to enable comparison of the state of the environment with and without a development. Although this may be desirable, it is problematical as Munn (1975) points out: "Fundamentally different alternatives are difficult to compare at the impact interpretation stage, since impacts differ in kind as well as in magnitude. Incremental alternatives have the advantages that the without action state is the same but the level of detail required to discriminate amongst different alternatives is often greater because differences are in degree and not type". Even the apparently straight-forward procedure of comparing the environment without the development with the post-development environment is not easy, as the environment is a dynamic process. However, reviewers seem agreed on an additional criterion: the need to consider fundamentally and incrementally different developments, as well as alternative sites. Munn (1975) states that "an ideal prediction system yields estimates of magnitude expressed in units required for the next stage of the assessment". The "ideal system" does not yet exist due to the difficulties of predicting all impacts in measurable units. For example, Winkle *et al.*, (1976) clarified the inherent difficulties in predicting the magnitude of ecological impacts. The difficulties increase as predictions are attempted for more complex circumstances; larger areas and longer timescales.

The evaluation of the significance of impacts is also difficult. Two aspects are involved; the importance of the impact itself and relative significance of different impacts. The evaluation procedure is subjective and controversial, because the significance may differ between social groups.

Since one impact rarely has a single ´social significance´ the importance of public involvement in the evaluation process is clear.

Six EIA methods were selected as being representative of the main types. In order to evaluate the utility of a number of EIA methods, their capabilities were assessed in relation to the neccessary criteria listed in the preceding section. A brief account of the methods chosen is presented below.

2.1. The McHarg overlay approach

This approach, proposed in Design with Nature (McHarg, 1969), relies on the ability to map an area and the impact constraints imposed by environmental and social land use factors. The different factors are drawn onto separate transparent maps of the area. For each factor, a shading system then indicates the degree of impact likely to be caused by the project under evaluation. The transparent maps are overlain on each other, to produce a visual display of all the mapped impacts. Using this composite map, it is possible to identify alternative project locations which would give rise to fewest environmental impacts. The process of producing a composite map and identifying sites of reduced environmental impact can best be undertaken on a computer (McHarg, 1969).

2.2. The Leopold Matrix

The Leopold matrix as proposed by Leopold et al., (1971) in U.S. Department of the Interior Geological Survey Circular 645, as A Procedure for Evaluating Environmental Impact, is well-known and has formed the basis of several other methods. The matrix consists of a list of human actions on the horizontal axis, and a list of environmental factors on the vertical axis. A visual display of any interactions considered to be significant can be produced by placing a diagonal line in the appropriate cell of the matrix. A numerical evaluation of the magnitude of the impact (on a 1-10 scale) can be placed in the upper right-hand corner of the cell, while an evaluation, also on a 1-10 scale, to signify the importance of the impact can be placed in the bottom left-hand corner. These values are not equatable. A written description of any significant impacts together with any columns or rows in which many interactions occur, must be presented. Such documents should also include description of the proposed and alternative developments, probable impacts, unavoidable adverse impacts, irreversible and irretrievable commitments of resources, and discussion of objections to the proposed development, as laid down in NEPA guidelines. A detailed discussion of this method and its variants can be found in the paper on methods by Bisset which is contained in this volume.

2.3. The Environmental Evaluation System

The Environmental Evaluation System (EES) was developed by Dee *et al.*, (1973) in the Battelle Laboratories for the U.S. Bureau of Land Reclamation. This method is based upon the development of value functions for seventy-eight environmental parameters. A measurement for each parameter is converted to a 0-1 scale of environmental quality, in which 0 represents bad and 1 represents good. Such value functions are constructed on the judgement of experts. Each expert assigns a weight to each parameter from a total of one thousand "parameter importance units", using a ranked pairwise comparison technique. The weighted list of each expert is then subject to the Delphi technique to obtain a consensus weighting from other experts. Then each environmental quality value is multiplied by its weighting value to produce the impact value. The final environmental impact value is defined as the difference between the sum of the impact values with and without the project. A discussion of this method can also be found in the methods paper by Bisset contained in this volume.

2.4. The Sorensen Matrix-network

The matrix-network developed by Sorensen (1971), integrates into one display a matrix for the identification of impacts, a network of consequent impacts, and proposals for actions to avoid or reduce adverse consequences. The resource uses are given on the horizontal axis of the matrix, while the vertical axis consists of a list of human actions (termed causal factors by Sorensen) involved in the proposed development. For each cell in which an interaction is identified, a causal chain is given in a table next to the matrix. The chain of impacts can be followed through to the desired level of detail until an effect is reached which creates a resource-use conflict. The number of impacts included in the chain will, therefore, depend upon whether the predicted impacts are considered to be significant in resourse-use conflict. This method is discussed in the paper by Wathern included in this volume.

2.5. Adaptive Environmental Assessment and Management

This method, proposed by Holling (1978), does not consider that the analysis and evaluation stages of impact assessment are separate; instead the aim is to produce a flexible, adaptive, management approach. Initially, a workshop comprising scientists, managers and policy-makers meets to consider and define the variables, management acts, objectives, indicators, time and spatial boundaries, and to determine alternative actions. The core group of experts then develops an explicit computer model of the system. Since the model is based on an understanding of the processes, as well as the components, it will mimic behavior over time and be responsive to different modes of management. The model is then subjected to validity testing, and is used to develop resilient policies to avoid adverse impacts (Holling, 1978).

2.6. The PADC Manual

The PADC Manual, prepared by Clark et al., (1976 and 1981), was produced in an attempt to resolve two problems encountered by U.K. planners when assessing development proposals. These two problems are those of obtaining sufficient detailed information from prospective developers, and the lack of a systematic procedure for the appraisal of proposals. This method comprises three activities; acquisition of information, identification of likely impacts and appraisal of these impacts. The method relies upon checklists and matrices. The PADC Manual is reviewed in the paper by Clark in this volume.

3. EVALUATION OF UTILITY

The six methods were examined to determine whether they satisfied any of the criteria listed in section 2. A table was constructed to compare the fulfilment of the criteria by the six methods (see table 1). Since many of the fifteen criteria listed are composites, construction of this table involved their subdivision to yield thirty individual criteria. In table 1, a star is used to indicate that the author of the method claims that criteria will be fulfilled. The table thus provides a concise picture of the relative strengths and weaknesses of the different methods. It does not provide an evaluation of the extent to which the author's claims may be justified in practice.

	McHarg Overlay	Leopold Matrix	Dee Environmental Evaluation System	Sorensen Matrix Network	Holling Adaptive Modelling	PADC Manual
Flexibility					*	*
Screening		*				*
Alternative Sites	*	*	*	*	*	
Fundamental Alternatives	*	*	*	*	*	
Incremental Alternatives	*	*	*	*	*	*
No Development		*	*		*	*
Area Considered		*	*	*	*	*
Geophysical and Meteorological	*	*	*	*	*	*
Ecological	*	*	*	*	*	*
Socio-Economic	*	*	*	*	*	*
Induced				*	*	*
Synergistic etc					*	*
Positive and Negative Effects		*	*	*	*	*
Reversibility, Resource Commitment		*			*	*
Magnitude	*	*	*	*	*	*
Signifiance	*	*	*	*	*	
Risk		*			*	*
Uncertainty			*	*	*	
Resiliance					*	
Monitoring				*	*	*
Validity of Results		R			*	
Construction		*	*			*
Operation	*	*	*		*	*
Long-Term		*				*
Post-Operational						
Concise Communication	*		*	*	*	*
Cost						
Time						1
Data						
Personnel						P

<u>Key</u>

Y Explicit Method Proposed or Statement made by Authors of Methods

R Reviewed

P Planning Department

1 1 Year

<u>Table 1: Comparison of Six EIA Methods</u>

3.1. Conclusions on a theoretical examination of utility

Table 1 suggests that the Holling method is capable of fulfilling most (21) of the thirty criteria; followed by the PADC method (20) and the Leopold method (18); only about half of the criteria are met by the Dee method (15) and the Sorensen approach (14); the McHarg approach satisfied the least number of criteria. Since this comparison is theoretical, and in reality different methods are more appropriate for different tasks, no recommendations of utility can be made as a result of table 1. There is a need, therefore, for an examination of the utility of different methods in practice.

4. EVALUATION OF UTILITY BY USE OF A CASE STUDY

The utility of the six EIA methods was examined by the use of a retrospective case study of a stormwater balancing lake, Willen Lake, situated in Milton Keynes (50 miles north of London). The six methods were compared with the existing *ad hoc* consultative procedures based on U.K. planning, development control and environmental legislation. The lake was designed in consultation with the Anglian Water Authority, engineering consultants, recreation managers and ecologists. The main objective of the lake was to control potential flooding caused by increased surface water runoff on to impervious Oxford clays following urban development. Linked to this objective was a desire to provide wildlife and recreational amenities.

4.1. Method of investigation

Investigation of the consultative decision-making process showed that no potential environmental consequences of Willen Lake were considered. Priority consideration was given to the following objectives:

(i) to control the release of stormwater into the River Ouse, in order to restrict downstream flooding;
(ii) to design the lake with minimum cost;
(iii) to limit oil pollution of the river;
(iv) to create a lakeside development site to attract hotel and office development;
(v) to use the lake and lakeside to provide facilities for residents; and

TABLE 2 : IDENTIFICATION OF SIGNIFICANT CONFLICTS BETWEEN ENVIRONMENTAL PROCESSES AND HUMAN OBJECTIVES

Method	flooding	erosion	oil (etc) pollution	Biological requirements of fish	macrophyte growth process	silting and colonisation	pollution conflict with ecology	recreation conflict with ecology	ecology conflict with recreation
Present	*	*	*	*	⊗	⊗	⊗	⊗	⊗
McHarg	⊗	⊗	⊗	⊗	⊗	⊗	⊗	⊗	⊗
Leopold	*	⊗	*	⊗	⊗	⊗	*	*	⊗
Dee	⊗	⊗	*	⊗	⊗	⊗	*	*	⊗
Sorensen	*	*	*	⊗	⊗	*	*	*	⊗
Holling	?	?	?	?	?	?	?	?	?
Clark	*	⊗	*	⊗	⊗	*	*	*	⊗

Key Impact Identification * Probable ? Possible ⊗ Unlikely

(vi) to create an environment suitable for those interested in wildlife and angling.

The study examined the performance of the existing, ad hoc decision-making process (hereafter referred to as the Willen approach) in terms of whether it achieved its objectives and of whether unpredicted impacts occurred. Interviews were held with personnel involved with the management and use of the Willen, and those affected by the balancing lake after its construction.

4.2. Results of the investigation

These interviews provided evidence to show that the Willen approach had satisfactorily fulfilled all the objectives listed above. The interviews however, revealed the existance of resource-use conflicts between objectives (v) and (vi), which had not been considered: some recreational uses of the lake conflicted with the environmental needs for the wildlife objectives. The interviews also provided evidence of problems arising from the management of the lake. No allowance had been made during the calculation of the lake storage volume for the growth of submerged aquatic plants. Excessive growth was controlled by the use of costly and degradable herbicides. Destruction of the plants led to their decomposition, causing odour problems for residents and recreation seekers, while also destroying aquatic life dependent on the plants; causing instability in the lake ecology, and threatening the basis for the desired fish and bird life. The growth of submerged plants also conflicted with recreational boating on the lake. Other problems included plant colonization of the lake margins and increased siltation caused by the plants.

4.3. Discussion

The problems outlined above were not predicted by the Willen approach, it was possible that some of the EIA methods would have done so, especially as some of the methods are explicitly designed to identify resource-use conflicts, and conflicts arising between environmental processes and human objectives (for example, Sorensen and Holling). Table 2 was therefore constructed in order to compare the effectiveness of the present ad hoc decision process, as illustrated by the Willen approach, with the potential effectiveness of the six EIA methods described earlier. The table indicates where the author of an EIA method specifies consideration of an environmental process which had been considered by the Willen approach, and those processes omitted.

5. CONCLUSIONS

Table 2 indicates that the McHarg method would not have been as useful as the Willen approach. The Dee and Leopold methods also failed to identify conflicts identified by the Willen approach. While the PADC Manual appeared to identify a greater number of problems missed by the Willen approach, the Sorensen approach seemed likely to identify most conflicts. Identification of specific conflicts by the Holling workshop method would depend upon the participants; it is not possible to be sure of the effectiveness of this method.

In conclusion, this retrospective case study reveals an area of inadequacy both in the existing consultative decision process and also in existing EIA methods. All inadequately consider conflicts between environmental processes and human objectives, and the limitations and costs which such processes can cause. Perhaps the limitations of reductionist science are partly responsible, in that systems are reduced to their component parts in order to understand and control them. While this approach is successful in terms of its ability to understand the behaviour of specified parts, it provides only a limited comprehension of the interactions within the system as a whole. This limited comprehension of interactions has led to unwanted impacts. Hence there is an increasing need for communication between specialists. Environmental science has developed partly to counter previous over-specialization in science and society. However a major area of inadequacy in environmental science is the limited understanding of ecological processes and the even more limited abilities to predict ecological consequences. Warner and Bromley (1974) after reviewing the methods proposed by Dee, Leopold and others, concluded that EIA has largely been concerned with structural aspects rather than functional aspects. This emphasis has been due, in part, to a lack of measurable parameters to accommodate system functions. Important functional concepts, such as food webs, energy pathways, diversity patterns and nutrient cycles, have been widely discussed but seldom included in EIA methods. This is partly due to the early stage of the science of ecology, in which predictive abilities are only now evolving.

REFERENCES

Beer, A.: 1977, 'Environmental Impact Assessment: A Review Article', Town Planning Review, 48 (4), 389-396.

Catlow, J. and C.G. Thirlwall: 1976, Environmental Impact Assessment, Department of Environment Research Report No.11, HMSO, London.

Clark, B.D., K. Chapman, R. Bisset and P. Wathern: 1976, The Assessment of Major Industrial Applications: A Manual, Department of Environment Research Report No. 13, HMSO, London.

Clark, B.D., K. Chapman, R. Bisset and P. Wathern: 1978, Environmental Impact Assessment in the USA: A Critical Review, Department of Environment Research Report No. 26, HMSO, London.

Clark, B.D., K. Chapman, R. Bisset and P. Wathern: 1981, Assessment of Major Industrial Applications: A Manual, HMSO, London.

Dee, N., J. Baker, N. Drobney and K. Duke: 1973, 'An Environmental Evaluation System for Water Resource Planning', Water Resources Research, 9 (3), 523-535.

Holling, C.S. (ed.): 1978, Adaptive Environmental Assessment and Management, International Series on Applied Systems Analysis 3, John Wiley and Sons, Chichester.

Leopold, C.B., F.E. Clarke, B.R. Hanshaw and J.R. Balsley: 1971, A Procedure For Evaluating Environmental Impact, Geological Survey Circular No. 645, U.S. Geological Survey, Washington D.C..

McHarg, I.: 1971, Design with Nature, Doubleday Natural History Press, New York.

Munn, R.E. (ed.): 1975, EIA: Principles and Procedures, SCOPE Report No. 5, International Council of Scientific Unions, Toronto.

Ortolano, L. and W.W. Hill: 1972, An Analysis of Environmental Statements for the Corps of Engineers' Water Projects, Stanford University, California.

Skutsh, M. McC. and R.T.N. Flowerdew: 1976, 'Measurement Techniques in Environmental Assessment', Environmental Conservation, 3 (3), 209-217.

Sorensen, J.: 1971, A Framework for Identification and Control of Resource Degradation and Conflict in the Multiple Use of the Coastal Zone, Department of Landscape Architecture, University of California, Berkeley.

Warner, M.L. and D.W. Bromley: 1974, Environmental Impact Assessment: A Review of Three Methodologies, Institute for Environmental Studies, University of Wisconsin, Madison.

Winkle, W.V., S.W. Christensen and J.S. Mattice: 1976, 'Two Roles of Ecologists in Defining and Determining the Acceptability of impacts', Journal of Environmental Studies, 9, 247-254.

AFFILIATION

Ros Atkins was formerly a member of the PADC research unit and has now turned to a new career of nursing in Leeds.

M. Nay Htun

DEVELOPMENT OF UNEP GUIDELINES FOR ASSESSING INDUSTRIAL ENVIRONMENTAL IMPACT AND ENVIRONMENTAL CRITERIA FOR THE SITING OF INDUSTRY

1. INTRODUCTION

The United Nations Environment Programme, UNEP, with the assistance of W.S. Atkins Research and Development, started work on developing guidelines for Assessing Industrial Environmental Impact and Environmental Criteria for the Siting of Industry in March 1978. At three ad hoc expert meetings it was agreed that the guidelines to be developed should not be of a cook book type of manual which would recommend a rigid step by step approach for undertaking environmental impact assessment. A rigid approach was considered not suitable since the guidelines would need to have global applications. Furthermore, as each industrial project would be somewhat different from another, and sited in different types of locations, a flexible approach would be needed to carry out impact assessments. The guidelines should therefore aim to provide a framework from which an appropriate environmental impact assessment process could be structured, taking into account the local conditions.

The draft document was sent to a large number of persons in many countries of the world for review. The comments were reflected and incorporated in the final Guidelines. Three regional workshops have been held in Mexico City, Tashkent and Kuala Lumpur to present the Guidelines.

The Guidelines consist of four major chapters. These deal with procedure, methodology, assessment of pollution and ecological effects, and assessment of social economic effects.

2. PROCEDURAL FRAMEWORK

This chapter offers suggestions to those charged with deciding whether or not to have procedures for the assessment of industrial environmental impact and the siting of industry. The

B. D. Clark et al. (eds.), Perspectives on Environmental Impact Assessment, 253–263.

aspects of procedure which are relevant to establishing a system for undertaking and reviewing the environmental aspects of industrial projects are discussed. Whereas environmental impact assessment can obviously be undertaken without a set of formal procedures, it is nevertheless important to consider the use of such procedures in order to ensure firstly that all relevant projects are subject to some form of organized study and secondly, that the best use is made of the ensuing results. The procedures define a structure for achieving a formal assessment of the environmental implications of industrial projects. They are of particular relevance at the feasibility study stage of industrial projects when the technical and commercial aspects are being examined and a site chosen. During this period, the procedures are concerned with the various enquiries and analyses to be undertaken and with the allocation of responsibility. Not all projects require a detailed environmental impact assessment and a comprehensive study is necessary only in a minority of cases. But a useful standard practice would be to include a preliminary environmental assessment as an integral part of all feasibility studies.

The chapter suggests the procedural framework that could be used in the following stages of a project development : strategic planning, feasibility studies, design and construction, and operation. Guidance on institutional procedures such as the roles of "regulatory agency", "review office", the public as well as procedures to be used when choosing a site, and the expansion of an existing project are also discussed.

The "regulatory agency" would establish administrative procedures for reviewing development procedures, vested with legal powers to request essential information and data, have the authority to impose conditions attached to the development and decide whether or not to allow the development to proceed and on what terms. The "agency" would also be responsible for establishing the machinery for environmental legislations, amongst which would be included standards, norms and criteria for industrial discharges, occupational health, products manufactured and environmental quality. The existence of such legislations will provide a reference for the scale and degree of mitigating measures which should be undertaken to ameliorate the adverse environmental impacts identified and predicted in the impact assessment study.

The "review office" suggests the responsibility of ensuring that the environmental impact assessment is properly carried out and evaluated. It will draw up the terms of reference for the assessment study and advise the "regulatory agency" on the significance of the environmental consequences and the suitable conditions to be applied to any permission to develop a site. The Guidelines also suggest some of the major functions of the "review office" and these are to :

(i) conduct preliminary environmental assessments;
(ii) request information and data relevant to environmental impacts, for which the office must be duly empowered;
(iii) decide the terms of reference and to initiate full environmental impact assessments;
(iv) ensure that the full environmental impact assessments have been adequately completed within the defined terms of reference;
(v) submit the completed, full impact assessment, together with any separate contributions from other organizations, with recommendations to the regulatory agency;
(vi) maintain records of completed studies and such other data and information as is relevant to environmental studies within its responsibilities;
(vii) serve as a channel and forum for the exchange of information and opinions on environmental matters, for example as a source for baseline data.

The project brief from the developer should also be submitted to the "review office". The brief should include the following items of information and data :

(i) name of the developer;
(ii) sources and quantities of raw materials needed for manufacturing;
(iii) types and quantities of products to be manufactured,
(iv) types of processes to be used;
(v) proposed methods for waste treatment;
(vi) proposed options for siting;
(vii) methods for transporting raw materials and finished products;
(viii) possible new infrastructure requirements;
(ix) expected new employment opportunities;
(x) timing of construction and initial operation.

Often, not all projects require a full environmental

impact assessment study. What would be required is an assembly of readily available information and data relevant to the potential environmental impact of the project. On the basis of this "preliminary" information, the "review office" will screen and decide the need for fuller studies including the terms of reference. The types of question or screening tests which the proposed project should be subjected to at the preliminary assessment stage would include :

(i) Are there significant physical and/or health hazards associated with the project ?
(ii) Will there be significant disturbances to existing communities ?
(iii) Will there be requirements for major physical infrastructure development ?
(iv) Is there potential conflict with other economic interests ?
(v) What is the size of the new employment opportunities in comparison to present local employment ?

Another element of the procedural framework, is the role of the public when carrying out environmental impact assessment. The reasons for public involvement are best described as being of a technical nature, since they are intended to improve the effectiveness of the environmental study by providing the concerns and perceptions of the community that would be affected by the development. There are however, other equally important reasons for widening the scope of involvement, in particular because of the potential conflicts of interest involved. The opportunity given for views to be represented and the exposure to public scrutiny of the assessment study process may serve to improve the acceptability of the final decision. Since the state of the environment has become a matter of growing concern to the public, the public wants to know more about the issues that have to be examined and the process used for examination. Various procedures that could be used for public involvement are discussed in the chapter.

3. METHODOLOGY FOR ENVIRONMENTAL IMPACT ASSESSMENT

This chapter is intended to be an introduction to the managerial and technical aspects involved in carrying out an environmental impact assessment and the preparation and presentation of the findings in an environmental assessment report. It is addressed primarily to administrators and especially those concerned with setting up and operating the "review office". The suggested elements in the Methodology include a study team, scoping processes that could be used to determine the scope and extent of the impacts that should be studied, tools that are available for carrying out impact studies, the methods of evaluating the impacts, and suggestions for the preparations of the study report.

With regard to the study team, a suggestion is made for a high calibre and dedicated core team co-ordinated by a person who has good managerial and communicating abilities. The extent of the subject matter is such that few, if any, individuals could adequately cover the whole range of expertise required. Generally, the bulk of the work could be undertaken by the core team and specialist advice and assistance acquired on a limited basis, to augment the team as and when needed. The team would be responsible for interpreting the terms of reference of the study, drawing up a programme for its implementation, deciding the sources for information and data, preparing plans for consultation, evaluation and reporting. An essential management function is that of defining the terms of reference for individual specialists and the collation of the views and findings of the specialists into a clear, concise and comprehensive statement of potential environmental impacts.

With regard to identifying the scope of the study, every attempt should be made to concentrate the assessment on those aspects of environmental impact which, it is anticipated, will be of significance in the final decision-making. This process is designed to focus the study on the most relevant investigations by establishing the boundries to the study area, the range of processes and siting alternatives to be considered, and so on. There should be enough flexibility in defining the scope of the assessment, at the beginning of the study, so that other important impacts which were not foreseen then could be taken into account subsequently should there be a need.

There are a wide number of techniques and tools that are being used to identify, evaluate, quantify and predict environmental impacts. The pros and cons, disadvantages and advantages of some of the major tools such as checklists, interacting matrix, overlays, adaptive modelling, etc., are briefly discussed in the chapter.

A major aspect of the impact study is the preparation and presentation of the study report. The content suggested would start with a statement of the terms of reference on which the study has been based, to be followed by a description of the sources of the environmental impacts and their potential consequences, as well as the identification of the groups and community interests which would be affected. It is strongly emphasized that the report must be prepared in a concise and succinct form understandable and meaningful to decision-makers as well as the community that will be affected by the project. It should not be written in highly technical and specialized terms, nor couched in jargons only understandable to the experts. The report could be supplemented with annexes containing such highly technical and scientific findings. The need to integrate the physical impact findings with the socio - economic findings are stressed in this chapter.

4. ASSESSMENT OF POLLUTION AND ECOLOGICAL EFFECTS

This chapter describes the necessary ecological inputs to assessing industrial environmental impact. It is directed primarily at the technologists and scientists charged with carrying out environmental impact assessment.

The objective assessment of the impact of industrial development projects upon the natural environment is enhanced by a sound data base. This can be established by obtaining relevant data about the nature and characteristics of the proposed sites and the surroundings, and by ensuring that every advantage is taken of available existing data. The acquisition of baseline information and data can be expensive. It is important to have a clear idea of what will be needed, how much of this will be needed, as well as the format it should be in, so as to facilitate the assessment. When the necessary baseline data is not available, it will have to be measured, taking into account any seasonal changes. The existence of baseline data at the beginning of the assessment study will enable the predicted impacts to be monitored and the net changes determined.

The chapter includes suggestions on the steps that could be taken for a preliminary ecological assessment and if this indicates serious potential impacts then it is to be followed by the steps necessary for a more detailed impact assessment study.

The guidelines suggest that each aspect of project construction and operation, beginning from raw material extraction, site preparation and construction process operations, to waste disposal and control, should be analysed for potential environmental impacts.

The chapter contains screening test tables for climate and air quality, water, geology, soils, ecology, environmentally sensitive areas, land use and land capability, noise and vibration, and visual quality. Each of these nine elements are given in more detailed sub-elements and presented in individual tables which identify the potential impacts, the information that is required and the possible sources of information. Augmenting the screening test tables are baseline summary tables corresponding to these nine elements and their sub-elements. Each individual table indicates the information/expertise required, the methodology for acquiring the baseline data, and type of findings/measurements that should result from these findings and measurements. Examples of the screening test tables and baseline summary tables are shown in Table 1.

5. ASSESSMENT OF SOCIO-ECONOMIC EFFECTS

The traditional ways of appraising projects have generally not fully covered the socio-economic impacts of developments. These impacts are very often inseparable from and are no less important than impacts on the physical environment. There are analogies to be drawn between assessment methodologies employed by the natural and physical sciences and those used by the social sciences. Although there is a parallel between the methodological frameworks suggested for physical impact assessments and that for the socio-economic studies, attention is drawn to the fact that social sciences are not characterized by the complete objectivity which is frequently claimed for the physical sciences. Although quantitative methods are often used by social scientists, value judgements are generally implicit in the findings. Socio-economic investigationsare inevitably addressed to matters which are extremely variable, complex and often controversial. Furthermore, the concepts of marginality and non-marginality need to be often employed to assess impacts of projects.

The chapter discusses the role and importance of assessing social and economic impacts from an industrial development

Table 1

EXAMPLE OF SCREENING TEST TABLE

Climate and Air Quality

Sub-element	Potential Impact(s)	Required Information	Sources of Information
Precipitation/ humidity	Will the project have an impact upon the local precipitation/humidity pattern? Will the project be sited in a "high risk" area?	Precipitation/ humidity data incl- uding unusual cond- itions - flash floods, etc.	Meteorological records; existing residents in area.
Temperature	Will the project have an impact upon the local temperature pattern?	Temperature data, including extremes.	Meteorological records.

EXAMPLE OF BASELINE SUMMARY TABLE

Sub-element	Objectives	Required Information/ Specialist(s)	Methodology	Findings/ Measurements
Precipitation/ humidity	Protection of life/human health	Precipitation/humidity data; unusual conditions - flash floods, very high rainfall, hail, fog, snow. Meteorolo- gist.	Location of site; estimate risk potential; measurement of micro-climate.	Risk of unusual conditions; nature and char- acter of micro- climate.
Temperature	Protection of life/human health.	Temperature data; unusual conditions - extremes in temperature; heat dome(s), frequency and extent of temperature inversions; valley downwash conditions; ventilation potential. Meteorologist.	Location of site; estimate risk potential; measurement of micro-climate.	Risk of unusual conditions; nature and char- acter of micro- climate.

project and describes the major parameters which could be considered in a study. These include population structure, population dynamics, land use in settlement patterns, labour supply and employment structure, economic production and distribution, and income distribution and consumption. The framework for carrying out socio-economic assessment is given and examples of techniques such as field work and surveys for gathering information and data are also discussed.

6. DISSEMINATION OF THE GUIDELINES AND SOME CONCLUSIONS

The Guidelines have been widely disseminated in all regions of the world and workshops have been conducted for the Latin American, European, and Asia and Pacific regions to present them. Some of the major conclusions and findings of the regional workshops are :

The assessment study must include both physical and socio-economic impacts as they are interacting and interconnected. Physical changes in the environmental quality result in socio-economic impacts and vice versa. With regard to the details and scope of the studies, this would be dependent upon the prevailing conditions.

As each project is different from the other and also the location in which the projects are sited will be different, there is a need to ensure that the environmental impact assessment procedures, methodologies, and tools and techniques that are used should be flexible in order to ensure that the assessment study would be appropriately carried out. A standardized and rigid approach was considered to be impractical.

The involvement of the public to provide information and data in the assessment study is considered important and necessary. Different means should be used to ensure that there is good public involvement. With regard to public participation, that is the public actually taking part in the decision-making process, this depends upon the institutional procedures established in individual countries.

It is necessary to integrate the environmental impact assessment process at the very early project planning and feasibility studies stage. This would augment the planning process, generating alternatives with regard to siting, manufacturing processes, waste disposal methods, as well as finished products, enabling resource and environmental implications of each alternative to be identified and assessed. The availability of alternatives, with their potential impacts identified, would enhance the decision-making process. Furthermore, integrating the environmental impact assessment process at the early stages of a project would facilitate design modifications and changes with minimal delays and costs. These have been the major criticisms of and concerns with environmental impact assessments which have been justified because the process was introduced later on in the project cycle when plant construction had started.

The procedures and methodologies for environmental impact assessment must not be cumbersome and bureaucratic. Furthermore it must not develop into a system which would be considered as a mechanism for stopping a project on non-environmental grounds. The report resulting from the study must be clearly and concisely presented. It must not be couched in technical jargon which will not be understandable or meaningful to the decision-maker and the community affected by the project.

The institutional procedures to be established should also include post assessment. This is important to ensure the recommendations made in the assessment report to prevent and ameliorate adverse environmental impacts are implemented. Also, this would be a mechanism to actually measure the net changes identified and predicted in the study when the project is in operation. The post assessment results would enable the impact assessment methodology to be improved accordingly.

There is one way in which a country can increasingly improve the methodology for environmental impact assessment and that is to increasingly use the process, adapting, modifying and improving with the experience gained.

An increasing number of countries are now either formally implementing procedures for, or strongly encouraging the use of environmental impact assessment. Furthermore, international funding agencies such as the World Bank, UNDP, the Regional Development Banks, with UNEP, had on 1 February 1980 signed a Declaration which included clauses promoting the use of assessment procedures in development projects. Furthermore, an increasing number of bilateral funding agencies are also requiring environmental assessments to be carried out before funds are made available to aid development. A significant support for using environmental impact assessment in development projects was also advocated in the Brandt report "North South - a programme for survival".

AFFILIATION

Nay Htun is currently the Director of UNEP's Regional Office for Asia and the Pacific, based in Bangkok, Thailand. Previously he was with the UNEP Industry and Environment Office, Paris, France, where amongst his responsibilities was the development of the UNEP Guidelines for Assessing Industrial Environmental Impact and Environmental Criteria for the Siting of Industry. He graduated with a Ph.D degree in Chemical Engineering from Imperial College, London University, in 1966.

Xaver Monbailliu

ASSESSMENT OF VISUAL IMPACT

1. INTRODUCTION

The increasing environmental awareness among the general public has led to greater care for areas of particular ecological and aesthetic importance. Thus there has been increased concern over the intrusion of developments into areas of landscape beauty. This has consequently stimulated the need to analyse the landscape characteristics and evaluate or classify landscape beauty. The resulting techniques can be broadly classified into those which are either subjective or objective in nature. The first group of techniques to be discussed, attempts to compare visual landscape characteristics by a ranking of landscape units and is essentially subjective. The second group attempts to incorporate statistical techniques with representative numbers of observers.

2. SUBJECTIVE EVALUATION TECHNIQUES

In the attempted comparison of visual landscape characteristics, the landscape units to be ranked are defined as:

(i) Grid squares, for example, a 1 km^2 grid;

(ii) Landscape tracts, units defined by relief or vegetation;

(iii) Large areas of land with homogeneous characteristics, for example, an agricultural plane.

Once the landscape units have been identified, a verbal description is usually the next stage, when various micro-features of the landscape are analysed. This stage then allows a grading system to be developed, on the basis of a qualitative or quantitative appreciation of the landscape. Depending upon the project, landscape appreciation studies, as used in development control, usually attempt to define one or more of the following:

(i) The visual scale of the overall landscape as observed from the road or any other transport corridor (canal, rail or footpath). These may

B. D. Clark et al. (eds.), Perspectives on Environmental Impact Assessment, 265–271.

be highly panoramic landscapes or confined views as found in narrow valleys.

(ii) The internal scale of the landscape, such as a hedgerow landscape with a historical and intimate land use pattern, or extensive areas of monoculture with few shelter belts.

(iii) The complexity rate of landscape texture or diversity rate. A rocky hillside can visually "absorb" transmission lines to a greater extent than a meadow landscape of even texture. The opposite may be true for an underground pipeline.

(iv) The degree of artificialization of the viewed landscape. It is more difficult to associate a high voltage line, a symbol of contemporary technology and political centralization, with an area of outstanding natural beauty, than with an industrialized landscape which spoils the original landscape unity.

(v) The degree of historical and cultural significance of a landscape. It is difficult to integrate certain forms of land use, for example, a power plant or camp site with an archaeological site.

(vi) The importance of rarity. Certain plant communities or geomorphological configurations are unique at a regional, national or international level.

(vii) The readability of the viewed landscape. The landscape pattern of an area with hedgerows can be more "readable" than, for example, fields under polyculture.

(viii) Weather conditions such as precipitation and fog in valleys may limit the landscape views.

(ix) The degree of landscape unity.

(x) Seasonal changes and effects of colours in the landscape.

(xi) The ability to view the landscape from the road or other transport corridor. Views may be limited by cuttings and other obstructions.

(xii) Linear elements, such as canals, roads, valleys encourage viewing in a particular direction. It is clear that the location of an artifact in the same field width, produces a noticeable visual impact.

(xiii) Some landscapes contain attractive or unattractive elements such as a waterfall or spoil heap. These visually dominating elements (visual detractors) transfer attention from the new artifact which will then have a reduced visual impact.

(xiv) Relief can sometimes determine the regional choice of an infrastructure.

(xv) People sense the landscape within their cultural background; Why is an electricity pylon viewed as ugly by most westerners, while it is exalted by the Chinese as modern technology? Ski lifts are acceptable visually in the winter as they are seen as being functional to the enjoyment of skiing, while in the summer they are regarded as being ugly. We must be fully aware of our cultural and personal affiliations with landscape scenery.

Most classical landscape analysis techniques, developed for specific projects such as the siting of power lines, pipelines, roads and any other large projects, involve the use of perspective drawings and sketches, models and photomontages. Photomontage techniques may include superimposed photography or projected images in order to present a better visual appreciation of the development scheme. More recently, computer mapping and graphics have been employed in order to ensure the best integration of the proposed development in the landscape. These techniques are mainly designed to enable the possible landscape intrusion to be visualized, but they do not justify the location of the artifact. It is of greater importance to study and determine the geographical location of the site of least visual impact. This site, may however, be unsuitable in relation to more importand factors such as the technical and financial constraints of the project itself. Ecological, socio-economic and cultural repercussions may also be important locational determinants; for example, a highway will be a major factor in the choice of the geographical position of future development. On the other hand, due to increasing opposition by local inhabitants to the visual intrusion of such projects as power lines and highways, the siting of these infrastructural facilities, especially in areas of outstanding natural beauty, is increasingly being determined by a visual evaluation of the landscape.

Once the study area has been deliniated, an initial regional analysis will determine the landscape units to be used in the later analysis and evaluation stages. If the grid square method is being used, then no preliminary survey will be required. However, due to the costs and time consuming nature of the method (up to a year may be required to obtain objective data), a technique of sub-dividing the study area is likely. Such a technique divides the study area into visual landscape tracts or into landscape areas of homogeneous physiognomic characteristics.

2.1 A Case Study

A visual evaluation technique has been developed for the Scottish Highlands (Countryside Commission for Scotland, 1971). This method consists of appraising "landscape elements" according to their effects on the scenery of the tract. If the artifact is inconspicuous, due to screening or a landscape of high complexity, a value of zero will be attributed. The nominal value, one, is scored when the artifact is conspicuous and two when it is very conspicuous. If a score of one or two has been attributed, one may ask if the landscape would be more or less attractive without that artifact, or if the landscape would not be modified. The following scores can then be attributed:

2 for a negative contribution
1 if poor
0 neutral
1 if good
2 for a positive contribution

This relatively rapid landscape assessment technique is accompanied by complementary description of the land use, relief, and landscape tract views. Homogeneous tracts can thus be compared. While this study was conceived as a desk study for the general appraisal of the landscape in the Scottish Highlands, the application of this technique is difficult in certain cases. Difficulties can be expected both in relation to relatively flat landscapes, since the landscape tracts are easier to use in mountainous terrain as seen from the valley, and also in impact studies related to the positioning of intrusive elements such as pylons, since no positive scores would be attributed to these elements. Indeed, every impact study requiring visual landscape assessments needs a specific approach, since the criterion of judgement applied, relate directly to the visual impact engendered by the artifact and to the type of landscape. Further, financial considerations and contractual deadlines, constitute the final choice of assessment method.

In two other studies, one by Hebblethwaite (1970) and the other by Tandy (1971), both British landscape consultants, the landscape view was given a quality rating of high or very high, depending on topography, vegetation and land use per quarter of one km grid square. Tandy also gave quantitative ratings to the landscape unit.

In a study to find possible sites for industry in the Clyde estuary, Professor Weddle (1969), analysed the local landscape on the assumption that a landscape is only of value when it is of some use to the community. He evaluated what he terms the "inherent values of a landscape" such as views, background scenery and foreground vegetation, and the acquired value by aggregating the importance to resident and visiting populations. Thus, following Weddle, the intrinsic landscape value and the acquired value equals the aggregate value of a landscape. Each land unit was given a one, two or three star rating. This study is of particular interest, since it integrates the frequency that the inhabitants view the landscape, with the more intrinsic landscape values.

3. OBJECTIVE EVALUATION TECHNIQUES

These evaluation techniques attempt to avoid the subjectivity of the observer and assessor by using a representative sample of observers viewing and statistically evaluating the landscape. It is clear, that this sophisticated approach is undertaken by regional authorities rather than by private consultants preparing an impact study for a proposed development. Fines (1968) in his project for East Sussex, asked forty-five people, the majority having considerable experience in planning and design, to grade pictures depicting the East Sussex landscape. The qualitative scores of the landscape beauty ranged from "spectacular" to "unsightly". The landscape views were evaluated <u>in situ</u>, while the landscape tracts were determined. In order to refine the ranking system of landscape beauty per tract the standard deviation of the results was taken. The consequences of this study have been applied in the regional plans, and have suggested that tracts well above average beauty could well be chosen as preservation zones. The twenty pictures by Fines, were perhaps not representative enough of the local landscape; indeed twenty is too small a number to represent the visual image of a region.

The most sophisticated landscape assessment technique used in the UK is the Coventry-Solihull-Warwickshire subregional plan undertaken by Warwickshire County Council (1970). The County lacks to a certain degree dramatic landscape, therefore, in a rather homogeneous countryside, great care was given to the choice of landscape elements (parameters). Twenty-three of these parameters were used, these included; relief, agriculture, woodland, parks, moorland, water, urbanized areas, industry, wasteland and a series of linear elements such as hedgerows, trees, watercourses, roads, power lines, railway tracks, farms, listed buildings, churches and windmills. The second stage involved the evaluation of one km grid squares in terms of beauty using a ranking system of twenty-six scores as seen by one observer. An original ranking system from zero to ten proved to be too small for the study area.

After the identification of land form, land use and land features per km^2, and the ranking of each grid in terms of beauty, a regression analysis was applied in order to determine the contribution of each factor to visual quality. In addition, an index of inter-visibility was also used for each grid square.

This method is important because of the correlation between landscape parameters and subjective quality scores. A high degree of correlation was proven to exist. Mirenovitcz (I.A.U.R.I.F., 1978) has recently used this technique in France, in a landscape planning study for an area north of Paris in the Departement du Val d'Oise. Mirenovitcz working for the Institut d'Aménagement et d'Urbanisme de la Région d'Ile-de-France (I.A.U.R.I.F.), has chosen grid squares of 500 x 500 m and increased the number of landscape parameters from twenty-three to thirty-two. He classified the various forms of landscape into fifty-one homogeneous groups after applying the landscape parameters to some seven hundred grid squares. As in the Warwickshire study, he used regression analysis in order to identify correlations within these landscape types. Mirenovitcz also attempted to reduce the subjectivity implicit in his personal value judgement on landscape beauty by asking sixteen observers (far too small a sample), to rank fifty-one pictures representing the landscape types of the study area. The visual quality of each landscape type was thus measured, scoring values between one and fifty-one. The standard deviations were also analysed. It is interesting to note that, after computer analysis, the Val d'Oise study concluded that eight percent of all the landscape types can be determined by only three landscape or land features, namely; woodland, electricity lines and topography.

4. CONCLUSION

In conclusion, it is obvious that a great deal of research is required in landscape assessment, and that it has only been possible to touch on a few of the techniques. The subjective and objective evaluation techniques represent two different avenues out of which, it is to be hoped that, appropriate assessment techniques will develop. These techniques will then allow a proper consideration of the visual implications of proposed developments.

REFERENCES

Countryside Commission for Scotland: 1971, A Planning Classification of Scottish Landscape Resources, CCS Occasional Paper No. 1, Perth, Scotland.

Fines, K.D.: 1968, 'Landscape Evaluation: A Research Project in East Sussex'. Regional Studies 2(1) pp. 41-55.

Hebblethwaite, R.L.: 1970, Landscape Assessment, Landscape Research Group Seminar, York University, UK.

Institut d'Aménagement et d'Urbanisme de la Région d'Ile-de-France: 1978, Methodes d'Evaluation des Caracteres Physiques et Humains du Paysage, Paris, I.A.U.R.I.F.

Tandy, C.V.R.: 1971, Landscape Evaluation Techniques, Land Use Consultants, Croydon, England.

Warwickshire County Council: 1971, Coventry-Solihull-Warwickshire, a strategy for the Sub-region, Supplementary Report No. 5, Warwick, England.

Weddle, A.E.: 1969, 'Landscape Evaluation: Case Study in the Clyde Estuary', Journal Royal Town Planning Institute, 55(9), pp. 387-389.

AFFILIATION

Xaver Monbailliu is an Environmental Consultant in Paris, France.

Peter Wathern

ECOLOGICAL IMPACT ASSESSMENT

1. INTRODUCTION

The current level of concern for natural, particularly biological, resources has been responsible for the imposition, in many countries, of a legal framework to ensure that decisions on development proposals are taken with a thorough understanding of their environmental consequences. The aspiration to reveal ecological impacts, however, has outstripped the technical capability of predicting changes within biological systems, with the result that projections of ecological impacts have often been incorrect in the past. As predictions are uncertain, the role of the ecologist in environmental impact assessment should be carefully appraised. So too should the expectations placed on ecologists by decision-makers and the public. It is important to appreciate the characteristics of biological systems which make prediction difficult in order to assess the prognosis of the ecologist. Initially, this should be limited to a statement of the likely changes, without an evaluation of whether these are "good" or "bad". This viewpoint can be formulated only against certain objectives which may involve aspects of applied ecology such as nature conservation or agricultural and forestry production.

2. ECOSYSTEM CHARACTERISTICS

The basic units of organization studied in ecology, ecosystems, result from "... biological units interacting with the physical environment ..." (Odum, 1975). The physico-chemical and biological components of an ecosystem are inexorably linked. The consequences of this inter-relationship is that impacts can result from the direct effects of development on biota, for example the destruction of vegetation on a building site by a bulldozer. Alternatively, impacts may be effected through changes in the physico-chemical environment (Figure 1). For example, construction of a jetty in an estuary may change sediment-

B. D. Clark et al. (eds.), Perspectives on Environmental Impact Assessment, 273–292.

ation patterns which are reflected in changes in the distribution of estuarine invertebrates. In many instances changes operate through a number of intermediaries and the inter-relationships between the biological and physico-chemical environment may be poorly understood. In addition, knowledge of how some systems respond to perturbations is unavailable.

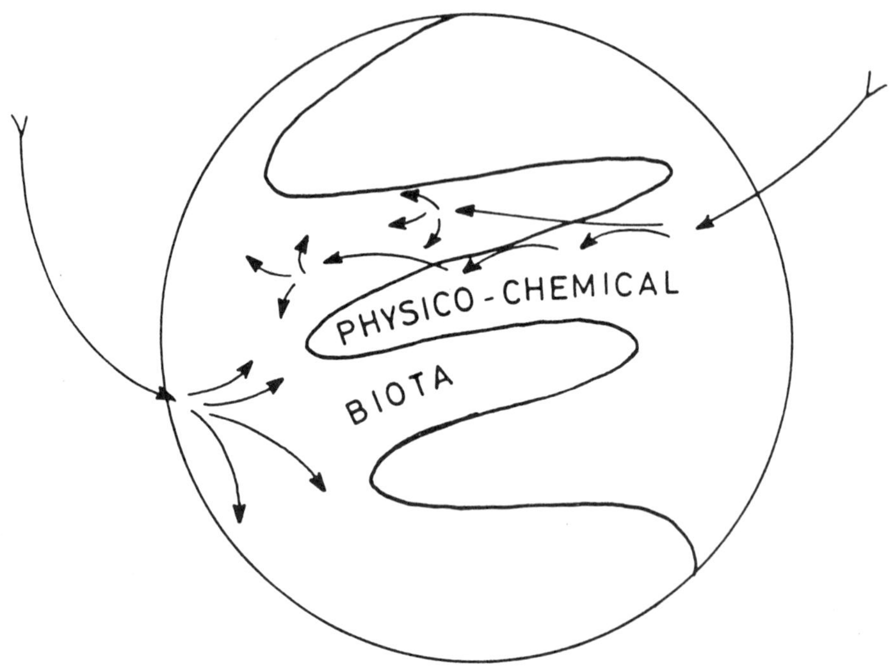

Figure 1 Direct and indirect impacts of development on ecosystems.

In order to predict impacts it is essential to know both the magnitude of the environmental change and the response of the system under consideration to changes. Ecosystems respond to changes in different ways. Wathern and Gilbert (1979) for example, have shown that grasslands with contrasting species composition respond in different ways to fertilizer applications. Within an ecosystem the response of different species to a particular environmental change varies. This phenomenon is well known in pollution monitoring where certain species are used as indicators of pollutants, because of their sensitivity. Hawksworth and Rose (1970) have shown that species of the lichen genus Lobaria are absent from trees in areas where the mean winter sulphur dioxide concentration exceeds 30μg/m^3, whereas Parmelia saxatilis and Hypogymnia physodes are present up to sulphur dioxide concentrations of 70μg/m^3. Even within a single species, ecotypic variation may lead to a variety of responses. Tolerant ecotypes of a number of different plant species have long been known to grow on mine spoil heaps contaminated by heavy metals such as copper, lead and zinc. The growth of non-tolerant ecotypes is severely retarded in heavy metal contaminated soils, while tolerant ecotypes are not affected (see for example Bradshaw et al. 1978).

Predicting ecological impacts depends, to a certain extent, upon the transfer of experience from one situation to another. This may be invalid, however, because of variations in response at the ecosystem, species and ecotype level. Holling (1978) has argued that it is not possible to postulate a generalized model of ecological impact in which there is a decline in impact with decreasing distance from the source of change, as this model does not accommodate these varying responses. In reality, no simple generalized model of impacts can be formulated to allow for the numerous potential interations between environmental parameters and ecosystems.

The components of an ecosystem respond to changes in environmental parameters in a number of ways (Figure 2). The most simple situation is one in which there is a linear relationship between change in the environmental parameter and the response of a system. When this type of response exists, impacts can be predicted readily. Responses, however, may deviate markedly from this simple linear relationship. Incremental changes in an environmental parameter may lead to progressively greater or smaller response in a non-linear relationship. The most difficult situation to consider involves a sigmoid relationship between the response of a system and changes in environmental parameters. Such situations are characterized by a threshold below which incremental changes produce little response and systems accommodate changes in environmental parameters. It is for this reason, for example, that waterbodies can absorb certain quantities of pollutants without adverse effects. Above the threshold, however, the response may be exponential. In the case of pollutants, this may lead to a rapid deterioration in environmental quality with only small changes in the level of discharge.

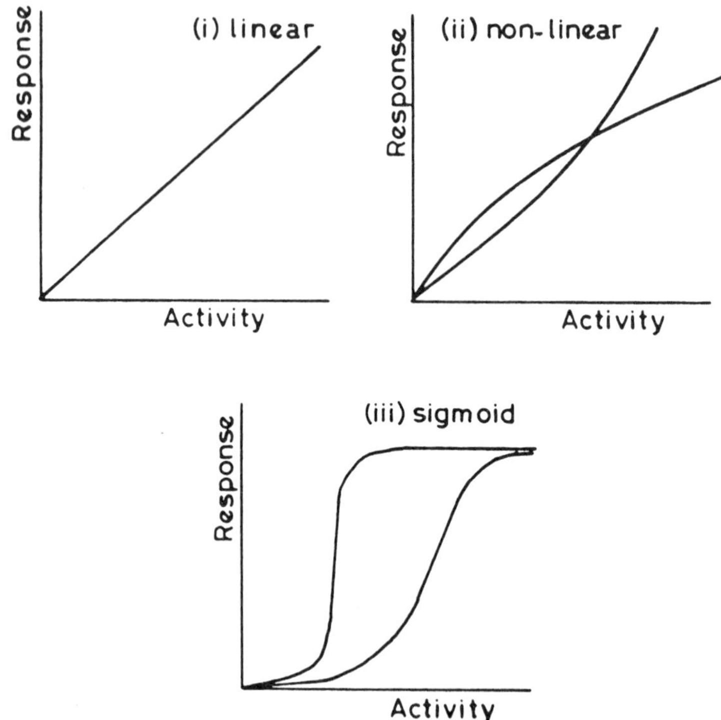

Figure 2 The response of ecosystems to change in environmental parameters.

Resilience of an ecosystem is an important factor in determining its response to change. Resilience, the probability of extinction, is a measure of the ability of a system to persist in the present of perturbations (Holling 1973). First, the system may recover and eventually return to the original equilibrium. For example, cessation of fur seal hunting in Antarctica since the early part of the century has led to a recovery of seal numbers, but not yet to their former levels. Secondly, a new equilibrium related to the degree of disturbance may be established. Herring were present originally at high density in the North Sea, but there has been no recovery of stocks following control of commercial fishing, instead, new low density equilibria have established.

Finally, the original perturbation may initiate a sequence of change which proceeds to extinction. This situation probably occurred in the United States with the passenger pigeon which became extinct despite protective legislation. It is likely that its populations fell below thresholds necessary for successful breeding.

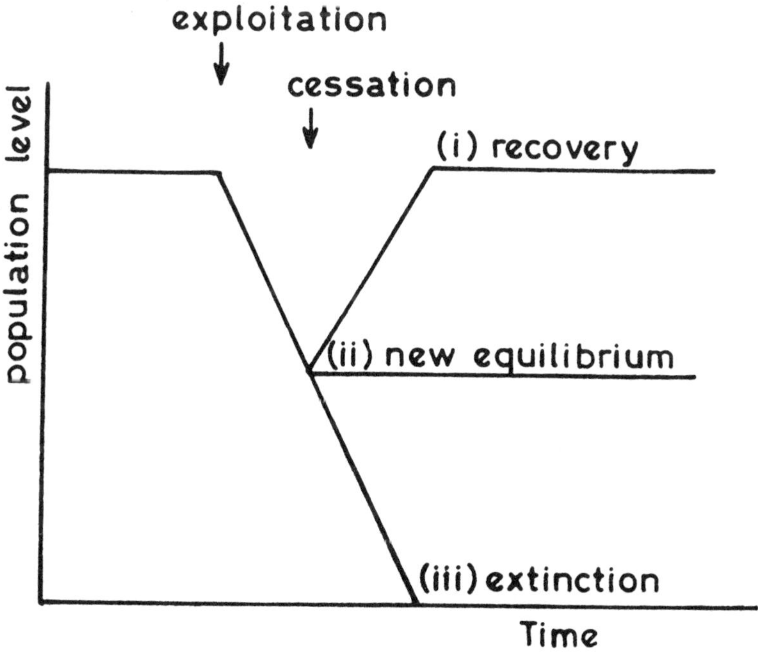

Figure 3 The response of populations to changes in environmental parameters.

3. A GENERALIZED THEORY OF DEVELOPMENT

To aid an understanding of ecological impacts, it is important to establish whether there is any pattern to the effects of development on ecosystems. Ecosystems are not static and changes over time can often be predicted at least in stochastic terms. Clement (1916) proposed a general theory of temporal change in ecosystems, the theory of succession, which has been elaborated by a number of workers. An account of the changes in twenty-two attributes related to the composition, characteristics, strategies and functioning of ecosystems over the course of succession has been presented by Odum (1969). As there is also a temporal component to the effects of development on ecosystems, Odum's model can be used to assess the ecological effects of developments, but current knowledge prevents such a detailed appraisal. The twenty-two categories of Odum, however, can be redefined and reduced to eight. Observations on residential development construction sites in the UK indicate that there is a pattern to the effects of this type of activity.

Many of the changes are a result of differences in species composition of sites before and after development. The size of individuals declines markedly as mature organisms are replaced by immature specimens. This situation is exacerbated, because large plant species such as trees are replaced by small herbs and grasses. In combination small individuals produce a small biomass on a site. The net community productivity, that is the amount of production compared with the biomass, increases during development as the small individuals have a high rate of growth. Table 1 shows the relative growth rates of species characteristic of reinstated immature and mature grasslands in the UK. Species characteristic of developed ecosystems are opportunistic and have a broad niche in that their tolerance to environmental variations is wide. In contrast, species from more mature ecosystems have very specific environmental requirements. In addition, species from developed systems have short life cycles and have a strategy based on rapid growth to maturity and high reproductive rates. These species correspond to the ruderal-competitive strategy (Pioneer strategists) of Grime (1969).

Development also affects certain process changes within ecosystems. A characteristic feature of developed systems is the free availability of materials, particularly nutrients. Chadwick (1973) has argued that materials within ecosystems

Species	Relative Growth Rate (R_{max})
Old (mature) grasslands	
Calluna vulgaris	0.35
Campanula rotundifolia	0.81
Nardus stricta	0.71
Thymus drucei	0.72
Viola riviniana	0.65
Argostis tenuis	1.36
Festuca ovina	1.00
Poa pratensis	1.26
Sieglingia decumbens	0.60
Festuca rubra	1.18
New (immature) grasslands	
Poa annua	2.70
Ranunculus repens	1.39
Rumex obtusifolius	1.49
Senecio squalidus	2.28
Trifolium repens	1.26
Lolium perenne	1.30
Poa trivialis	1.40
Dactylis glomerata	1.31
Holcus lanatus	2.01

Table 1 The relative growth rates of selected species from contrasted grasslands. (R_{max} values from Grime and Hunt 1975).

can be grouped into three categories (Figure 4). First, there are stored materials which are in an insoluble unavailable form. Secondly, materials can be fixed within biological systems. Finally, there is a pool of soluble materials which are available to biological systems. In a mature undeveloped ecosystem these components are in dynamic equilibrium with the small pool of available materials thus limiting the rate of many biological processes. Developed or perturbed ecosystems are characterized by greatly increased pools of available materials. This situation, for example, applies to nutrients and accounts for the eutrophic nature of developed soils and water bodies. The distribution of synthetic pollutants produced by man is similar. The cycling of nutrients in mature ecosystems is characterized by the low availability of free nutrients, particularly nitrate and phosphate, compared with the total accumulated within the system. In tropical rain forest, for example, fifty eight per cent of the total nitrogen is fixed in the living biomass (Ovington 1962). Nutrients are released from dead organic matter by the action of micro-organisms and are rapidly taken up by living organisms. Thus, nutrient cycles in mature ecosystems are closed.

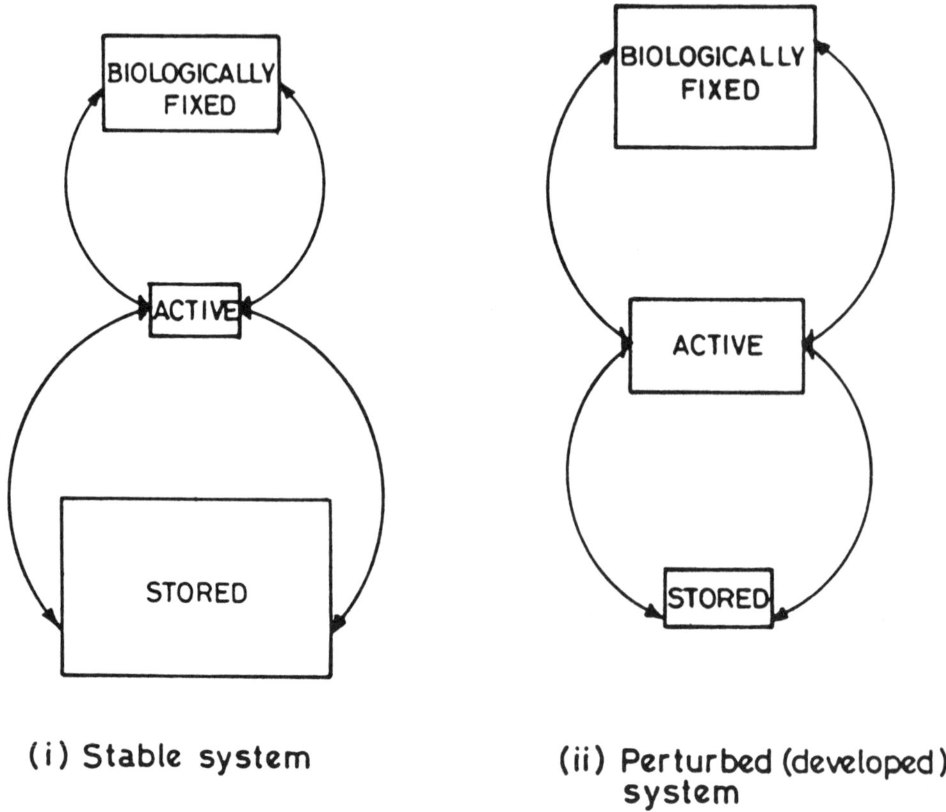

Figure 4 Distribution of nutrients in mature and developed ecosystems. The size of the boxes indicates the relative proportions of materials (after Chadwick 1973).

Elton and Miller (1954) have partitioned ecosystems into four structural units (Figure 5). Between each of these structural units is a steep gradient of changing environmental conditions, referred to as ecotones. Ecotones, transitional areas combining features of either extremity as well as attributes unique to the ecotone, are diverse elements in an ecosystem. The replacement of woodlands, for example, by more simple systems reduces the complexity of community structure. The pattern within mature ecosystems is complex with much structural heterogeneity. Complexity is reduced further as small habitats are removed and uniform systems are recreated (Figure 6).

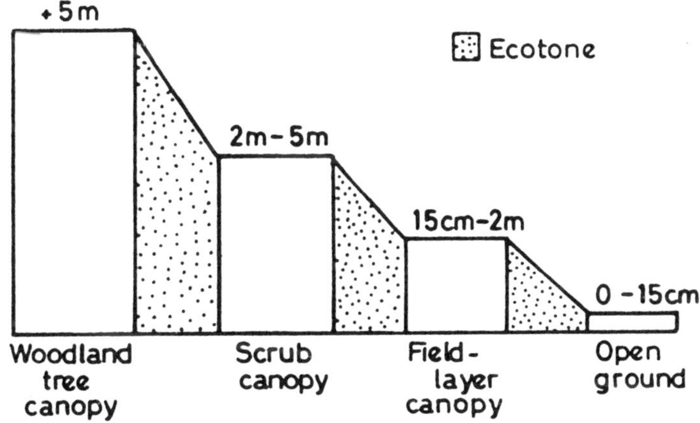

Figure 5 Structural differentiation of habitats (after Elton and Miller 1954).

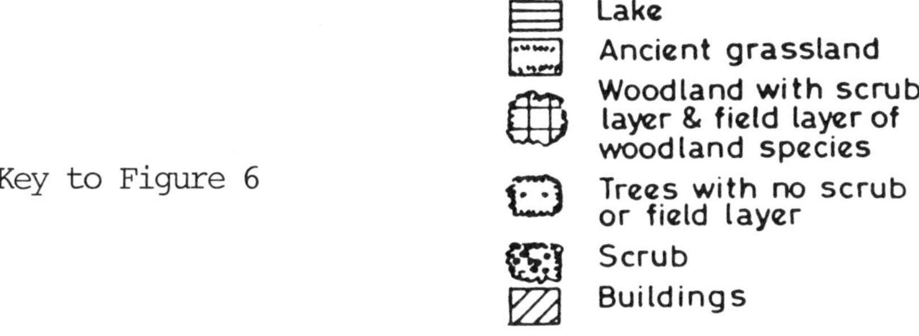

Key to Figure 6

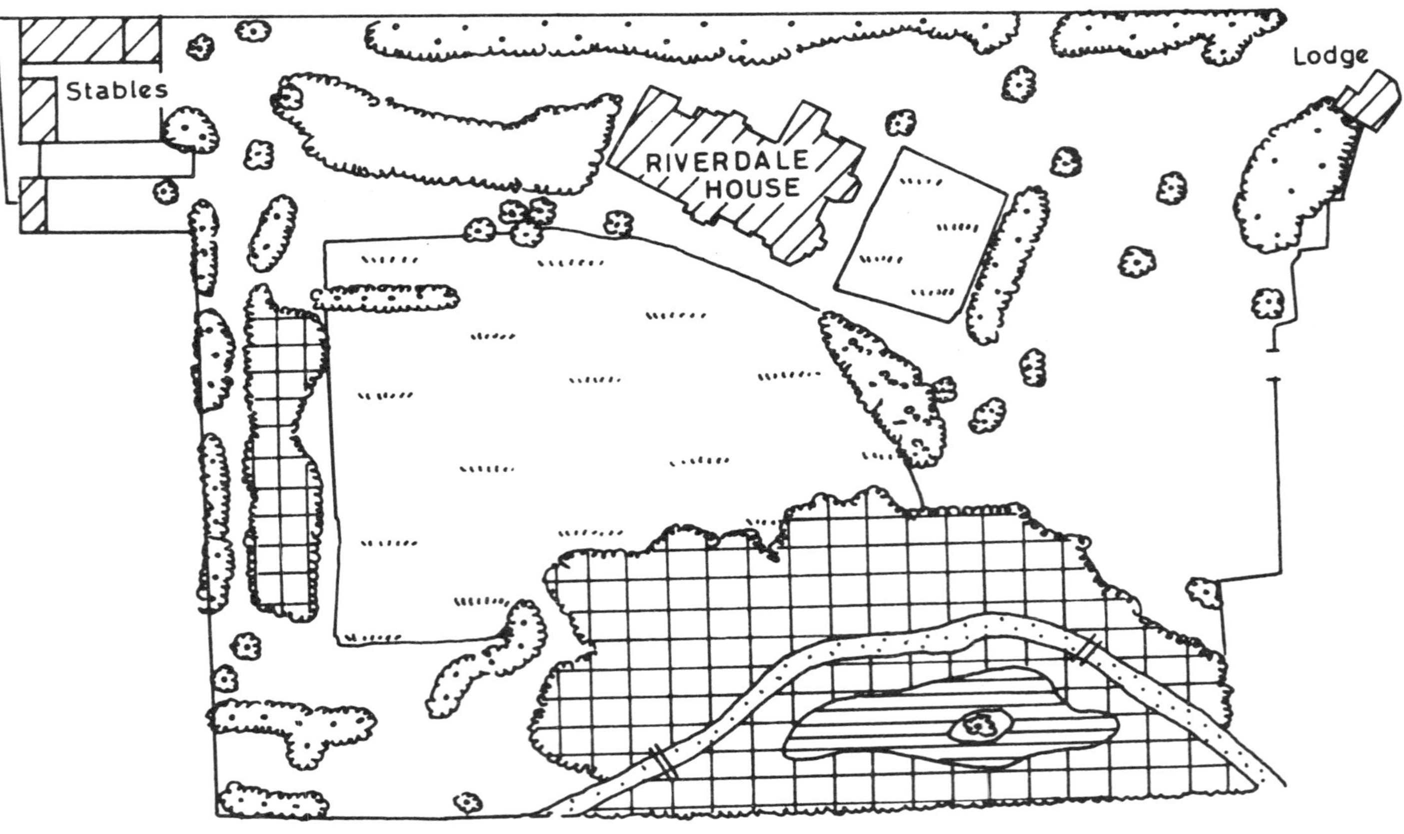

Figure 6(a) The structural complexity of habitats on a residential development before construction.

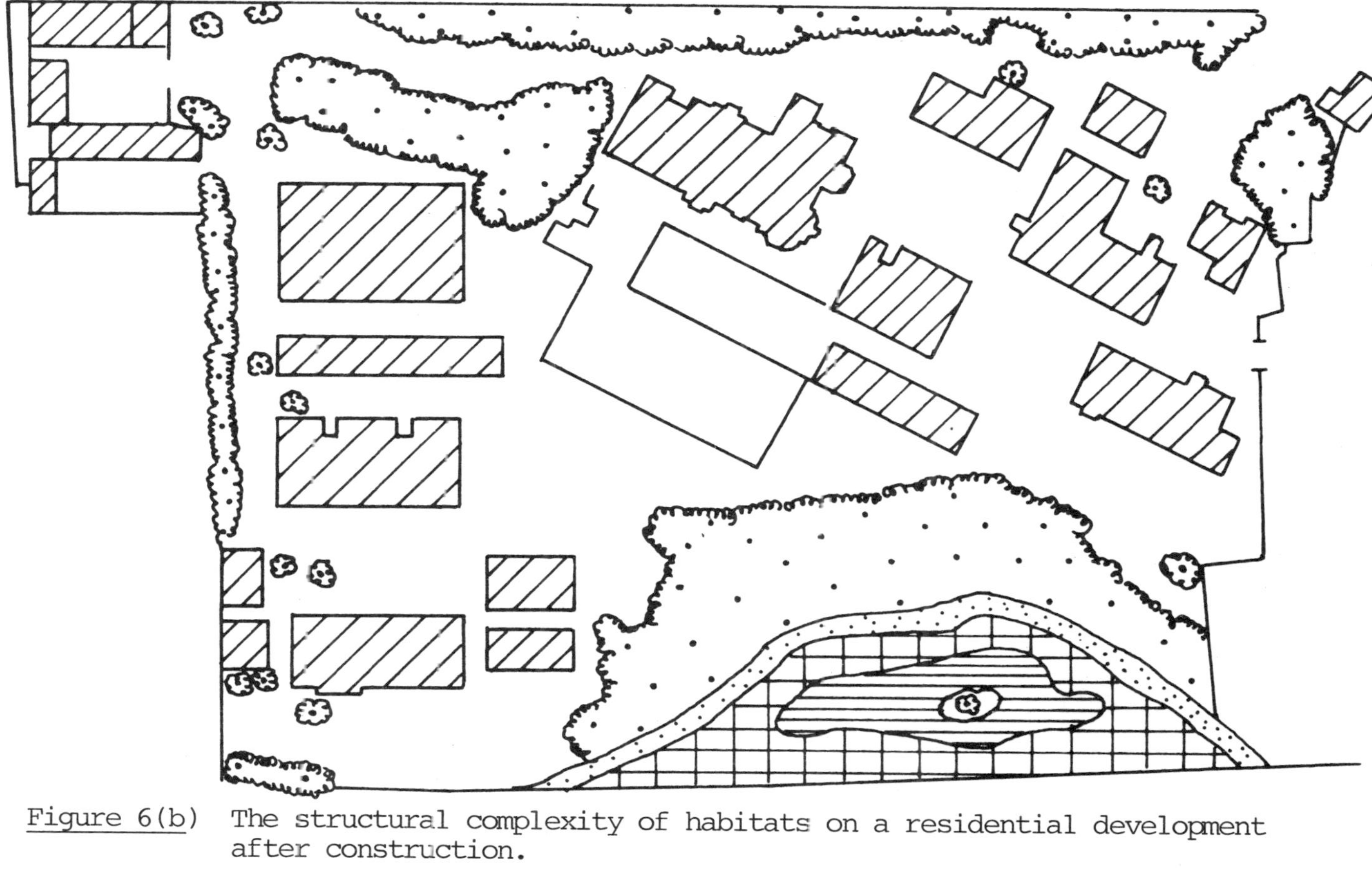

Figure 6(b) The structural complexity of habitats on a residential development after construction.

The final feature in Table 2, overall homeostasis, is supported by only inconclusive observations. Elton (1958) has argued that agricultural intensification, a form of "development", results in systems characterized by an increased frequency of epidemic outbreaks of crop pests, as natural checks to population increase are removed. On development sites in the UK adventive plant species such as rose bay willow herb (Chamaenerion angustifolium) and Japanese Knotgrass (Polygonum cuspidatum) exploit disturbed areas and, lacking effective control, spread rapidly.

ATTRIBUTE	SUCCESSION	DEVELOPMENT
1. Organism size	Increases	Decreases
2. Biomass	Increases	Decreases
3. Net productivity	Decreases	Increases
4. Niche specialization	Narrows	Widens
5. Life histories	Lengthen	Shorten
6. Nutrient cycles	Close	Open
7. Community structure	Increasing Complexity	Simplification
8. Overall homeostasis	Improves	Degenerates

Table 2 A comparison of the ecological attributes of development and succession. (Source Wathern 1976).

Of the eight types of change induced by development,

the effect on overall homeostasis is the most speculative. It would appear, however, that the ecological effects of development follow a pattern and can be formulated into a general theory of development. Each of the eight trends occurring during succession is reversed during development (Table 2). Constructional activity, therefore, should be regarded as a retrograde succession. This theory, originally postulated with respect to construction sites for housing development in the UK (Wathern 1976), needs to be tested for other types of development.

The analogy between succession and development can be pursued. Successional changes are not immutable. Management can divert the course of succession, arrest it at a particular stage or even reverse the direction of change by returning systems to an earlier successional stage. Similarly, the adverse ecological consequences of development are not irrecusable. Management of development in order to conserve the major features of interest and sympathetic design of the new landscape will mitigate adverse impacts.

4. DETERMINING CARRYING CAPACITY

In considering development proposals, it is important to establish whether the proposed action is above a threshold which would lead to rapid deterioration of the environment. The ability of an area to accommodate change without deterioration is defined as the carrying capacity. Clearly, carrying capacity varies not only with the area under consideration, but also with the type of development proposed. Thus, for example, sand dune systems and woodlands respond to development in different ways, whilst residential and heavy industrial development have differing impacts upon the local environment. Consequently, models of carrying capacity must be formulated with respect to a particular type of area or to a development activity. Odum (1975) identified the difficulties of defining carrying capacities as one of the greatest impediments to effective regional planning and ecosystem management.

Use of predefined development capacities for project planning has been adopted experimentally in the United States (Hall, 1977). Master Environmental Impact Reports (MEIRS) Have been used to establish the limits of development within an area. The characteristics of the area are used in formulating these limits. This is a "balance sheet" approach to project planning in which incremental development is permitted until the predefined limits are reached. A moratorium on further development is imposed unless resources are made available to overcome the constraint, for example increased waste water treatment to meet water quality standards. Whereas federal standards can be used to set, for example the limits to air and water quality, many other attributes must be defined arbitrarily.

Carrying capacity can be defined with respect to a particular activity. Recreational impacts, particularly the effects of trampling, have been studies in a number of contrasted ecosystems, see for example the review of Liddle (1975a). Liddle (1975b) has postulated that there is a logarithmic relationship between resistence to trampling and the primary productivity of vegetation cover to 50% and the log of the primary productivity of the vegetation. Thus, given this model of recreational impact it would be possible to determine the carrying capacity of an area or establish whether a proposed recreational development would induce deterioration of the surrounding vegetation. Such mathematical models of carrying capacity, however, are rare and this limitation is a significant restriction on the objective assessment of impacts.

Pollution tolerance may be regarded as a form of carrying capacity, since the maintenance of a particular assemblage of organisms at a particular site will be dependent upon restricting pollution to levels below critical thresholds. With point source and area wide pollution models it is possible to project pollution patterns and regulate levels through planning and pollution control. For many organisms, the level of tolerance is unknown even amongst groups which have been studied in detail, for example, the sensitivity of lichens to sulphur dioxide pollution. Accommodating non-pollution aspects, particularly the effects of habitat modification and disturbance on wildlife, is a recurrent problem in defining carrying capacity. Marquiss, Newton and Ratcliffe (1978) for example have shown a decline in raven (Corvus corax) populations, with increasing afforestation in the border region of southern Scotland and northern England. The importance of range integrity in ravens and large raptors such as buzzards (Buteo buteo) and eagles (Aquila chrysaetos) is not known. Helliwell (1971) has argued that afforestation leads to increased bird diversity, but his work shows this to be the result of an influx of widespread woodland generalist bird species exceeding the decline in more restricted low density moorland species. Thus, defining a carrying capacity of uplands for afforestation even with respect to a restricted group of organisms, birds, is difficult.

5. AN APPROACH TO ASSESSING ECOLOGICAL IMPACTS

For much of its history, ecology has been a descriptive science, but it is undergoing a transformation with the increasing capability of modelling ecosystems. This is providing the potential to predict changes resulting from specific development proposals, but this ability can be applied only when there is adequate knowledge on the development proposals and the functioning of the ecosystem is understood. The further one departs from this ideal the less valid predictions become. From the above discussion it is evident that detailed guidance on assessing ecological impacts which is applicable in all situations cannot be given. However, an approach to assessing ecological impacts has been provided by Clark et al. (1976). There are three main phases to this approach; describing the site and its environs, identifying the major determining factors and finally projecting likely changes.

A recurrent criticism of environmental impact statements produced in the United States in the early 1970s has been the encyclopaedic approach to site description. A site description is essential and the major difficulty lies in determining the level of detail necessary to provide a satisfactory analysis. Compiling a complete species list for a site is a waste of resources. As far as possible existing inventories should be used, but in most situations this data will not be available and site surveys will be required. These should be rapid surveys designed to establish the affinities of the site with respect to known habitat types and indicate attributes of particular interest, for example any rare or unique features that may be present. This information should be used to establish the status of the site, whether designated or not, particularly against any existing objectively derived criteria. For example, sites containing one per cent of any Western European Wader species are defined as sites of international importance (Smart 1976). It is important that sites of importance for nature conservation which have not been designated should be identified. Areas of comparable quality to designated sites should be retained in the landscape so that isolated reserves do not remain the only surviving examples of certain habitat or vegetation types. Successful nature conservation, is ultimately dependent upon management of the whole landscape rather than protection of a few isolated reserves which may be dependent upon the surrounding mosaic of habitats for survival.

The survey should be extended selectively into the surrounding area to identify components which may be adversely affected by the proposed development. The areas to be considered will vary with each proposal. For example, if development would reduce water tables, it would be necessary to consider wetlands in the surrounding area, whilst pollution sensitive organisms would be considered if pollution levels would increase. Similarly, if the development would result in greatly increased vehicular movements, vegetation along the road network could be assessed.

It is important to establish the major factors which determine the nature of the ecosystems under consideration. These factors may be physical, chemical or the result of management practices. For example topography may be an important factor in the development of wetlands in a valley or flood plain. The chemical composition of the groundwater and detailed hydrology will determine the composition of the vegetation which develops. Oligotrophic lakes are characterized by low nutrient levels; similarly the composition of many grasslands is determined by base and nutrient status. Continuity of established management patterns, in some instances extending through historic times, in the form of particular grazing or burning regimes, may be necessary to maintain the composition of certain types of vegetation. It is only through an understanding of such determining factors that likely impacts can be projected.

Detailed impacts are site specific and depend on both the nature of the site and the proposed development. Appraisal should consider both on-site and off-site changes, but a different approach is required to assess both types of change. It must be accepted that over much of a development site existing habitats will be intentionally or inadvertantly destroyed by constructional activity. The quality of the habitats on the site should be the major concern at this stage as well as identifying ways in which habitats could be protected during the construction and operational phases. Parts of the site not required for construction should be delimited and the opportunities not only for conserving existing habitats, but also for creating new habitats should be appraised.

To assess off-site changes it is necessary to establish whether development would induce changes in environmental parameters, such as water levels, water quality and atmospheric quality, beyond the site boundary. Models of, for example, pollution patterns can be used to show spatial aspects of environmental change likely to result from the development. The surrounding area should then be appraised systematically to establish whether there are any habitats sensitive to change in a particular parameter. When the sensitivity of target organisms is known it may be possible to make detailed projections of change. Usually, it will be necessary for the ecologist to give an indication of the changes in more general terms. This situation is likely to apply when the functioning of ecosystems is poorly understood particularly in developing countries. Whenever possible, however, these projections should be couched in numerical terms with an estimate of the credence which can be placed in the assessment. This could take the form of an estimated probability of occurrence.

In many instances the affects of development are not known in detail. Large-scale proposals, however, frequently have a long lead time during which it may be possible to initiate investigations. Ecology is an experimental science and this characteristic should be exploited even within environmental impact assessment. A combination of field and laboratory investigations may reveal impacts of the development during the period of appraisal. It is essential, therefore, that the ecologist is involved in project planning from the outset. This is particularly important in areas, especially in developing countries where the functioning of ecosystems is inadequately understood and the existing data base is small. To date, the experimental approach to impact assessment has been largely ignored.

6. CONCLUSION

A review of ecosystem characteristics reveals that the prediction of ecological impacts is difficult because of the variability in responses of ecosystems, species and ecotypes. Although the general effects of development on ecosystems are understood and can be explained in ecological terms, individual ecosystems may respond in an unpredictable manner. This should not deter ecologists from becoming involved in environmental impact assessment. Their role is crucial and an abrogation from involvement will mean only that ecological aspects will be ignored by decision-makers. The ecologist has a responsibility to ensure that, even though projections may be uncertain, potential impacts should be determined and an indication of the reliability of the estimated change given. Similarly, decision-makers should be aware of the limitations within which an ecologist must make projections.

Not all development results in adverse impacts upon the environment. A massive civil engineering project in the Netherlands, Zuidelijk Flevoland for example, has produced habitats attractive for a wide range of wetland birds in large numbers (Wathern 1977). Gravel pit development in south eastern England has created important bird habitats (Harrison 1972, Catchpole and Tydeman 1975). Increased research on the recreation of natural vegetation should provide the capability to create new habitats to replace those destroyed during development. Techniques have been developed in the United Kingdom to create moorland vegetation (Gilbert and Wathern 1976) and semi-natural grasslands (Wathern and Gilbert 1978) rapidly after development. Throughout impact assessment, ecologists should look for opportunities to reduce the adverse effects of development and to create new habitats which will become valued resources in their own right in the future.

REFERENCES

Bradshaw, A.D., M.O. Humphreys, M.S. Johnson: 1978, 'The value of heavy metal tolerance in the vegetation of metalliferous mine wastes', in G.T. Goodman and M.J. Chadwick (eds.) Environmental Management of Mineral Wastes, Sijthoff and Noordhoff, Alphen aan den Rijn, pp. 311-334.

Catchpole, C.K. and C.F. Tydeman: 1975, 'Gravel pits as new wetland habitat for the conservation of breeding bird communities'. Biological Conservation 8, pp. 47-59.

Chadwick, M.J.: 1973, 'The cycling of materials in disturbed environment'. In M.J. Chadwick and G.T. Goodman (eds.), The Ecology of Resource Degradation and Renewal, Blackwell, Oxford, pp. 3-16.

Clark, B.D., K. Chapman, R. Bisset and P. Wathern: 1976, The Assessment of Major Industrial Applications: A Manual, DOE Research Report No. 13, Department of the Environment, London.

Clement, F.E.: 1916, Plant Succession: An Analysis of the Development of Vegetation, Carnegie Institute Washington Publication No. 212, Carnegie Institute, Washington.

Elton, C.S.: 1958, The Ecology of Invasions by Plants and Animals. Chapman and Hall, London.
Elton, C.S. and R.S. Miller: 1954, 'Ecological survey of animal communities with a practical system of classifying habitats by structural characteristics'. Journal of Ecology, 43, pp. 460-496.
Gilbert, O.L. and Wathern, P.: 1976, Towards the production of extensive Calluna swards. Landscape Design, 114, 35.
Grime, J.P.: 1979, Plant Strategies and Vegetation Process, Wiley, Chichester.
Grime, J.P. and R. Hunt: 1975, 'Relative growth rate: its range and adaptive significance in a local flora'. Journal of Ecology, 63, pp. 393-422.
Hall, R.C.: 1977, 'MEIRS - a method for evaluating the environmental impacts of general plans'. Water, Air and Soil Pollution, 7, pp. 251-260.
Hawksworth, D. and F. Rose: 1970, 'Qualitative scale for estimating sulphur dioxide air pollution in England and Wales using epiphytic lichens'. Nature Lond., 227, pp. 145-148.
Harrison, J.: 1972, A Gravel Pit Wildfowl Reserve, Claxton and Holmsdale, Sevenoaks.
Helliwell, D.R.: 1971, 'Changes in flora and fauna associated with afforestation of a Scottish moor'. Merlewood R & D Paper No. 24, Institute of Terrestrial Ecology.
Holling, C.S.: 1973, 'Resilience and stability of ecological systems' Annual Review of Ecological Systems, 4, pp. 1-23.
Holling, C.S.: 1978, Adaptive Environmental Assessment and Management, Wiley, Chichester.
Liddle, M.J.: 1975a, 'A selective review of the ecological effects of human trampling on natural ecosystems'. Biological Conservation, 8, pp. 17-36.
Liddle, M.J.: 1975b, 'A theoretical relationship between the primary productivity of vegetation and its ability to tolerate trampling', Biological Conservation, 8, pp. 251-255.
Marquiss, M., I. Newton and D.A. Ratcliffe: 1978, 'The decline of the raven Corvus corax in relation to afforestation in southern Scotland and northern England'. Journal of Applied Ecology, 15, pp. 129-144.
Odum, E.P.: 1969, 'A strategy of ecosystem development'. Science, 164, pp. 263-270.
Odum, E.P.: 1975, Ecology, second edition, Holt, Rinehart and Winston, New York.

Ovington, J.D.: 1962, 'Quantitative ecology and the woodland ecosystem concept'. Advances in Ecological Research, 1, pp. 103-192.

Smart, M.: 1976, Proceedings of the Fifth International Conference on Wetlands and Waterfowl, Heilig enhafen.

Wathern, P.: 1976, The Ecology of Development Sites, Unpublished Ph.D. thesis, University of Sheffield.

Wathern, P.: 1977, 'Some aspects of habitat building in the Dutch polderland'. Landscape Design, 118, pp. 20-21.

Wathern, P. and O.L. Gilbert: 1978, 'Artificial diversification of grassland with native herbs'. Journal of Environmental Management, 7, pp. 29-42.

Wathern, P. and O.L. Gilbert: 1979, 'The production of grassland on subsoil'. Journal of Environmental Management, 8, pp. 269-275.

AFFILIATION

Dr Peter Wathern is a Lecturer in the Department of Botany and Microbiology, University College of Wales, Aberystwyth, Wales.

Robert Johnston

WATER POLLUTION IMPACT ASSESSMENT - EXAMPLES FOR STUDY

General theoretical discussion of the techniques for assessing the effects of new constructions or operations on the aquatic environment would not greatly help those needing practical guidance or assistance with managing an EIA team. Some real-life examples may better illustrate the extraordinary variety of problems that arise and the awkward management situations that can happen. The chosen examples relate to the sea but the scientific principles and administrative situations are applicable to rivers and lakes.

1. AN INTER-ISLAND CAUSEWAY

1.1 Scenario

The participants were the local authority who wanted to replace an ageing 78-span bridge, 1.2 km long, between two offshore islands, by a road built on a solid barrier. The civil engineering company sought permission from the central government licensing authority to undertake constructional work on the sea-bed below high-water mark (London Dumping Convention). Fishermen's representatives were concerned about possible damage and limitation of access. The channel between the islands is mainly shallow with a clean sandy bottom and exposed sandbanks at low tide (Fig. 1). A narrow passage near the north shore allows small craft access to both sides at all states of tide.

1.2 Possible impacts

There are no industrial discharges and very little sewage. The island communities live mainly in scattered small crofts. Farming and fishing on a small scale for demersal fish species, crabs and lobsters are the usual occupations. Some potential exists for exploiting the abundant cockles (an edible bivalve mollusc) for mainland and European markets should the economics become more favourable. To preserve this possible resource it is vital not to disrupt

B. D. Clark et al. (eds.), Perspectives on Environmental Impact Assessment, 293–301.

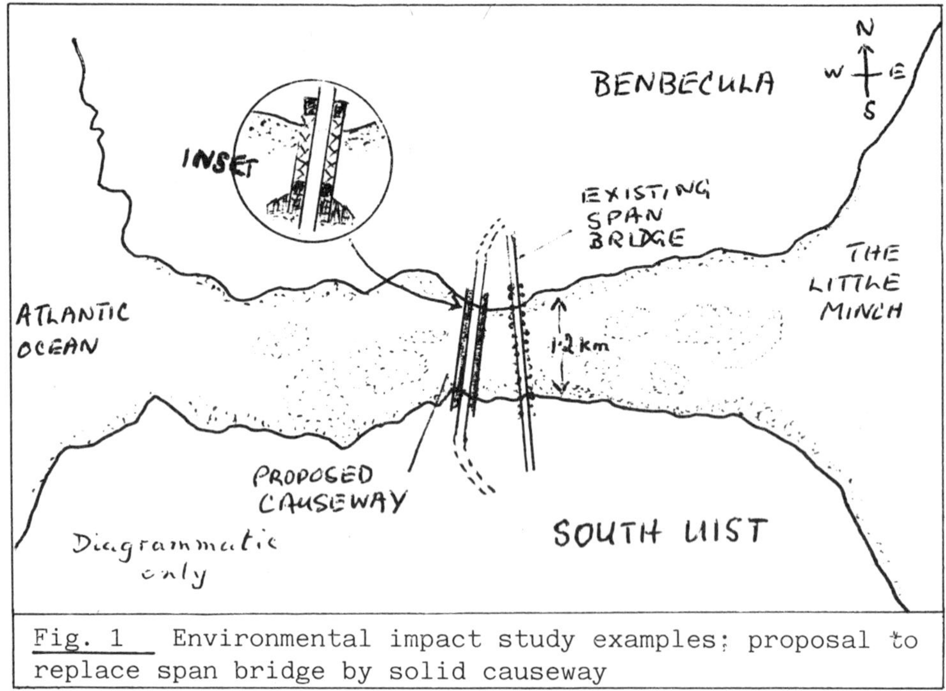

Fig. 1 Environmental impact study examples: proposal to replace span bridge by solid causeway

the existing pattern of sandy beaches and sandbanks.

1.3 Research

The developers undertook to engage an expert to measure water movements, tidal heights, wave pressures and the magnitude of any likely changes in sand distribution. The local authority and fishermen agreed to discuss the necessity for a passage for small boats.

1.4 Evaluation

The driving force for water movements in the channel was identified as the out-of-phase tidal cycles to the east and west. Open Atlantic and Minch wave regimes can be extremely violent but the long shallow sandy channel effectively dampens them. Provided the sand distribution remains unchanged there would be no threat to the causeway. Experience during construction of other inter-island barriers revealed possible serious extra costs created as the barrier approached closure. Accelerated tidal currents in the narrowing gap drags away the sand and foundations necessitating expensive additional reinforcement. It was also decided that the passage would be useful for fishing boats and small pleasure craft and, if the need arose, for marine rescue services.

1.5 Conclusion

It was mutually agreed to amend the plan for the causeway to include a short bridge section near the north shore. This would preserve access and also by building this section first the problems of closing the gap would be avoided. Open ducts would be built into the barrier to retain free flow of water. It was anticipated that the extra cost of the bridge section could be met from savings in establishing foundations across the gap.

2. COMMISSIONING THE WORLD'S LARGEST UNDERSEA NATURAL GAS PIPELINE

2.1 Scenario

Unlike the previous example which involved no recognised pollutants, this case encountered very large volumes of discharges containing toxic chemicals.

In commissioning the natural gas pipeline, 0.9m diameter, 160 nautical miles in length, three separate phases have to be considered, namely maintenance and pressure testing, cleaning, final cleaning and drying. There were many participants but the multinational oil company acting on behalf of a consortium of developers, the central government energy department and the pollution control authority acting as their advisers took the principal roles. Others included specialist chemical suppliers, experts on pipeline cleaning, local authorities, fishermen's representatives (including salmon fisheries) and inshore water pollution management. To retain commercial confidentiality the specialist firms and their products are not named. Anyone seeking detailed information should first approach the UK Department of Energy, Pipelines Engineering Division, Thames House South, Millbank, London.

2.2 Environmental impact assessment

Outline steps in the programme for commissioning the pipeline are listed in Table 1. What is briefly described

Table 1 Commissioning of the Brent gas pipeline

Stage	Action	Quantity
1 Initial filling	Inhibited sea water and fluorescent dye	–
Pressure testing	Pump in additional liquid (as above)	–
Relieve pressure	Release inhibited dyed sea water	6000 m^3 (inshore)
	INTERVAL	
2 Clean pipeline	Displace contents (as below)	165000 m^3 (offshore)
Separator gel	Add non-blending separator gel with inhibitors etc.	750 m^3 (offshore)
Cleaning gel	Add miscible cleaning gel with inhibitors	750 m^3 + 300t scale and silt
Fill pipeline	Add new inhibited freshwater	–
	INTERVAL	
3 Final cleaning	Displace contents (as below)	165000 m^3 (offshore)
Separator gel	Add non-blending separator gel with inhibitors etc.	750 m^3 (offshore)
Cleaning gel	Add miscible cleaning gel with inhibitors	750 m^3 and residual rust
Dewatering	Add batches of methanol	1000 m^3 (offshore)
Drying	Add methanol and treatment chemicals	525 m^3 (offshore)
Commiss- ioning	Displace seawards with natural gas	–

was an operation lasting several years. Individual steps took up to several months to evaluate and work out appropriate conditions for the release of the various discharges.

Phase 1 is designed to hold the pipeline with minimal corrosion and free from iron bacteria while the work on the production platforms and wellheads was completed. Clearance of grit, welding residues and scale is the second phase. Finally, any traces of residual loose particles are cleared, the pipeline dried, given its final preventative treatments and filled with natural gas under pressure ready for start-up.

Because much of this work was novel the effluent discharge arrangements were subject to change and in the light of experience gained, the treatment chemicals (oxygen scavengers, corrosion inhibitors, bactericides) altered. Inevitably many toxicity tests were needed - made complex by partially anaerobic conditions, insoluble precipitates and over the long term, progressive biodegradation of some of the biocides. In addition the hazards posed by the release of toxic gels had to be appraised in relation to behaviour on discharge and possible entry into fish and clogging of fishing nets.

2.3 Research

A major issue was the capacity of the water offshore to dilute and de-toxify the large volumes (about 50 million US gallons) of liquids and gels and just as critical the inshore discharge of some of the dyed inhibited water. Toxicity tests were made using shrimps and juvenile plaice. Practical experiments were conducted on breaking up the gels into small pieces that would aid loss of toxic chemicals and promote natural decay. Tests were made on the ability of trawl netting to capture gel of various sizes and attempts were made to feed gel to commercial fish species.

2.4 Evaluation

Both the oil industry research team and the control agency had particular problems to solve in relation to specific domestic responsibilities. On many of the problems and their solutions, the scientists on both sides worked closely in consultation. The safe discharge of inhibited waters was achieved by agreeing to certain pumping speeds which matched dispersion in the neighbouring sea area. The gels were smashed up as they emerged from the pipeline outlet using high pressure water jets. Fish would seldom approach particles of gel and if a particle was taken into the mouth it was speedily rejected. Follow up surveys revealed no demonstrable effects in the marine environment though the dyed effluent looked very impressive.

2.5 Conclusions

The lesson from this huge project was the value of free face-to-face discussion among those concerned throughout the various episodes. The many participants gained confidence from the evident intent shown by the project managers, administrators and scientists to ensure that each unit process was fully appraised, discussed with parties affected and brought to a conclusion with every possible safeguard.

3. DREDGING AND OYSTERS

In many ways this was the most difficult of the environmental impact examples to evaluate for three main reasons: (a) the large scale of the dredging programme relative to the small enclosed sea inlet (b) the presence of several sensitive fishery targets and the special case of Scotland's only commercial oyster fishery (c) the problems of establishing neutral decision-making.

3.1 Scenario

A nationally-owned ferry service sought to enlarge a ferry terminal, turning area and shipping channel in a small, muddy, shallow sea inlet. The national licensing authority had the responsibility of deciding on spoil dumping grounds and of evaluating any environmental damage arising from the dredging. The oyster company wished to protect their statutory right and obtain damages for oysters removed by dredging and harmed or killed by siltation caused by the dredging.

3.2 Environmental impact assessment

Environmental damage can arise from the removal of bottom sediments during dredging, from the clouds of disturbed silt, by the deposition of spoil at the dump ground and the spread of the plume of suspended solids. Outwith the sea inlet the area is known to be stormy. There are significant trawl and seine net fisheries and within the inlet a salmon farm. Not far seawards is an extremely valuable herring spawning ground. A major oyster fishery, established by legal decree at the beginning of the century, occupies the inner part of the inlet.

3.3 Research

In brief, after consultations, the control authority identified a spoil dumping ground just at the mouth of the inlet. Very tight restrictions were made to secure the even spread of spoil in relation to the depth of water; all large boulders were separately dumped in a rocky area off the shipping route.

The other main issue was to determine the pattern of fallout of silt around the dredger and find a way of measuring its effects on the oysters. Earlier oyster surveys had already determined the distribution of the oysters in the inner area and the characteristics of the juvenile and adult populations.

The ferry company engaged expert biologists to advise them and also had the benefit of advice from dredging technologists. The oyster owner did not engage scientific advisers.

It was not difficult to determine the pattern of suspended silt by conventional methods. This was established for varying wind conditions and results agreed closely between the university and control scientists.

What was a great deal more difficult to assess was the effect of prolonged exposure of the oysters *in situ* to long periods of enhanced silt deposition. In the absence of any other body with appropriate resources, the scientists of the controlling agency undertook a major site investigation using native oysters set out in plastic-coated stout wire cages on heavy moorings. The caged oysters were observed at the end of a period of field investigations (some 2 to 3 weeks after laying) with no losses and no mortalities. After some 5 months the cages were raised and the oysters examined. In general, where oysters occurred naturally survival in the cages was very good with only a few mortalities among the original 15 specimens. Where oysters were absent from the environment all died or only one or two survived. Near the dredging operation one cage was clogged with mud with no survivors but this also happened to one cage remote from the dredging. Some cages were missing particularly close to the dredging and oyster harvesting activity.

3.4 Evaluation

Despite strenuous efforts by the control authority to hold a demonstrably neutral position, the outcome of the field tests (largely because of missing cages) was debateable since various interpretations appealed to different parties. In the end the result completely satisfied no one.

3.5 Conclusion

The lesson learned here is most important. Where the control authority is placed in a situation between a huge nationalised industry and a small private owner it can only be deemed unwise for it to attempt to act as a neutral arbiter even, as in this case, when there may be no other local body adequately equipped. In these circumstances it is essential that an agreed method of reaching a decision on damage be hammered out between the developer and the party claiming damages using, if need be, the experience of the central control agency. The field work and interpretation must be totally independent. If the results are indecisive they will most likely serve to limit the degree of uncertainty. Decision by arbitration should not be too difficult.

OVERALL CONCLUSION

These examples emphasize (1) the care that may be needed in situations where a development may not involve poisonous substances (2) the great complexity of some major undertakings and the advisability of close consultations if speedy and harmonious solutions must be found (3) the care that must be exercised to demonstrate fair and equitable solutions.

It is clear that any environmental control body must have a team leader of wide experience together with administrators that are alert to the status of all the participants. Ideally the lead scientist must have access to experts in physical, chemical and biological aspects of pollution. Often major industries have their own expert environmental advisers and by working with the control team at a scientific level, many difficult issues can be anticipated and resolved on these grounds.

At the administrative level the post holds problems related to statutory issues, consultation with interested bodies, legal responsibilities and so on. It is most important that the framework of any problem encountered, and measures aimed at resolving crucial differences, do not place the control authority in a vulnerable position with regard to demonstrating total impartiality.

AFFILIATION

Dr. R.Johnston is a retired marine scientist, formerly of the Marine Laboratory, Banchory, Department of Agriculture and Fisheries for Scotland.

William Ritchie

PHYSICAL ENVIRONMENTAL ASSESSMENT: SOME PARTICULAR PROBLEMS OF THE COASTAL ZONE

From the Arctic to the Equator the coastal zone is probably under greater and more varied forms of pressure for development than any other land surface. These pressures are not new but there is considerable evidence to suggest that development may be accelerating as the less industrialized nations strive for commercial and industrial growth, and the sophisticated industrial and post-industrial nations find an increasing number of relatively affluent people moving to the coast, either temporarily as tourists or permanently as residents. In general, attention is focused on low, unconsolidated coastal landforms. Partly this is due to the need for relatively level terrain; partly it is related to ease of excavation for harbours, canals and water-based leisure complexes; partly it is related to the prime tourist resource of sand beaches; partly it is related to the need of large industrial and power plants to be near tidewater for cooling water and effluent discharge and partly it is due to lower costs associated with ease of transhipment of raw materials and finished products. A special but increasing form of pressure is now generated by the need to bring the products of the bed and sub-surface of the continental shelves ashore, normally by pipeline.

These many forms of coastal development, whether of large or small scale, pose special problems for anyone engaged in impact assessment. Environmental impact assessment of the coastal zone is not only related to such routine elements as landforms, soil, drainage, vegetation, micro-climate etc. (although these may have some special problems relating to proximity to the sea) but is a product of the dynamic, unbounded system of the marine environment. To understand the special difficulties it is necessary to summarise the main factors that produce and modify typical unconsolidated coastlines viz.

B. D. Clark et al. (eds.), Perspectives on Environmental Impact Assessment, 303–309.

1. GEOLOGICAL SETTING

e.g. rock platform, coral reef, rock headlands, glaciated lowland plain etc.

This is essentially a static factor, creating the framework which contains the mobile elements. It is also of direct relevance to sediment supply and possibly exerts a direct influence on surface form. It might also be a factor in the stability of the coastal zone e.g. in volcanic areas.

2. LAND PROCESSES

e.g. drainage, landslips, freeze and thaw effects, chemical weathering etc.

These processes are not unique to the coastal zone and are as varied as there are bioclimatic zones which control the nature and rate of these geomorphological processes.

3. SEDIMENT SUPPLY

e.g. river and stream deposition, shells, coral fragments, glacial deposits etc.

This is an important factor since, by definition, unconsolidated coastlines are created and maintained by marine processes transporting and depositing those materials onshore. At the critical contact zone of the beach, the saltmarsh, the delta - the rate and amount of this supply determines whether the coastline is in a state of dynamic balance or net accretion or net erosion.

4. BIOLOGICAL ACTIVITY

e.g. coral growth, vegetation growth etc.

The importance of this factor varies but is never wholly absent. Rock platforms in tropical areas, for example, have been shown to be substantially reduced by the activities of shellfish and other organisms in the intertidal zone. The saltmarsh and the sand dune are two landforms that are created by vegetation. In some areas dead vegetation and lifeforms become a significant part of the sediment budget e.g. shell sand beaches, deltaic marshes.

5. MARINE PROCESSES

These may be divided into chemical/biological effects such as the creation of lime-rich coastal materials, and physical effects that relate to changes in water elevation, waves and currents. It is this factor that is of critical importance in coastal zone assessment.

6. EXISTING LAND USE

Existing land uses such as those relating to recreation, amenity space, agriculture, forestry etc. are important controls of coastline development. The existing land use (or uses) can effect one or all of the five groups of factors as listed above. Agriculture, for example, will alter drainage and vegetation and thereby effect the natural development of the coastal zone. The proposed new development will obviously eliminate or alter the existing land use.

These six groups of factors should be used as a check-list for environmental evaluation work along any coastline. The relative importance of each will vary but standard matrix approaches will identify those factors that are of particular importance. The matrix approach might allow attention to be focused on a single or a few factors which lead logically into the inevitable web or series of interactions and feedbacks that characterise physical as well as biological systems.

SPECIAL PROBLEMS OF THE ASSESSMENT OF THE PHYSICAL ENVIRONMENT OF COASTAL ZONES

A survey of recent literature reveals that coastal zone assessment work tends to adopt a biological emphasis, summarized by the phrase (Clark, 1971) "the identification of ecosystem hazards that are consequent upon a specific project and undertaking" while this is certainly a valid approach it tends to reduce the importance of physical change, direct or indirect, to the landforms and waterbodies that constitute the coastal zone. These physical entities are the habitats for the ecosystems and it seems logical to place at least equal emphasis on the physical environmental conditions and processes as on the biological systems.

Another reason for emphasising physical factors is the specialised nature of coastal engineering work which cannot avoid the necessity of evaluating the effects of tides, waves and currents. Most developments at the coast include design elements that relate to the possibility of alterations in sediment budget and the statistical probability of short or long term elevations in sea level.

In spite of the manifest need to study such process factors as onshore and alongshore movements, both of which are a direct function of incident wave energy, it is rarely possible to do so. Such is the nature of the coastal zone that assessment work usually requires the use of indicators to summarize complex interactions; there is rarely time for the direct measurement of such process variables as wave action or sediment movements. Indicators are invariably morphological and in this sense the appraisal of the physical environment is similar to the use of key species in biological assessment. This use of static evidence to deduce dynamic change is dangerous but unavoidable. Nevertheless the developer and planner alike must recognize the uncertainties of an assessment method that uses indirect evidence, but the alternative approach, based on measurement and field recording, is probably prohibitively time consuming, expensive, and on occasion, impossible. Some process information is nevertheless available in those situations where a national agency or scientific programme has been monitoring and recording relevant information and data as part of a national or regional service. To take some obvious examples tidal ranges, meteorological statistics, river flow statistics, quality of coastal water and some other factors are normally available from published sources. Part of the assessment procedure is therefore to collate existing data. It is often remarkable how much information is readily available if the correct sources can be located. Frequently a much neglected source is maps and aerial photographs. The gaps in knowledge that remain must obviously be filled by field work.

Some of the special difficulties of environmental assessment of coastlines are related not only to the fact that it is the junction, even the collision zone, of two environments, the land and the sea, but also to the unbounded nature of the marine system. Normally the effects of waves, currents, tides and sediment movements cannot be restricted to the specific area of development. The coastal geomorphologist recognises two main types of unconsolidated coastline: the coastlines of impeded and unimpeded transport. The impeded coastline as exemplified by a semi-circular bay is the closest approximation to a closed system where linkages and feedbacks can be identified and adverse environmental effects might be more easily controlled. As the coastline opens-out, unimpeded transport means that the effects of a particular development might extend for great distances in both directions along the coast. Environmental effects are therefore more difficult to predict and minimise.

Many aspects of environmental assessment work at the coast are similar to assessment procedures elsewhere but may require special attention. Direct engineering work for example poses special problems due to the need for protection against the forces of the sea. Drainage and water supply, site access and the disposal of waste and excavated material may provide particular problems and greater cause for concern when the development is at the coast.

In environmental assessment some variables can be calculated or computed but many variables are unquantifiable; scenic value, disturbance to wildlife, loss of amenity. Such problems are not unique to the coastal zone but appear to have particular significance when so many of the coastlines in many nations are regarded as scenic, or as part of wildlife protection and conservation areas, or are under actual or potential recreational pressure from both the landwards and seawards directions.

As a result of this complexity, the dynamic linkages, the lack of hard process information and the open-ended nature of many coastal systems it is easy for a development to be stopped on the basis of insufficient information. Similarly unquantifiable objections such as excessive disturbance or reduction in scenic quality are value judgements that can be pressed-home with relative ease. The coastal zone typifies the central problem of some elements of environmental assessment work, i.e. the apparent need to quantify so many issues that are essentially qualitative. Since for the most part this is impossible, environmental assessment at the coastline consists of two elements, a semi-quantitative appraisal within the limits of existing knowledge and a qualitative estimation which must be more than mere opinion but should be based on comparative judgement. This comparative judgement may take various forms. After the first reconnaissance stage, it should be possible to state how the area under development compares with similar areas in the wider regional setting; if the investigator has sufficient experience it should be possible to cast the net relatively widely and venture a reasonable opinion on how this particular area compares with other similar areas elsewhere. Similarly, comparative knowledge may be used as a predictive device in relation to the probable physical development of the area. This is, in effect, the rationale of the morphological approach to landforms. The ability to state that a coastline is eroding and unstable is rarely made on the direct observation of that process but more on certain surface characteristics and signs which suggest that a particular process is active. This decision may lie in the individual experience of the investigator or related to his knowledge of current literature and research work.

The assessment of the probable effects of a proposed development on the physical environment of an unconsolidated coastline contains all the elements of impact assessment work elsewhere plus some special factors which add complexity and a degree of uncertainty to the impact analysis procedures. These difficulties result from the dynamic nature of the junction of two environments, one of which is essentially fluid and unbounded. Side effects and repercussions on relatively distant areas also require additional attention. The importance of the coastal zone for recreation, amenity, conservation and wildlife frequently means that these elements must receive greater than average attention from all parties - developer, investigator and planning authority - involved in the enterprise. Finally all parties must accept the need to recognise the unpredictable and sometimes hazardous nature of inserting fixed structures (and almost all coastal developments involve the construction of some form of fixed structure) into a zone which is intrinsically and essentially mobile. Unless special care is taken to retain a degree of mobility and flexibility, and to design for absorption rather than resistance, then the inevitable consequence is a total alteration in geomorphological character. This point might be illustrated by a single simple example as follows:

visualize a development where a sand beach and a line of dunes are replaced by a sea wall. In geomorphological terms, what has happened is that one has introduced a static, concrete cliff into a previously totally mobile and dynamic, unconsolidated coastal system; the amount of consequent environmental change should not therefore come as a surprise. Whether or not this is acceptable is not the point at issue but this example illustrates the difficulty of assessment, in that the entire environmental frame of reference as well as the process-form relationships of the geomorphological system which it contains, has altered completely.

The crux of the special difficulties of physical environmental assessment procedures in unconsolidated coastlines is therefore the need to recognise the highly dynamic and mutable nature of these landforms. Even moderate amounts of man-induced change produce far-reaching consequences, especially in open, high energy coastlines of unimpeded sediment transport, and these changes are usually unpredictable in their extent and duration. Many coastal zones epitomize the central problem of the development of fragile landforms and ecosystems: man's technology has provided unquestioned capacity to produce change but his knowledge of the wider consequences of such changes remains largely unknown or misunderstood.

REFERENCE

Clark, J.: 1971, Coastal Ecosystems
The Conservation Foundation, Washington

AFFILIATION

William Ritchie is Professor of Physical Geography and now head of the Department of Geography at the University of Aberdeen, Scotland.

Alastair Gemmell

SOLID AND LIQUID WASTES AND THE IMPACTS OF THEIR DISPOSAL

1. INTRODUCTION

Wastes are unwanted or undesirable products of life. They range in character from organic materials such as human and animal excreta to metallic, plastic and chemical byproducts of manufacturing industry. The term 'waste' can also include artifacts once considered useful by their owners but now obsolete or worn out and hence abandoned.

Wastes are usually only a problem when they occur in locations or in concentrations which may adversely affect the standard of life of the inhabitants of that area. Most wastes are perfectly tolerable at concentrations below a critical threshold but once that threshold is breached, environmental degradation will ensue unless preventative measures are initiated. Generally speaking, the problem of waste increases as the number of people increases and as their degree of mobility decreases. A nomadic group in a desert area will not consider themselves affected by a significant waste dispoal problem, for they can simply move and leave their wastes behind. In contrast sedentary life styles, dominant over most of the globe, result in waste continually being generated at fixed locations. To prevent accumulation of wastes continuing to the point at which a hazard is presented, some form of waste disposal system is required.

The feasibility of developing a waste disposal scheme which may be of universal application is very low. Disposal techniques vary greatly in cost, as a consequence of which some waste management agencies may be highly restricted in their ability to implement a scheme, purely on economic grounds. They may have to settle for less efficient but cheaper techniques. Secondly, the types of waste generated by a society vary greatly. In a non-industrialised country, waste products are largely organic (biodegradeable) materials, together with ashes derived from such activities as cooking (Pickford, 1977) and small-scale metal-working. As

B. D. Clark et al. (eds.), Perspectives on Environmental Impact Assessment, 311–340.

society becomes more industrial and consumer-oriented, so the character of the waste products and the techniques necessary for their safe disposal change. Lastly, because differing wastes pose individual problems of safe disposal, it may become necessary to have a range of disposal facilities available, some of which are of a highly specialised nature. The treatment and disposal of radioactive waste is a particular example of this.

The planner requires to be aware of the impact on the environment of the various techniques of waste disposal. This paper is intended to outline particular impacts related to the main waste disposal techniques, and to provide guidelines on methods of assessment and containment of these impacts.

2. PRINCIPLES OF DISPOSAL

There are two basic principles or philosophies governing disposal of waste. The first principle can be summarised as 'dilute and disperse'. Waste material is deposited in dilute form in a position where it can be dispersed by natural processes such as wind or streamflow, with a concomitant decrease in levels of concentration. In this way, it is intended that the concentration of wastes be kept below the critical threshold at which it becomes environmentally harmful.

The second principle can be summed up in the maxim 'concentrate and contain'. This principle is especially widely applied to hazardous wastes. Wastes are collected and concentrated so as to minimise the area contaminated by their presence. At such concentrations, these materials are often highly toxic, and so strictly enforced techniques of containment may be required to prevent leakage and environmental damage. In some countries, the techniques of confinement have evolved into legally enforced codes of practice (eg Department of Environment, 1976).

In reality, practices have often evolved which contain elements of both principles. Many industrial and domestic wastes are disposed of by initial collection for treatment at some centralised facility prior either to dispersal or to permanent containment. Although collection techniques are not really the concern of this paper, they do have significance in terms of assessing the likely impact of disposing of certain wastes. This is because the concentration of wastes at a treatment or disposal centre will almost always require transportation of 'raw' wastes. Cheremisinoff et al. (1979) note that wastes are moved by road, rail, ship and pipeline to treatment and disposal sites. Should safeguards prove inadequate and spillage take place in transit, the impact related to the disposal of that material will be much more widespread than might have been anticipated in relation to the site itself. A particular example of this problem is the transport of nuclear waste across Britain to a designated disposal site. Here, the mere knowledge of the temporary proximity of such waste material is sufficient to have an impact in terms of acute public anxiety in the vicinity of the designated routeway.

Hazardous wastes in general require special plans, and should be considered separately in any waste disposal scheme. Fields and Lindsey (1975) identify five categories of hazardous waste:-

(i) toxic chemical wastes,
(ii) radioactive wastes,
(iii) flammable wastes,
(iv) explosive wastes,
(v) biological wastes.

Careful analysis of the composition and characteristics of such wastes may be necessary in order to effect their safe disposal (Anon., 1978).

Nuclear wastes provide particular problems (Campbell, 1978). Current research is concentrating on ways of encapsulating such material in an inert shell prior to its disposal. Materials of particular interest as encapsulants are cement, glass and various ceramics. Once encapsulation is complete, then disposal can take place, usually in a deep burial site (eg mine shaft) or by dumping in the ocean. Such techniques are not however suitable for highly radioactive waste, for which a suitable means of disposal is still being urgently sought. As radioactive waste is a problem of relatively recent origin, those disposal techniques which have been used must be regarded as unproven with regard to long-term environmental impact.

The general question of hazardous waste disposal is one in which the impacts to be evaluated are not just physical, but also psychological. As implied in an earlier paragraph, there may be a significant mental impact on a populace faced with the possibility of 'nuclear dumping' in their vicinity. In such circumstances, the planner may find himself having to design waste disposal strategies which are partly shaped by non-technical constraints.

The selection of the most appropriate plan for the disposal of wastes cannot be undertaken successfully by the planner unless he possesses a knowledge of the potential environmental impacts of each of the alternative techniques of disposal he may have under consideration. How these impacts may be anticipated, avoided, or monitored will form the substance of this paper.

Before considering the individual characteristics of particular disposal techniques, it is worth making some more general observations. In attempting to produce a waste disposal strategy, the planner needs to consider not just techniques of disposal, but also the likely character of the wastes to be processed. Points of particular concern include:-

(i) the composition of the waste, remembering that this may not stay constant over time - a certain flexibility should be built into any scheme to allow for this;

(ii) the rate at which waste is generated within the catchment of a waste disposal facility, for if it is below the capacity of that facility then the plant may prove uneconomic to run;

(iii) the ability of any waste collection system to balance the objectives of a) not leaving waste any longer than absolutely necessary at the point of generation, and b) not feeding waste into the disposal facilities at a rate greater than those facilities can handle;

(iv) the fact that over time there is likely to be a steady enlargement of the area in which the impact of a waste disposal scheme is felt, the rate of diffusion of the impact being a function partly of the disposal technique, partly of the composition of the waste, and partly of geological and topographical factors at the disposal site.

3. LANDFILL DISPOSAL

Because of its simplicity and widespread applicability, landfill disposal is probably the oldest and by far the most common recognised waste disposal technique, being dominant even in industrial nations. Porteous (1977) quotes figures to demonstrate that no less than 89% of U.K. refuse is disposed of by landfill techniques, while U.S. figures indicate a similar preponderance in North America. Accordingly, the problems of impact assessment related to landfill disposal practices are those the planner is most likely to encounter.

Landfill disposal is a technique based primarily on the principle of 'concentrate and contain'. Waste is tipped at a designated site, often an old gravel pit or quarry, or into specially dug trenches. Such a practice would give rise to problems of flies, odour and vermin infestation unless it were carefully managed along the lines suggested below. Even in its crudest form it localises problems of environmental deterioration and is therefore preferable to the random deposition of waste on the land. It also allows a greater degree of control over the impact of disposal than would be the case were wastes to be deposited directly into uncontrolled, and largely uncontrollable, media such as rivers or the sea.

3.1 Principles of landfill management

The environmental impact of landfill disposal of waste can be greatly reduced by careful management to produce what is referred to as a 'sanitary landfill', ie a landfill in which the tipping surface is regularly sealed to prevent problems with odour and pests. This is viewed as being so desirable that some countries have introduced legal codes of practice laying down the procedures by which sites shall be managed.

The British code of practice (Department of the Environment, 1976) is typical of those dealing with sanitary landfills. For a landfill to be deemed sanitary, all waste must be compacted as soon as it is received at the site. On-site compaction is carried out, using special plant (eg Anon., 1980), at least once a day. Refuse should be placed within the landfill either by bulldozing over the end of the area in such a way as to result in the creation of a face with a gradient of no more than 1 in 3, or by tipping in front of the face, then bulldozing the waste onto the face from below. This latter method is believed to give better compaction of the wastes (Tasell, 1979). At the end of each working day, or after a layer of refuse has reached a thickness of 2.5 m, whichever is the sooner, all working surfaces are to be covered with inert material to a depth of 15 cm. Soil is usually used for this purpose. Clay soils should be used with caution, for they have a tendency to shrink and crack when dry, producing ruptures in the sealing layer. Before applying the sealing layer, any large items such as refrigerators or abandoned cars should be crushed or broken so that they do not protrude through the cover material. When the tip is full, or has ended its useful life, it should be sealed with a final layer of soil at least 1 metre in thickness, in which there is a subsoil layer and a topsoil layer.

The maintenance of surface drainage routes may prove a problem in the management of a landfill site. The U.K. code of practice requires that all cover material be laid to a gradient to assist the drainage of any precipitation which might fall on the site. Note that flowing water has the power to erode the cover material and expose underlying wastes. This problem can be reduced to some degree by ensuring that gradients are kept as low as possible though still permitting active drainage. It may also prove necessary to divert or to culvert watercourses which cross the site. If the landfill lies in a natural hollow there is a risk of inflow of surface water from the surrounding area, potentially leading to erosion of cover material. In such cases protective dykes or bunds should be built around the upslope flanks of the site.

The successful management of a landfill site is greatly enhanced by the keeping of precise records detailing both what materials are being deposited there, and where within the actual site the different materials are being stored. Mistakes might lead to harmful chemical interactions, problems which may be avoided by careful segregation of dumping areas within a landfill. The Department of the Environment (1976) gives details of such potential interactions.

3.2 Potential impacts of landfill disposal

3.2.1 Groundwater contamination

Wastes, both solid and liquid, undergo changes during storage in a landfill. Some of these changes raise the possibility of groundwater contamination. One such agency is heat. Fungaroli and Steiner (1971) demonstrated that the temperatures within a layer of waste rise rapidly soon after the layer is sealed, followed by a lowering of the heat level. At the same time, the initial aerobic conditions within the layer will tend to be superceded by anaerobic ones. Under such circumstances bacterial decomposition of the wastes takes place, forming new compounds within the tip. The combination of heat and chemical change gives rise to 'sweating' within the tip. Liquids generated by this process, together with any wastes deposited in liquid form, will gradually percolate downwards through the tip. Such percolation is likely to accelerate if the site is inadequately sealed against infiltration by rain, surface water flow or throughflow. This raises the possibility of some wastes being dissolved during the passage of water through the landfill.

Liquid emanating from the base or sides of a landfill is known as leachate. It will probably contain a variety of chemical and bacteriological pollutants, some of which could be injurious to health were they to enter drinking water supplies. Experiments suggest that the appearance of leachate will be delayed for a time after the deposition of waste in the ground, the length of the lag period varying according to the composition of the waste and the amount of liquid it contains. As the dumped material becomes increasingly saturated, so leachate will begin to appear (Ministry of Housing and Local Government, 1961). Once the tipped material has reached field capacity, the emission of leachate will roughly balance the amount of liquid that enters the tip.

The initial leachate from a tip is usually highly concentrated (Ministry of Housing and Local Government, 1961; Farquhar and Rovers, 1975). Its concentration of pollutants will reduce with time (Farquhar et al., 1972), but the total elimination of pollutants may be a lengthy process (Golwer et al., 1977). If the concentrated leachate reaches groundwater, then the water will become polluted. If the base of a landfill actually lies below the local water table, leachate can be absorbed directly into the groundwater flow. It is somewhat less of a problem if the base of the site is well above the groundwater table so that leachate has to percolate through sediment or rock before entering groundwater. Studies have demonstrated that leachate strength is reduced during its passage through sediments, both vertically downwards and laterally away from its source (Zanoni, 1972).

The rate of attenuation of leachate strength is related to the nature of the materials through which it passes on its way to the watertable. During the passage of the leachate, bacteriological and chemical pollutants are reduced in concentration by a series of processes including mechanical filtration, gaseous and chemical exchanges, sorption and microbiological activity (Farquhar and Rovers, 1975).

Fine-grained sediments, eg silts and clays, transmit

leachate at a very slow rate, and so afford the maximum opportunity for attenuation of pollutants during flow. Although the efficiency of this natural filtering decreases with increasing coarseness of sediment, even a fine sand bed will reduce bacterial levels in leachate by two orders of magnitude in less than 2 m of travel (Ministry of Housing and Local Government, 1961). The effectiveness of the sediment as a filter does however reduce with time, for it will itself become polluted.

It has often been assumed that a separation of only a few metres between the base of a landfill and the water table will allow adequate filtering of any leachate. The actual distance deemed 'safe' is a matter of some disagreement. Apgar and Langmuir (1971) suggest that leachate can be present in groundwater in detectable amounts 12-15 m below the base of a landfill. In contrast, the Ministry of Housing and Local Government (1961) suggested that as little as 24 ft (c. 7 m) be regarded as the minimum distance from a tip at which water could safely be extracted for domestic purposes. Other researchers are considerably more cautious. Anderson and Dornbush (1967) found that a distance of some 1200 ft (c. 350 m) of travel was required before groundwater quality was back to base levels. A more extreme example was cited by Newton (1979) in which polluted water was found to be appearing at a spring some 600 m from its parent tip. A possible explanation for these variations is that the distance over which significant levels of pollution are detectable downflow from a landfill site will reduce with increasing age of the tip, though Farquhar _et al_. (1972) note that leachate from one site which had lain 'fallow' for seven years was still detectable 700 m away.

It will be apparent from the evidence cited that landfill disposal of wastes will almost inevitably result in some pollution of groundwater supplies, particularly in areas with a high water table. Golwer _et al_. (1977) suggest that in such areas waste should be dumped in mounds rather than placed in pits. While this would help to increase the distance between the base of the refuse and the water table, it would considerably increase the visual impact of the site. The planner might thus be required to balance one impact against another when trying to select a disposal technique.

The ideal site is one in which there is a thick layer of fairly fine-grained sediment below the landfill, and in which the base of the landfill is substantially above the water table. In some locations the waste may be deposited directly onto bedrock, with no intervening layers of sediment. Worries have been expressed about this situation (Apgar and Langmuir, 1971) particularly if the bedrock is permeable or fissured. The validity of these fears is challenged by some workers (eg Waterton, 1969) who consider certain permeable rocks, limestone in particular, to be natural filters and to have the same properties for attenuating pollution as sands and silts.

As indicated above, the ideal site for a landfill is one in which leachate will have difficulty reaching groundwater level. Where there are only permeable sediments, attempts have been made to line landfill pits with relatively impermeable materials such as clays, asphalt, bentonite and concrete (Geswein, 1975). Ideally both the floor and the sides of a pit should be so treated, to avoid the problem of contamination of throughflow. Unfortunately, if leachate cannot escape into the ground beneath a landfill, it will accumulate within the fill, and may require secondary treatment prior to being pumped away. A particular hazard here is that some liner materials may react chemically with the ponded leachate (Fields and Lindsey, 1975), and possibly lose their effectiveness as a result. Careful management is of paramount importance in the avoidance of this problem.

3.2.2 Odours and gases

Bacterial decomposition of waste, besides leading to the 'sweating' detailed above, frequently produced gases - mainly carbon dioxide and methane with trace amounts of hydrogen sulphide (McGarry, 1977). The particular problem here lies with the methane, which is flammable and potentially explosive. Carbon dioxide is a lesser worry, but can dissolve in water to form a weak acid capable of dissolving some mineral matter. The character of leachate emanating from the site may be modified as a consequence. The risk of explosion is especially acute if pockets of methane accumulate within the landfill. Should the gas explode, not only would the sealing layers of inert material be ruptured, but there would be a risk that the careful sealing and segregation of certain wastes would be rendered ineffective, leading to the greatly increased risk of further impact on the environment.

The installation of a system of gas venting and control is the best way to avoid problems with methane and other gases. Cheremisinoff et al. (1979) suggest that gas movement be controlled by providing layers of permeable material such as gravel within the landfill. These materials provide easy routes for the movement of gases within the landfill, so preventing the accumulation of gas pockets. Such routes should open at the surface of the tip so that gases can be vented into the atmosphere. In the event of a landfill not having been constructed in this manner, it may be necessary to install venting pipes and burners in the waste so that the methane can be flared off. There should be little environmental impact from the venting of gases if it is done carefully. Vents should not be sited near buildings, but if such a safeguard is not feasible then discharge should be at an elevation above roof level.

3.2.3 Visual and audible impacts

The operation of a landfill disposal site, even if well-managed, will still be a source of noise and a visible intrusion into the landscape. Even when the landfill is sited well away from habitation, it will have an impact caused by the noise of waste collection vehicles moving to and from the site. Unless new roads are to be constructed to serve the site, there is little that can be done about this impact. In extreme cases it may prove necessary to provide grants to assist local householders with the sound-proofing of their dwellings.

The visual impact of a landfill tends to be less than might be suspected because of the practice of using hollows or holes in the ground for disposal. Even in cases where a site is widely visible, simple measures such as the planting of a screen of trees can do a lot to reduce the impact.

In dry countries there may be a problem with dust from the site. The use of sprinklers will help to keep the dust down. However it must be remembered that sprinklers will add to the problem of leachate generation, and their use should be avoided in sites where groundwater pollution is a possible impact.

Even in a well-managed landfill, there will still be a potential impact from the presence of paper and other light materials being blown about the countryside. This problem is reduced by rapid sealing of tipped material, but papers may be taken by the wind during the tipping act itself. A trap of some sort is the best way of reducing this problem. Suitably located screen fencing around the site should be a help here, and will have the secondary benefit of helping to hide the site from general view.

3.3 Selecting a Site for Landfill Operations

Although the selection of a landfill site will generally be undertaken with a view to minimising the resultant environmental impact, it has to be recognised that such a selection will have to be based on compromise. It would be most unusual to find a site which satisfied all the criteria the planner might desire. It may instead be necessary to 'improve' a conveniently located site by artificial means such as the installation of liners. This is especially important if hazardous wastes are to be disposed of there, though some of the constraints may be relaxed if the waste is to be composed of less toxic materials.

The planner needs to be far-sighted, particularly if there is any possibility of a future change in the character of material to be dumped. It is advisable that sites be selected with a view to the successful disposal of the most toxic materials that one can realistically imagine being deposited there during the working life of the site.

In highly-industrialised and developed regions like Britain, sites suitable for landfill disposal are becoming increasingly difficult to find. The choice of site is often limited by economic factors, to former pits and quarries. Although the extraction rates of minerals are rather greater than the rates at which wastes are being generated (Porteous, 1977), mineral workings are often sited well away from any urban centres. Under such circumstances new landfill sites may have to be created at more convenient locations. When choosing a new site, or selecting from a range of potential sites, the planner will need certain data which given an insight into the likely environmental impact of landfill disposal at that site.

The first essential is a knowledge of the hydrogeology of the site. This is critical if groundwater pollution is seen as a potential hazard. All possible sites should be surveyed to determine the depth to groundwater, the degree to which the water table rises and falls with the seasons, and the speed, direction and likely distribution of groundwater flow. The permeability and mineral composition of the sediments beneath the site should also be investigated. This gives some idea of the likely rate of leachate attenuation which might be expected if a disposal site were

CLASS	DESCRIPTION	NOTES
Class 1	Significant element of containment of liquids	Usually fine-grained compact rocks, including clays, shales and mudstones. These sites will have only slow percolation of leachate, particularly if the fine stratum is at least 6 m thick. Such sites are not recommended for large volumes of liquid, because of problems of ponding.
Class 2	Slow leachate migration and considerable attentuation	Usually in fine sand or silt, and not in contact with any saturated layer. Such sites may be used for liquid wastes.
Class 3	Rapid leachate migration and insignificant attentuation	Suitable only for the deposition of inert materials.

Table 1: Classes of Landfill Sites (Based on Department of the Environment, 1976).

established there. On the basis of such data, it is possible to classify sites with regard to the types of waste that can safely be stored there.

The Department of the Environment (1976) distinguishes three classes of site on hydrogeological grounds (Table 1). Class 1 sites will require constant monitoring of liquid levels within the tip, for the possibility exists that ponded leachate will spill out of the site and pollute the ground and any nearby surface waters. Newton (1979) suggests that this problem can be solved by pumping of leachate away from the site and spraying it onto nearby vegetation. It will then infiltrate the soil, and be rendered less toxic by attentuation. This practice does however imply that there will be an impact in an area beyond the immediate confines of the tip. Class 1 sites should therefore not be used for the storage of soluble hazardous wastes in regions of high precipitation. Class 2 sites can be regarded as the most generally useful sites, though again there is the problem of potential ponding of leachate in wet regions. Class 3 sites do not suffer from ponding problems, but pose a greater risk of leachate pollution of groundwater. These sites may require to be lined in some fashion before that can be regarded as suitable for the disposal of liquid wastes.

The second important factor to be considered when selecting a landfill site is the local geology, particularly the proximity to faults. This may be critical if hazardous wastes are to be deposited at the landfill. Where such wastes are stored, there should be no possibility of groundwater contamination. No leaks or spills can be allowed. Many hazardous wastes are in liquid form, and barriers against percolation, leakage or spillage of wastes will be required for safety. Any factor increasing the risk of rupture of containing structures will be very important. For this reason, hazardous wastes should not be deposited in areas where there is a risk of earthquake or landslip. A geological survey may often be required to assess the chances of such events occurring. As an extra safeguard, planning constraints noted by Fields and Lindsey (1975) emphasise the necessity of accurate record keeping and careful management so that materials which might react, should the barriers separating them be breached, are stored as far apart as possible within the tip or even are kept at separate tips.

The third factor is topography (Table 2). When

TOPOGRAPHY	CHARACTERISTICS	LIKELY ENVIRONMENTAL IMPACTS
Floodplain sites	Flat, poorly drained, often close to or at water table. Tendency to be in fine-grained material. Often Class 1 or Class 2 sites.	Danger of rapid production of leachate, and often liable to flood and pollute surface water. Should never be used for hazardous waste disposal.
Valley-side slopes	Sloping, usually well-drained, above level of water table. May be in solid rock or soils.	Free from flood danger, quite well drained usually, infiltration rates low because water runs over surface. May present a visual impact, but usually easily screened.
Ridges and Summits	Sloping, likely to be well above water table. Usually in solid rocks.	No flood danger, usually well drained. Likely to present a strong visual impact and to suffer from wind dispersal of dust and paper. Not easily accessible sites in many environments, possibly necessitating new road construction.

Table 2: Topographical Location and Environmental Impact of Landfill Disposal Sites

selecting a site, a simple evaluation of the topographic situation and its implications will help the planner greatly. Topography directs and channels certain impacts. As an example, access routes to a landfill tend to follow low ground whenever possible. This means that high-level disposal sites, although possessing some hydrological advantages, may require the construction of access roads, a procedure which generates its own level of impact. Here again, the planner will need to consider these various impacts before deciding on a site.

The last major factor to be considered in site selection is climate. Precipitation levels are of significance in determining the likelihood of leachate reaching groundwater. The critical amount is not known, indeed will probably vary as other climatic inputs change, but Cheremisinoff et al. (1979) note one study which suggests that in areas where sludge is deposited on the land, a precipitation level of as little as 760 mm per annum will produce leachate capable of polluting groundwater beyond acceptable quality standards.

Sites for the disposal of hazardous wastes are best located in dry areas. Campbell (1978) suggests that such sites ideally should allow evaporation of any precipitation that falls on the site. In this way, problems of leachate production, ponding and overspill will be greatly reduced.

On the basis of the above guidelines, the planner should be able to select landfill disposal sites which will minimise environmental impacts, and direct them along predictable and (hopefully) controllable channels.

3.4 Monitoring the Impact of Existing Landfills

In situations where a landfill site is already in operation, it may often be a relevant exercise to monitor the impact that site is having on the environment. The resultant data serve two functions. First, they provide an indication of the contaminant attentuation capabilities of the site. This information will aid decision-making regarding the types of waste which might safely be disposed of at that site and whether the site is suitable for the types of waste presently being dumped there. Secondly, a determination of current levels of impact will be necessary if steps are to be taken to contain, reduce or avoid that impact as far as possible.

In monitoring existing landfills, the principal concern must be for the maintenance of groundwater quality, especially if that water is used for domestic purposes. Should groundwater pollution occur, it then becomes necessary to ascertain the geographical extent of the contamination so that the drawing of polluted water for human use can be prevented.

A system of piezometers should be installed around the landfill, to monitor variations in water pressure. From such data, predictions can be made as to the direction and pattern of groundwater movement a critical element in the design of a water-quality monitoring system. To monitor the actual spread of pollution from a landfill, water sampling wells should be sunk in locations determined from the piezometric data. This will allow the determination of levels of contamination and geographical variations in those levels. One row of wells needs to be sunk just upflow from the landfill, so that baselevel composition of groundwater can be determined. On the downflow side of the landfill, wells should be located so as to tap the likely line of travel of the main pollution plume. As Hoeks (1977) notes, the siting of wells should be considered with the utmost case. A landfill if a point source of pollution, and the route which the contaminated groundwater will follow will be quite narrow, especially close to the source. Wells only a few metres from the main pollution flow may contain safe water. A test bore may indicate clean water, although a well sunk only a few metres away may still be tapping water which is unsafe for consumption. Careful and detailed monitoring is the only way to avoid such a situation.

Some researchers believe that no matter how rapid the attenuation of pollution by passage through fine-grained sediment, water known to have had contact with a landfill site must always be used with caution. Johnson (1979) points out that there are no known cases in which the passage of polluted water through sediment has cleaned it to the extent where it is fit to drink, though it may prove quite acceptable for irrigation purposes. If this is true, then it may be more important to determine the geographical extent of the pollution plume from a landfill site that it is to determine its rate of attenuation.

The sinking of a large number of monitoring wells is likely to be both costly and potentially unsightly. A monitoring technique which offers greater flexibility and reduced cost would appear to be highly desirable. Geophysical methods, particularly electrical resistivity techniques (Cartwright and McComas, 1968) look promising. Resistivity values are affected by the composition of water within the soil, and can be used to assess that composition. As they are also affected by the character of the sediment itself, great care and skill may be required in their interpretation. Cartwright and McComas conclude that resistivity is viable where the earth materials are uniform and the depth of the water table is constant. If these conditions do not obtain, then especial care is needed in the interpretation of the results.

3.5 Impacts where Landfills are to be used for Land Reclamation

The majority of landfill sites are returned to agricultural use or are built upon once the useful life of the site is ended and restoration has taken place. Some are even used as parkland. Although such sites should not present a hazard if the original landfill was well managed, problems of compaction and settling of wastes may cause drainage lines to be interrupted or broken, with resultant development of ponding (Aplet and Conn, 1977). For similar reasons, the installation of a gas venting system must be regarded as essential if a landfill is to be restored for subsequent use.

The production of leachate within the landfill is likely to continue for a considerable time after active deposition of wastes has ceased. In cases where the waste was being used directly for land reclamation, deposited into marshes or coastal areas with a view to extending the acreage of useful land, it must be borne in mind that such leachate will prove very hard to control. The planner would be advised in such cases to minimise any such impact before it happens by the simple expedient of keeping a very tight control over the types of material dumped at the site. In this way, any resultant leachate will be relatively innocuous, and the subsequent dilution and dispersion will ensure that any impact is kept to very low levels.

4. LAND APPLICATION OF WASTE WATER

The careful treatment of liquid wastes prior to disposal is advisable for a number of reasons. Many liquid wastes, particularly industrial wastes, are highly toxic and could cause great damage to the environment were they to escape or spill. Being liquids they are highly mobile and once spilt are effectively 'out of control' and cannot be contained (Johnson, 1979). It is therefore unsafe to dispose of them in landfills other than Class 1 sites. The safe disposal of liquids should be a multi-stage operation involving pretreatment of the waste followed by disposal under carefully controlled conditions.

The term 'wastewater' covers a variety of liquid wastes ranging in character from sewage to industrial byproducts. Many such wastes are originally treated by settling in oxidation ponds and lagoons, a process facilitating segregation of the waste into liquid and sludge. The latter is highly concentrated waste which may be disposed of by incineration, land application as a fertiliser in some cases, or in landfills. Some liquid wastes may require chemical treatment prior to disposal, while biodegradation of sewage and other wastes is quite a prevalent treatment in hot climates (Mara, 1977).

The direct discharge of untreated liquid wastes into landfills, rivers or the sea carries a strong risk of pollution and damage to the local ecology. Paper milling is one industry known to have discharged large amounts of liquid wastes directly into rivers, killing fish and other water creatures for considerable distance downstream from the point of injection. Stanley (1970) reported that the problem could be substantially reduced by the use of settling ponds coupled with screening filters to remove suspended pollutants such as wood fibres. Experiment showed that a four-day detention period in an aerated lagoon would reduce the BOD of the waste liquids by some 70%. This is not a complete solution to the problem, but a simple and effective first step towards the reduction of environmental impact associated with the production of paper.

Given the attenuating properties of many sediments, dilute wastewater should be cleaned quite effectively by passing through a thickness of soil. Land application of waste liquids utilises and augments this process by using growing plants to aid the cleansing of the waters (Wright and Rovey, 1979). In Australia, U.S.S.R. and U.S.A. the concept has been expanded so that wastewaters are used to irrigate crops. Such a system will provide an economic return in addition to purification of wastewater, but this reward is obtainable only after installation of expensive equipment.

If wastewater is to be used economically for irrigation, the site chosen must have access to a constant supply of wastewater throughout the growing season of whatever crops are being raised. For this reason the technique is most effective in the vicinity of an urban centre. The use of pipelines will increase the distance wastewater can be transported effectively, but will also increase the initial investment required.

4.1 Techniques of Wastewater Application

The majority of wastewaters will require pretreatment before being applied to the land. At the most fundamental level, physical screening to remove solids will avoid possible clogging of spray nozzles. Where the wastewater contains undesirable pollutants, particularly if it is to be used for the production of foodstuffs, some chemical and bacteriological pretreatments may be required.

The basic techniques for the application of wastewater to land are:-

(i) Overland flow. Here the water is released at the top of a slope. The water then flows downslope over the surface (which is often terraced), and collected at the foot of the slope in runoff collection ditches. It is then either recycled for further treatment or released. The wastewater is cleaned by a combination of biological action and physical filtering by vegetation, especially grasses. Overland flow is not a recommended technique where the soil is highly permeable, and is not possible when the temperature is far below freezing point.

(ii) Rapid Infiltration. Waste waters are ponded and allowed to infiltrate the soil, the objectives being to use the soil properties to clean the wastewater, to recharge groundwater levels and to allow the withdrawal of renovated water (Cheremisinoff et al., 1979). The technique has the advantage that it requires little space. It requires permeable soils and should not be used where there is a high water table as this greatly increases the risk of groundwater pollution.

(iii) Irrigation. Wastewaters have been used successfully for the irrigation of crops, including foodstuffs and timber. Any water used to irrigate foodstuffs will normally undergo a course of treatment prior to land application, lest the crop be unfit for human consumption. At some sites sewage sludge is also applied to the land, but this is not recommended if the crop is to be eaten raw unless the sludge had been allowed to stand for at least a year prior to application. Wastewaters have been applied to the land by any of a variety of techniques, including sprays, sprinklers and buried pipe distribution systems (Cheremisinoff et al., 1979). In terms of environmental impact there is little to choose between them.

(iv) Integrated Projects. An approach to wastewater treatment currently being pioneered in Michigan, U.S.A., is the integrated use of a system of lakes, marshes and irrigation systems for wastewater cleansing (Downs, 1978). Treated wastewater from a sewage plant is passed down a chain of lakes, four in all, with three marshes between the third and fourth lakes. Water can be pumped from any lake for irrigation purposes. The lakes not only facilitate settling out of pollutants, but water plants and algae flourish in the nitrogen- and phosphorus-rich waters of the upper lakes. The organisms remove these nutrients from the waters so that the liquid leaving the bottom lake is of very good quality. The water plants can be harvested for silage, while the water extracted for irrigation is further filtered and purified in the growing of corn, alfalfa, grass and some species of tree. Another benefit of the system is that fish thrive in the clearer waters of the lowest lake in the chain.

The experiment sounds very promising, but required a large amount of space and particular drainage characteristics in order to be feasible. Its application must therefore be somewhat limited despite its considerable efficiency.

4.2 Impacts of Land Application of Wastewater

Dissatisfaction has been expressed concerning potential impacts resulting from land application of wastewaters (Johnson, 1979). Not enough is known of the overall effects of land application techniques to enable them to be pursued with complete confidence. It is known, for example, that should wastewaters containing bacterial or viral components enter either groundwater or surface water flows, contamination of domestic water supplies is very likely to occur. Johnson cites cases in which material leaking from sewage lagoons has entered an aquifer to produce severe outbreaks of diseases such as gastro-enteritis.

The main body of opinion on the topic is rather less pessimistic. Land application of wastewater is supported with the proviso that it be carefully monitored and controlled. Points of particular importance in making an assessment of likely impacts include:-

(i) The hydrogeology of the site. This should be thoroughly surveyed prior to the implementation of any scheme. It will provide information about the most suitable mode of application of the wastewater, and the rate of application that the site can withstand without groundwater contamination becoming inevitable.

(ii) The topography of the site. This will influence the effectiveness of irrigation. If the land is flat there will be a danger of ponding and stagnation of water, especially if the soil is impermeable. On the other hand, if the slopes are too steep, wastewater runoff may enter surface streams without having been adequately filtered.

(iii) The climate. This will determine the length of the growing season. Wastewater application outside this period will merely increase the chance of groundwater pollution. In areas with a short growing season it may prove necessary to hold wastewater in storage ponds until it can be utilised. This adds to the cost of such schemes. The use of spray irrigation may not be recommended for some parts of the world, as it has been found to provide ideal breeding conditions for a variety of insects, including mosquitos (Downs, 1978). In hot climates, the use of irrigation techniques may have some unfortunate side-effects. High evaporation rates may result in the precipitation of pollutants in the surface layers of the soil. This may result in loss of soil quality and in extreme cases the crops may be contaminated.

Careless use of the technique raises the spectre of uncontainable pollution of aquifers and surface water flows. However, if practised under carefully controlled conditions, land impact of wastewaters can be a beneficial procedure with positive impacts.

5. INCINERATION AND PYROLYSIS

Incineration has long been a favourite method of waste disposal. If performed in a controlled environment such as an incineration plant the burning of rubbish is a very effective method of waste treatment. The volume of the waste will be reduced by some 90% (Pickford, 1977), leaving an ash which is most commonly disposed of by landfill tipping.

Pyrolysis is a technique which has evolved from incineration. The refuse is burnt in the absence of air, by a process called 'destructive distillation'. This not only considerably reduces the volume of wastes, but allows the reclamation of oils and gases from rubbers and plastics (Porteous, 1977). These products can then be sold to help offset the cost of the process.

5.1 Impacts of Incineration and Pyrolysis

Where incineration is carried out efficiently it may prove to be the safest method of disposing of chemical wastes such as PCB's (polychlorinated biphenyls), though the smoke from the incinerator may itself by a pollutant and the ash will still be a toxic hazard (Fields and Lindsey, 1975) requiring disposal by landfill.

Both pyrolysis and incineration plants produce a visual impact in the landscape, and are also expensive to construct. In addition the running costs, particularly of incineration, are high. Pyrolysis is now considered to be more economical in the long term by virture of the saleable properties of the reclaimed materials.

The cost of incineration is so high that the Waste Management Advisory Council (1976) believe that it will soon be completely replaced by pyrolysis. Attempts have been made to reduce the economic burden of incineration by mixing other materials, particularly sewage sludge, with the domestic wastes prior to burning. This helps to reduce the problem of sewage sludge disposal. A second potential aid to economic operation of incineration facilities is the selling of the heat emitted by the process. This can be marketed directly for heating purposes as for example near Nottingham where it is used in a district heating scheme, or it can be utilised to generate electricity which may then be sold. For all its economic drawbacks, incineration is of great use where landfill tipping space is at a premium. The reduction in volume of waste is such that the useful life of landfill sites is greatly extended. Greater Manchester is one area which is utilising this facility (Millbank, 1978).

As incineration reduces rather than eliminates wastes, it reduces rather than eliminates the need for new landfill disposal sites. On economic grounds it is not recommended for use in environments where suitable landfill sites are freely available.

6. COMPOSTING

Composting is the technique of processing organic wastes to give a humus-like product capable of being used as a top dressing for soil. As with incineration and pyrolysis, composting requires an initial investment in plant machinery to separate materials, especially metals, necessary to enable the process to work (Porteous, 1977). The efficiency of the process is being improved and new techniques of composting are being developed which greatly increase the speed of the process, and which give a nutrient-rich compost as compared with the previous rather inert product (Anon., 1982).

Composting has little environmental impact other than that of the construction and operation of the plant itself, though the building of the necessary roadways and other accoutrements of any refuse collection scheme need to be taken into account. The compost itself is often of low nitrogenous content, and can have little impact other than to absorb the force of rainfall and thereby reduce the risk of soil erosion. The compost is rather more useful in arid countries by virtue of its ability to retain soil moisture when applied as a top-dressing. Composting may thus prove of greater long-term economic benefit in some parts of the world than in others.

7. RECLAMATION OF MATERIALS

Much of the waste generated in the wealthier and industrialised nations is capable of being recycled. This has a number of benefits. It can particularly lessen the demand on scarce raw materials and at the same time relieve the pressure on waste disposal plants by reducing the volume of waste. Materials such as glass, paper, tin and ferrous metals can be reclaimed leaving only irredeemable wastes to be disposed of, usually by landfill techniques.

In some parts of the world, hand separation of

materials for recycling is practised. In other areas techniques like magnetic sorting and gravity sorting are utilised. Such techniques require the construction of automated plants, entailing considerable expenditure. The economic impact of instigating a policy of reclamation of wastes is hard to judge. It is expensive, but the sale of the reclaimed material will theoretically offset that expense, but is very dependent on the current local and wider markets for the particular materials. As an example, the sorting and recycling of paper is regarded as uneconomic at present in some areas, but not in others.

Reclamation has a beneficial environmental impact, in that it reduces the volume of waste requiring safe disposal. Its suitability for implementation may in practice rest largely on local economic factors.

8. BIOGAS

Methane gas, generated during the anaerobic fermentation of organic wastes, is a common byproduct of landfill disposal techniques. In many landfills pipes and burners are used to flare off the gas. This is seen by some as a waste of a valuable asset. Collection systems are now employed in some pits and the gas sold commercially. In this way not only is the risk of explosion reduced but there is a financial return to be set against the cost of waste disposal.

New techniques are being developed with the deliberate objective of producing methane from waste, particularly domestic and agricultural wastes. The anaerobic decomposition of such wastes under controlled conditions can give gas for heating and power generation, at the same time reducing the demands on other methods of waste disposal. In areas like India, small plants for the production of this 'biogas' were set up on a number of villages to provide power for heating and lighting. Although a good idea in theory, the scheme has been something of a failure because of economic problems and particularly the difficulty of maintenance of the facilities (McGarry, 1977).

Greater promise is shown in the use of biogas production plants as means to disposing of the great quantities of organic agricultural wastes resulting from the relatively intensive farming practices of Europe and North America. In Britain, five times as much agricultural slurry as sewage sludge is generated each year (Anon., 1982). This is far more than can be absorbed by the land, and presents a considerable problem in waste disposal. One solution is the erection of small biogas production units on individual farms or at co-operative centres. These plants can produce sufficient gas to drive generators or to heat steadings, while the residual digested material can be used as fertiliser.

The economics of biogas production are not especially favourable at present for considerable investment is required to set up a plant. In most parts of the world, biogas production will only be economically viable with government assistance, and without such aid small communities will almost certainly have to rely on other methods of waste disposal.

Where it is practised, biogas production has an indirect environmental impact. By reducing the need to spread raw slurry on the land, it limits the chance of groundwater pollution caused by leachate emanating from that slurry.

9. MISCELLANEOUS TECHNIQUES OF WASTE DISPOSAL

The search for safe ways of disposing of hazardous or very bulky wastes is continuing in a variety of directions. Some very bulky wastes, such as colliery spoil and mine tailings, are frequently disposed of by dumping on the land surface, forming unsightly spoil heaps. Apart from their visual impact, such tips can collapse should they become saturated with water, such as happended at Aberfan in Wales, when 144 people were killed in the collapse of a colliery spoil heap in 1966. Such impacts can be avoided best by careful management of spoil heaps. Spoil should be regraded and covered with topsoil to produce a new and relatively pleasant topography where before there was only ugliness. Appropriate techniques are detailed by Browne (1971) and Clouston (1977).

Ocean dumping has long been seen as an easy method of waste disposal and is unfortunately still prevalent in many parts of the world. Even within Western Europe many cities discharge raw sewage directly into the sea. The resultant pollution upsets the ecological balance of the sea, often to the extent of rendering it unsafe to bathe in. In extreme cases,it may result in fish being unfit for human consumption. Such levels of pollution can simply and effectively be reduced by the use of settling ponds in which liquid wastes are stored prior to discharge.

Impact	Waste Disposal Technique
Surface Water Pollution	Landfill, Wastewater application, Ocean dumping Spoil heaps and mine tailings.
Ground Water Pollution	Landfill, Wastewater application, Deep injection of wastes, Spoil heaps and mine tailings.
Atmospheric Pollution	Landfill (gases), Incineration and pyrolysis.
Visual Impact of Plant	Incineration and pyrolysis, Composting, Spoil heaps, Reclamation.
Drainage Disruption	Landfill, Spoil heaps, Wastewater application.
Landslip Hazard	Wastewater application, Spoil heaps.
Increased Tectonic Activity	Deep injection of waste.
Odour	Landfill, Incineration, Wastewater applications (occasionally), Spoil heaps.
Noise	Landfill, Reclamation, Incineration and pyrolysis.
Insect and Vermin Infestation	Landfill, Wastewater application.

Table 3: Summary of Impacts Associated with Waste Disposal Techniques

The oceans are also being used for the disposal of hazardous wastes. Current practice is to encase the waste in some inert material and cast it into drums for storage and disposal. The drums are shipped out to sea for dumping, or may be placed into deep wells or mineshafts. Such disposal techniques are really forms of waste storage rather than processing. They appear to have a very restricted impact, at least in the short term, but the long-term consequences of such actions are completely unknown.

The dumping of hazardous liquid wastes directly into the ocean is generally recognised as having a very severe impact, and cannot be recommended under any circumstances. Liquids of this type have been disposed of by injection into deep holes in the Earth's crust, such as mine shafts. Experience in the United States has shown that this form of disposal may lead to increased levels of tectonic activity as a result of the resultant lubrication of fault slip-lines. As the liquid is effectively out of control once injected, this again is a technique which cannot be recommended.

10. SUMMARY

Alternatives to landfill disposal of waste are constantly being developed. It would be true to say, however, that the majority of new developments will not eliminate landfill, but will rather reduce the amount of material requiring disposal in this fashion. It is likely that the future will continue the trend towards multi-mode waste disposal strategies. The environmental impact resulting from such strategies will be complex combinations of those detailed in Table 3. There will thus be an increase in the range of impacts which will affect the environment, though the magnitude of individual impacts is likely to be reduced.

As yet there is no technique for waste disposal which can be truly described as impact-free. The planner is required to perform a balancing act between cost on one hand and the type and degree of likely impact on the other. At present the building of complex and expensive plants will certainly ameliorate the demand for landfill disposal though simultaneously introducing a fresh set of impacts to an area.

Landfill will continue to be the dominant waste disposal technique throughout the world, and the development of an environmental protection strategy will almost inevitably have at its hub the minimisation of impacts associated with landfill disposal techniques.

REFERENCES

Anonymous: 1978, 'Co-operative programme of research on the behaviour of hazardous wastes in landfill sites', Solid Wastes 68, 272-6.
Anonymous: 1980, 'Cambridgeshire masters its trash', Surveyor 155(4597), 24.
Anonymous: 1982, Environmental Data Services Ltd. (ENDS), Report 88, May 1982.
Anderson, J.R. and J.N. Dornbush: 1967, 'Influence of sanitary landfill on groundwater quality', Journal of the American Water Works Association 59, 457-470.
Apgar, M.A. and D. Langmuir: 1971, 'Groundwater pollution potential of a landfill above the water table', Ground-water 9, 76-96.
Aplet, J.A.H. and W.D. Conn: 1977, 'The uses of completed landfills', Conservation and Recycling 1, 237-246.
Browne, L.J. (ed.): 1971, Landscape Reclamation (Vol. 1), IPC Science and Technology Press Ltd., Guildford, England.
Campbell, A.: 1978, 'Geological investigation of disposal sites for hazardous waste', California Geology 31. 229-233.
Cartwright, K. and R. McComas: 1968, 'Geophysical surveys in the vicinity of sanitary landfills in northeastern Illinois', Ground Water 6, 23-30.
Cheremisinoff, N.P., P.N. Cheremisinoff, P. Ellerbusch and A.J. Perna: 1979, Industrial and Hazardous Waste Impoundment, Ann Arbor Science, distributed by John Wiley.
Clouston, B.: 1977, 'How Britain is cleaning up its mine ruins', Landscape Architecture 67, 253-262.
Department of the Environment: 1976, Waste Management Paper No. 4. The Licensing of Waste Disposal Sites. H.M.S.O. London.
Downs, C.R.: 1978, 'Wastewater: a biological approach', Water Spectrum 10, 39-46.
Farquhar, G.J., R.N. Farvolden, H.M. Hill and F.A. Rovers: 1972, Sanitary Landfill Study Final Report, Vol. 1, Waterloo Research Institute, Canada.
Farquhar, G.J. and F.A. Rovers: 1975, 'Guidelines to landfill location and management for water pollution control', Sanitary Landfill Study 4, Waterloo Research Institute, Canada.

Fields, T. and A.W. Lindsey: 1975, Landfill disposal of hazardous wastes: a review of literature and known approaches, U.S. Environmental Protection Agency, Washington D.C. (PB-261 079).

Fungaroli, A.A. and R.L. Steiner: 1971, 'Laboratory study of the behaviour of a sanitary landfill', Journal of the Water Pollution Control Federation 43, 252-267.

Geswein, A.J.: 1975, 'Liners for land disposal sites: an assessment', U.S. Environmental Protection Agency Report SW-137.

Golwer, A., G. Matthes and W. Schneider: 1977, 'Groundwater contamination by waste deposits and the implication for methods of refuse disposal', in Effects of urbanization and industrialization on the hydrological and on water quality, IAHS-AISH Publication 123, 365-365.

Hoeks, J.: 1977, 'Mobility of pollutants in soil and groundwater near waste disposal sites', in Effects of urbanization and industrialization on the hydrological regime and on water quality, IAHS-AISH Publication 123, 380-388.

Johnson, C.C. Jr.: 1979, 'Land application of waste - an accident waiting to happen', Groundwater 17, 69-72.

McGarry, M.G.: 1977, 'Domestic wastes as an economic resource: biogas and fish culture', in R. Feachem, M. McGarry and D. Mara (eds.), Water, wastes and health in hot climates, pp. 347-364, Wiley-Interscience.

Mara, D.D.: 1977, 'Wastewater treatment in hot countries', in R. Feachem, M.M. McGarry and D. Mara (eds.), Water, wastes and health in hot climates, pp. 264-283, Wiley-Interscience.

Millbank, P.: 1978, 'Tackling waste in Greater Manchester', Surveyor 151 (4475), 9-16.

Ministry of Housing and Local Government: 1961, 'Pollution of water by tipped refuse', Report of the Technical Committee on the Experimental Disposal of House Refuse in Wet and Dry Pits, H.M.S.O. London.

Newton, J.W.: 1979, 'Leachate treatment on landfill sites', Surveyor 153 (4532), 33-35.

Pickford, J.: 1977, 'Solid waste in hot climates', in R. Feachem, M.M. McGarry and D. Mara (eds.), Water wastes and health in hot climates, pp. 320-344, Wiley-Interscience.

Porteous, A.: 1977, Recycling resources refuse, Longman, London.

Stanley, D.R.: 1970, 'Liquid waste in the pulp and paper industry', in M.A. Ward (ed.), Man and his Environment,

Volume 1. Proceedings of the first Banff Conference on Pollution, May 16-17, 1968.

Tasell, M.G.: 1979, 'The disposal of controlled waste - a county council approach', Water Pollution Control 78, 106-114.

Waste Management Advisory Council: 1976, 1st Report, H.M.S.O. London.

Waterton, T.: 1969, 'The effect of tipped domestic refuse on groundwater quality', Water Treatment and Examination 18, 15-17.

Wright, T.R. and C.K. Rovey: 1979, 'Land application of waste - state of the art', Groundwater 17, 47-61.

Zanoni, A.E.: 1972, 'Groundwater pollution and sanitary landfills. A critical review', Groundwater 10, 3-13.

AFFILIATION

Dr Alastair Gemmell is a Lecturer in the Department of Geography at University of Aberdeen, Scotland.

A. H. Harris

THE ECONOMIC INPUT INTO ENVIRONMENTAL IMPACT ASSESSMENT

1. INTRODUCTION

There are a number of approaches which economists have adopted when considering the impact of a project on society. The traditional approach has been to attempt to quantify the costs and benefits of the project in a way which allows their comparison. The aim here is to assess a project and its ranking among competing mutually exclusive projects using net benefit as a gauge. In order to obtain a single measure a numeraire for all social costs and benefits is used - money.

The valuation of the net benefit of a project is the aim of the technique, cost-benefit analysis. Its usefulness lies in its supposed ability to generate a single value by which a project's desirability can be measured. It is seemingly a tool which decision makers would relish providing as it does an apparent objective measure of the relative worth of projects taking into account not only their relative costs but also their relative benefits over a time period.

There are a great many problems with this technique many of which are discussed in the literature and textbooks such as Mishan (1971) or Pearce and Nash (1981). With reference to projects which involve environmental costs and benefits the particular problems are:

1) The technique of CBA requires a measurement of the change in net real income of those affected by the project. This is done by placing a valuation on the commodities consumed by those individuals as well as their direct and indirect income from employment. By measuring the expected change in prices and money incomes a figure is obtained for the net change in aggregate income for the affected group as a whole.
   A major problem here is that many aspects of the environment and natural resources have either no

B. D. Clark et al. (eds.), Perspectives on Environmental Impact Assessment, 341–363.

price attached to them or the market fails to price them at a level which reflects their worth to society. This provides the economist with the task of allocating a price to those commodities and services which is a reflection of that worth.

2) A second set of problems is the more general question of the significance of market prices in a world of uncertainty, imperfect information, and an unequal distribution of income, wealth and power. The use of money as a numeraire may be legitimate in practice even if we have some problems with valuing clean air, life itself, or the risk of death, but the use of market prices may not be as objective as it may seem at first. This however is not the place to go into a detailed discussion of the normative significance of market prices.

How economists have tackled these problems will be examined later. The purpose of such an examination is that CBA may provide some assistance at the post-EIA stage when decisions about competing projects are to be made. CBA, by summarising information, and making comparison easier, may ease the decision making process. It may be argued that it over-simplifies but that is a question of the use to which the results are put.

Even if the whole technique of CBA is rejected there are some parts which EIA may be able to use. The economist who assesses the benefits of a project must first quantify the effects on product and labour markets as well as the physical effects of a project. The techniques used to do this may be of use to those concerned with EIA.

The methodology used by economists when considering the effects on markets of a project is to start from the *ex ante* position of an equilibrium set of prices and quantities demanded by consumers and supplied by producers. The project upsets this balance perhaps by increasing demand for goods and services, by affecting the supply of raw materials, and thereby affecting prices and incomes. The economist who wishes to measure the change in real income associated with the project may examine the expected new equilibrium at the end of the project and quantify the different levels of income, prices, quantities and employment. Although the use of equilibrium concepts will not capture the full dynamic effects of the project this approach might be regarded as a first step in a full CBA and EIA.

In order to undertake this analysis the economist must

first build a model of the economy in question. This may be a country, a region, a city or a locality depending on the nature of the project. This model can then be used to simulate the introduction of a project which affects the economy and the environment. In a later section various attempts to do this will be discussed with particular reference to environmental programmes, and such projects as the building of a nuclear power station.

2. VALUING POLLUTION DAMAGE

There exists a large number of studies of the costs of air and water and noise pollution. Often the studies are framed in terms of the benefits of regulating such pollution but the methodology is identical. Two recent reviews have comprehensively surveyed the field. Freeman (1979) is a report prepared for the United States Council on Environmental Quality, and Pearce (1982) is a survey prepared for the OECD Environment Directorate which covers similar ground and in addition comments on Freeman's attempts to synthesise the results of various studies and present "best judgement" estimates of air and water control benefits.

At the outset this literature takes as given the physical relationships between rates and times of discharge, environmental quality, and the flow of environmental services to individuals. The economic input into this relationship occurs when consideration is given to the functional relationship between environmental services and economic welfare.

The economist (with certain notable exceptions) takes as given the physical effects of pollution and pollution control on vegetation, property, health, visibility etc. The economist is thus reliant on accurate information on these aspects and a lack of knowledge may in some instances be a major barrier to empirical estimates.

There are 3 basic approaches to determining the values that individuals place on improvements in environmental quality. 1) Direct questioning 2) Voting 3) Surrogate markets. These techniques are described in Freeman (1979) and are used to discuss the damage caused to health, soiling and cleaning, recreation, productivity, amenity, the risk to workers of accident and ill health due to occupational hazards.

Table 1: Estimates of the Elasticity of Mortality with Respect to Air Pollution - From Lave and Seskin

	Combined Elasticity		
Model	Air Pollutants (Combined)	Unadjusted Mortality Rate	Age-sex-race adjusted Mortality Rate
1960 Annual Cross-Section - 117 SMSA's	Sulfates & particulates	.094	.096
1969 Annual Cross-Section - 112 SMSA's	Sulfates & particulates	.116	.100
1969 Annual Cross-Section - 69 SMSA's	Sulfates & particulates	.106	.096
1969 Annual Cross-Section - 69 SMSA's	Sulfur di-oxide and particulates	.126	.110
1960-69 Annual Cross-sectional time-series - 26 SMSA's	Sulfates & particulates	.094	.102
1960-68 Annual Cross-sectional time-series - 15 SMSA's	Sulfates & particulates	.118	.126
1962-68 Annual Cross-sectional time-series - 15 SMSA's	Sulfur di-oxide and particulates	.106	.114
1963-64 Daily time-series - Chicago	Sulfur dioxide	.108	

Source: Lave and Seskin (1977) Table 10.1, p. 218.

2.1 The Damage from Air Pollution

2.1.1 Health Damage

The first of these, the damage to health has used a number of valuation techniques. The principal difficulty has been in estimating the physical relationship between pollution concentration and health status. A large number of studies have looked at this problem some using long term individual controlled case studies while others used cross-section aggregate statistical methods. One recent review of the air pollution - health relationship is Lipfert (1978).

Most of the economic studies of the health benefits of pollution control have used a framework of analysis pioneered by Lave and Seskin (1977). This has involved obtaining data on mortality rates in a number of areas and using multiple regression techniques estimating a relationship between these rates and air pollution taking into account other explanatory variables such as socio-economic differences, climatic variance as well as other important variables thought to influence health. In this way an estimated relationship is built up which allows the calculation of the responsiveness of the mortality rate to the level of pollution other things being equal. This is called the elasticity of mortality with respect to air pollution.

Table 1 below gives an example of this from Lave and Seskin. This shows that a one per cent change in the mean concentration of air pollution is estimated to result in between a 0.094 and 0.126 per cent change in the mortality rate from its mean value.

Unfortunately there is by no means agreement among analysts on whether these effects exist let alone on their strength. However let us for the moment assume that these effects exist. What now does the economist do?

The last step for the economist is to use a monetary measure of willingness to pay to assign values to the predicted changes in mortality and morbidity. Fortunately EIA does not use this stage of the analysis and it will not be discussed here which is as well because estimates of willingness to pay to avoid the certainty equivalent of death vary by over an order of magnitude. A more interesting approach is to look at the implied valuation of life of various projects found to be acceptable.

An example of this process is Freeman's synthesis. He first accepts that the elasticity of mortality with respect to stationary source air pollution in the U.S. is about 0.05. This is used along with figures on the reduction in air pollution in the U.S. from 1970 to 1978 of about 20% and a figure of $1m per life saved to give the results below in Table 2.

Table 2: Mortality Benefits for Stationary Source Air Pollution Control

ASSUMPTIONS:	Reduction in Pollution =	20%
	Elasticity of mortality with respect to pollution (most reasonable point estimate = .05)	.01-.1
CONDITIONS:	U.S. Mortality in 1978	1,900,000
	Percent of U.S. Population living in metropolitan areas	73% in 1976
COMPUTATIONS:	U.S. Urban Mortality	1,390,000
	20% reduction in pollution reduces urban mortality by (most reasonable point estimate = 1%)	.2%-2%
	Mortality avoided (most reasonable point estimate = 13,900)	2,780-27,800 deaths
BENEFITS:	A. At $1 million per death avoided (most reasonable point estimate = $13.9 billion)	$2.8-$27.8 billion/yr.
	B. For illustrative purposes if the assumed value of life is $500,000, benefits are (most reasonable point estimate = $7.0 billion	$1.4-13.9 billion

Source: Freeman (1979) p.67

This type of analysis is also carried out for morbidity and for mobile sources.

2.1.2 Soiling and cleaning

A number of studies have looked at inter city differences in cleaning practices of households and related them to differences in total suspended particulate levels. Unit costs for cleaning tasks were used to compute the change in total cleaning costs due to the change in cleaning frequency as the level of suspended particulates was reduced. One example of this type of study is Liu and Yu (1976). Their conclusion was that the total benefits for reducing suspended particulates from 1970 levels to 45 $\mu g/m^3$ was \$5.0 b per year.

Freeman in his survey article concludes that the benefits of a 20% reduction from 1970 levels of all air pollutants lies in the range of \$0.5 b - \$5.0 b per annum at 1978 prices.

2.1.3 Vegetation

Damages to ornamental and commercial crops may be caused by ambient concentrations of oxidants, sulphur compounds and fluorides. The first attempt to estimate the national damages to vegetation due to oxidant pollution was made by Benedict, Moore and Smith (1973). A later less crude study by Heintz, Hershaft and Horak (1976) uses published field experiment data on crop losses due to oxidants. Freeman uses this study to estimate the total loss due to photochemical oxidants at 1978 levels is between \$1.2 b and \$12.2 billion per annum.

There is little hard data on vegetation losses due to sulphur compound pollution in spite of increasing concern about the effects of "acid rain" on crops and soil fertility.

2.1.4 Material damage

There are a great many studies of the damage to various materials and equipment due to air pollution and these cannot be summarised here. A very useful review is provided by Waddell (1974). Freeman estimates the total damage for 1970 is \$4.8 b in 1978 dollars. Assuming then a 20% reduction in levels the realised benefits from 1970 to 1978 would be between \$0.5 and 1.4 billion per year.

2.1.5 Aesthetics and visibility

There are two ways of estimating the value of the aesthetic damage done by air pollution. The first is by use of surveys, asking people what they would be willing to pay for the improvements. The second involves the use of a proxy market such as the housing market to extract the valuation of environmental amenity.

There is a large and growing literature in each of these areas. The problems with the first are the obvious ones of survey reliability especially given the nature of the good in question. The second approach uses the idea that the locational decisions on the part of households will be influenced by perceived environmental characteristics of the house and neighbourhood. This, it is argued, will be reflected in the price differential of houses at varying locations with varying levels of environmental amenity. Techniques have been developed to extract from house price information that portion of the price attributable to environmental amenity. Two main problems are, first the techniques are not wholly successful in involving certain technical assumptions which limit their accuracy. Second it is not clear that people's perceptions of environmental amenity correspond with the analyst's data on environmental quality or with the objective risks of the neighbourhood environment.

These two problems make interpretation of house price differentials difficult, including as they do, not only aesthetic considerations but also some element of perceived health risk, material damage, soiling and cleaning.

Freeman offers a range of estimated damage from stationary and mobile air pollution sources for a 20% reduction from 1970 levels of between $1.1 and 8.9 billion at 1978 prices.

Although it is too complex to add these figures to form some measure of the total damage from air pollution a number of points emerge. First about three quarters of the total benefits are estimated to be in the health category although this depends on the uncertainty surrounding the mortality-air pollution relationship and the value placed on life. Second, benefits as estimated are for one place, one time, and at one level of pollution. They do not necessarily apply to other areas with a different experience of pollution as well as other socio-economic characteristics. These results can not be transferred without a great deal of care.

2.3 The Damage from Water Pollution

There are a number of categories of damage from water pollution, recreation, non-user benefits, direct water use (domestic and commercial) and fisheries. Each of these uses requires techniques to evaluate the damage similar to those for air pollution and for this reason they will not be reviewed here.

Freeman (1979) estimates the annual benefit of meeting the United States water quality objectives set out in the Federal Water Pollution Control Act amendments of 1972. This law calls for all industrial dischargers to have controlled their discharges using the "best available technologies" by 1983. These estimates are reproduced below in Table 3.

Table 3: Summary of National Benefits of Meeting 1985 Water Quality Objectives By Category (in billions of 1978 dollars per year)

	Range	Most Likely Point Estimate
Recreation	\$4.1-\$14.1	\$6.7
Non-User Benefits:		
aesthetics, ecology, property values	1.0- 5.0	2.0
Diversion Uses		
Drinking Water-Health	0.0- 2.0	1.0
Municipal Treatment	0.6- 1.2	0.9
Households	0.1- 0.5	0.3
Industrial Supplies	0.4- 0.8	0.6
Commercial Fisheries	0.4- 1.2	0.8
Total	\$6.6-\$24.8	\$12.3

Source: Freeman (1979)

3. THE MACRO ECONOMIC EFFECTS OF ENVIRONMENTAL POLICIES

At the opposite extreme from the work of Freeman and others on evaluating the welfare effects of pollution damage with reference to individuals income changes, there exists some

Table 4: Economic Consequences of Additional Environmental Measures 1978-1985 (Alternative 1) (% changes)

	With compensation in wages								
After	1979	1980	1981	1982	1983	1984	1985	1986	1987
Volumes									
Private consumption	-0.1	-0.2	-0.3	-0.4	-0.5	-0.6	- 0.7	- 0.7	- 0.7
Investments	-0.9	-0.8	-0.3	-0.9	-1.3	-0.7	- 1.2	- 0.2	- 0.4
Exports	-0.1	-0.3	-0.5	-0.7	-0.9	-1.0	- 1.1	- 1.1	- 1.1
Imports	0.3	0.5	0.5	0.4	0.4	0.4	0.5	0.2	0.2
Production volume	0.1	-0.1	-0.2	-0.2	-0.4	-0.4	- 0.6	- 0.7	- 0.7
Production capacity	-0.1	-0.1	-0.2	-0.3	-0.3	-0.4	- 0.5	- 0.5	- 0.5
Prices									
Private consumption	0.2	0.4	0.5	0.6	0.7	0.7	0.8	0.8	0.7
Exports	0.2	0.3	0.4	0.5	0.6	0.7	0.7	0.7	0.7
Investments	0.1	0.2	0.3	0.3	0.3	0.4	0.4	0.4	0.3
Employment (1)	2.1	-0.5	-3.6	-5.9	-8.0	-9.0	-10.3	-13.3	-12.5
Unemployment (1) (in 1000 man years)	-1.4	0.3	2.4	3.9	5.3	6.0	6.9	8.9	8.4
Export balance (min fl.)	-0.4	-0.7	-0.9	-1.0	-1.1	-1.4	- 1.7	- 1.4	- 1.5

(1) The line 'Employment' refers to the net impact of the environmental programme in terms of the number of jobs created and foregone. The next line 'Unemployment' refers to the impacts of the programme on unemployment levels. The absolute figures in the two lines differ on account of those affected by the employment changes not necessarily having been or becoming registered as unemployed due to the different participation rates in the various sectors of the labour market.

Source: OECD (1982)

literature on the effects of complete programs of environmental control on the national economy.

The Environment Committee Group of Economic Experts to OECD (1982) provides a review of some of this work. These studies are done using econometric macro-models to assess the net overall effect of environmental expenditures upon the economy. Key economic aggregates of current concern include inflation, unemployment, productivity, balance of payments and general economic growth.

These studies are limited in general and specifically for the purposes of EIA, by the features inherent in the macro models. For example they are often too aggregated to be sensitive enough for the purposes of EIA. What they do however is point to a number of areas of concern for EIA if it wishes to consider the effects on the economy as a whole. These include:

<u>Direct Effects</u>

Additional demand for goods & services from enterprises and local authorities
Additional jobs for plant operation
Replacement of "productive" investment by pollution control investment (financial constraint)
Increased production costs and selling prices (polluting industries)

<u>Indirect Effects</u>

Multiplier effects on available income & private consumption
Speeding-up effect of domestic production on productive investment
Effects of additional demand on imports & available capacity
Effects of product and consumer price increases on private consumption
Effects of producer price increases on exports
Effects of capital goods price increases on capital utilisation costs
Effects of additional expenditure on financial and money markets and on rate of interest
Changes in business profitability conditions
Heavier tax burden to finance local authority expenditure

Two examples of the results of such models are shown in Tables 4 and 5. The first is from a recently published study for the Government of the Netherlands[(1)] which in the

Table 5: Impact of water protection investment programme on various national income account items, relative volume changes (%) and the impact on private consumption prices, the current account, the demand for labour, State finances and local tax scales (net pollution control costs)

	1976	1977	1978	1979	1980	1981	1982
GDP at market prices	0.3	0.3	0.5	0.5	0.5	0.6	0.6
Imports	0.4	0.4	0.8	0.8	0.7	0.9	0.9
Total Supply	0.3	0.3	0.6	0.6	0.6	0.7	0.7
Exports	-	-	-	-	-	-	-
Private Consumption	-	-	-	-	-	-	-
Public Consumption	-	-	-	-	-	-	-
Investment	1.1	1.2	2.2	2.4	2.3	2.7	2.6
- private	0.8	1.0	2.0	2.3	2.2	2.6	2.2
- public	1.9	1.9	2.8	2.8	2.7	2.8	2.8
Total Demand	0.3	0.3	0.6	0.6	0.6	0.7	0.7
Private Consump. Price Index (%)	0.2	0.2	0.2	0.2	0.2	0.2	0.2
Current Account mill.Mk. 1975 Prices	-150	-130	-280	-320	-360	-430	-490
Demand for labour 1000 persons	3.5	4.0	6.0	6.5	7.0	7.5	7.5
State budgetary surplus mill.Mk, current prices	0	20	60	60	50	70	70
Rise in local tax scales, increase in Finnish pennies	0.1	0.2	0.2	0.2	0.2	0.2	0.2

Source: OECD (1982)

case presented assumes no price increases from environmental programmes in other countries. The second is a Finnish study of the economic effects of a water protection programme covering a period 1976-82[2]. Some details of these studies are available in the OECD report as well as in the original reports.

What is important for EIA is the relevance of these studies to more localised programmes. Not only might macro-economic modelling provide some information on the effects of large projects but also on the costs of pollution abatement and their more general effects on the economy. Having said that it is dubious whether these models can be sufficiently dis-aggregated to provide relevant information on any but large scale programmes. It is necessary, therefore, to turn away from the macro-economics of the whole economy to a more local or regional level of analysis.

4. MULTIPLIER ANALYSIS

More often an EIA will be concerned with the effects of a project on the local or regional social and natural environment. One tool which economists have developed may be useful in this context - the regional economic multiplier.

Originally the Keynesian multiplier was developed as a means of demonstrating how variations in national income may result from a change in the level of injections of expenditure into the economy. The usual example is an increase in investment flows. The rise in investment adds to incomes which in turn are partly spent on other goods and services. In turn this means that those who produced the goods have also enjoyed a rise in income, and they subsequently spend part of this, and so the chain of events continues with a smaller sum of income being passed on at each stage. Clearly the extent of the eventual increase in income is determined by how much income is passed on at each stage i.e. by the propensity to consume of the parties in this sequence. The greater the propensity to consume out of income the higher will be the final level of national income. Conversely the higher the propensity to withdraw income from the economy (by savings, imports or taxes) the lower will be the value of the multiplier. The concept suggests that idle resources are available for the purpose of generating this extra income at a given set of prices.

The multiplier has been used in a number of studies to examine the effects of a project on regional economy. One study by Greig (1971) illustrates the simple use of the technique. Greig's objective is to investigate the income and employment multiplier effects of a large project on a sub-region, the Highlands of Scotland. The project was the creation of a £15 m pulp and paper mill at Fort William.

Briefly the procedure is as follows. The opening of the mill will increase the incomes of those employed. The direct effect of the project on employment and income is therefore relatively straightforward to calculate. These incomes will be spent and will create further income and employment in shops, services etc. Calculation of these rounds of indirect income effects depends on the estimation of the withdrawals or 'leakages' from the income flow by means of saving, imports from other regions, taxation etc. The calculation of the rounds of employment effects depends on estimating the relationship between income changes and employment not only in the private sector but also in the public service sector (schools, health care etc.).

Looking just at the first round, Greig defines the following for the region,

E = total employment
E_d = direct employment
w_d = average earnings of direct employees
w_p = average earnings of public sector employees
L = increase in local value added necessary to create one extra job in the services trade
θ = ratio of public service employees to other employees
s = average propensity to save
t = average tax rate
m = average propensity to import
v = proportion of an increase in income which is local value added
i.e. $v = (1-s-t)(1-m)$
ΔV = increase in local value added created by direct employees
i.e. $\Delta V = E_d w_d v$

The increase in indirect employment (ΔE_a) at the completion of the first round is

$$\Delta E_a = E_d\theta + \frac{\Delta V}{L}(1+\theta)$$

The first round multiplier (k_a) is

$$k_a = 1 + v + \frac{(E_d + \frac{\Delta V}{L})\ \theta w_p}{E_d w_d}$$

Greig then calculates the subsequent round multiplier and then multiplying the 1st round and subsequent round multipliers estimates the overall regional multiplier k_r. As a matter of interest the estimates are reproduced below in Table 6 with some sensitivity analysis.

Table 6: The Income and Employment Multiplier (From Greig (1971))

Estimate	k_a	k_b	k_r	k_E
Upper	1.4	1.35	1.54	2.66
Lower	1.35	1.25	1.44	1.90

This means for example that the mill with a direct employment of 850 people and another 500 in transport and forestry whose total earnings were £1.5 m were predicted to generate a total income of at least £2.1 m and employment of 2,565 jobs.

There are a number of problems with this work. The most obvious is the way in which the first round multiplier is calculated as the multiplied effects of the personal income of direct employees. No attempt is made to quantify the effects of the initial capital expenditure on the mill and any induced investment created by the increased income except public services.

Brownrigg (1973) in a study of the multiplier effects of the building of the University of Stirling has modified the model above to take account of the longer period of construction involved in that project as well as the induced investment in both private and public sectors to extend existing capacity. More recent studies in this area include Ashcroft and Swales (1982) and McGuire (1983).

These models are extremely useful in analysing the effects of new developments which are clearly recognisable and which create identifiable employment opportunities either for the existing unemployed workforce or for immigrants. It allows the economic consequences to be traced.

However where the programme is not confined to one enterprise or specific job creating development the approach is less useful, and one method for dealing with this situation is input-output analysis.

5. INPUT-OUTPUT ANALYSIS

An input-output table is both a descriptive framework for showing the relationship between industries and sectors and between inputs and outputs as well as an analytical tool for measuring the impact of autonomous disturbances on an economy's output and income.

Its use as the latter has been seen in a number of studies concerned with the effects of environmental projects on the wider regional and national economy[3]. Although suffering from a number of limiting assumptions it has the distinct advantage of being extremely practical in tracing out the direct and indirect effect on an economy of an injection of or reduction in expenditure in a more precise manner than the Keynesian income multiplier.

As a model it is very simple. Its key variables are the outputs of sectors into which the economy is divided. Each sector's output is allocated to its sales to other sectors and to final demand (consumption, exports, government, investment). Equilibrium is achieved when the output of each sector equals total purchases from that sector. The analytical model then allows us to vary the level of final demand (say by postulating £x m to be spent by government on a project) and to see the effect on all sectors creating a new equilibrium. A simple 3 sector input-output table is reproduced below.

To From	1	2	3	Final Demand	Gross Output
1	X_{11}	X_{12}	X_{13}	Y_1	X_1
2	X_{21}	X_{22}	X_{23}	Y_2	X_2
3	X_{31}	X_{32}	X_{33}	Y_3	X_3
V	V_1	V_2	-	-	V
Gross outlay	X_1	X_2	X_3	Y	X

X_{11} is the amount that ind_1 sells to itself
X_{12} is the amount that ind_1 sells to ind_2
X_{13} is the amount that ind_1 sells to ind_3
Y_1 is the amount that ind_1 sells to final consumption
X_1 is the sum of these amounts = gross output of $industry_1$
so: $X_1 - X_{11} - X_{12} - X_{13} = Y_1$

if we assume that when ind_2 purchases X_{12} it does so in some proportion to its own output $\frac{X_{12}}{X_2}$ and this is true for all industries we can define the direct input coefficients as

$$a_{11} = \frac{X_{11}}{X_1}, \quad a_{12} = \frac{X_{12}}{X_2}, \quad a_{13} = \frac{X_{13}}{X_3}$$

and rewrite the above table as

To / From	1	2	3	Final Demand	Gross Output
1	$a_{11}X_1$	$a_{12}X_2$	$a_{13}X_3$	Y_1	X_1
2	$a_{21}X_1$	$a_{22}X_2$	$a_{23}X_3$	Y_2	X_2
3	$a_{31}X_1$	$a_{32}X_2$	$a_{33}X_3$	Y_3	X_3
	⋮	⋮	⋮	⋮	⋮

This can be written for each industry such as industry 1 as

$$X_1 - a_{11}\, X_1 - a_{12}\, Y_2 - a_{13}Y_3 = Y_1$$

in matrix terms this is

$$X - AX = Y$$

where X and Y are column vectors of gross output and final demand respectively and A is the matrix of <u>direct input coefficients</u> a_{ij} this can be rewritten $X = (I - A)^{-1} Y$ where $(I - A)^{-1}$ is termed the <u>Leontief inverse matrix</u>. Each entry in this matrix is called an <u>interdependency coefficient</u> and represents the direct and indirect requirements of sector i per unit of final demand for the output of sector j. It is often written as X = BY.

The matrix B is useful in that it allows us to vary the size and composition of final demand (e.g. government expenditure) and see the levels of gross output of each sector computable with the change. In other words it allows an examination of the total impact on the economy of exogenous disturbances.

The simplest multiplier used in input-output analysis to discuss economic impact analysis is the output multiplier. This simply measures the sum of direct and indirect requirements from all sectors needed to deliver one additional £ of output of j to final demand. It is derived by summing the entries in the column under industry j in the Leontief inverse matrix table, i.e.

$$\sum_{i=1}^{n} b_{ij} \quad i = (1,\ldots n).$$

where b_{ij} = inverse matrix coefficients

h_{Ri} = entry in the row vector of household coefficients

The employment multiplier is obtained by first obtaining estimates of a relationship between employment and output and then converting the output or income multipliers into employment terms.

A numerical example 2 industries

To / From	1	2	Households	Other final demand	Gross Output
1	20	45	30	5	100
2	40	15	30	65	150
Households	20	60	10	10	100
Other Value Added + Imports	20	30	30		80
Gross Outlay	100	150	100	80	430

Direct input coefficient
Matrix A

$$\begin{bmatrix} 0.2 & 0.3 \\ 0.4 & 0.1 \end{bmatrix}$$

This shows the direct purchases from industry i (at the left) per £ of the output of industry j (at the top).

If we subtract this matrix from the identity matrix $\begin{bmatrix} 1 & 0 \\ 0 & 1 \end{bmatrix}$ and then inverting the result we get

$$B = \begin{bmatrix} 1.5 & 0.5 \\ 0.67 & 1.33 \end{bmatrix}$$

This matrix shows us the direct and indirect purchases of industry j (at the top) from industries i (to the left) per £ of final demand from the product of j.

Multipliers obtained can now be examined. Simply by adding up the columns of the inverse matrix we get the output multiplier i.e. the total value of industry requirements per unit of final demand for the industry in question. This gives us 2.17 for industry 1 and 1.83 for industry 2.

However if we are interested in the income multiplier we must consider the effect of the industry's purchases on the income of households. This is done by multiplying the column entries of the inverse matrix by the household coefficients for the supplying industries. The latter come from the original table and are 20/100 or 0.2 and 60/150 or 0.4. This gives the direct and indirect income changes per £ change in the final demand for industry 1 as (1.5 x 0.2) + (0.67 x 0.4) 0.568. This is sometimes called the regional income coefficient for industry 1. In order to find the multiplier we divide this coefficient by the household coefficient for that industry 0.2 giving 2.84.

A change in final demand for the products of industry 1 by £10 m would lead to an increase in income of £5.68 m. In other words a change in the income of those employed in industry 1 of £2 m would increase income by £5.68 m.

6. EXTENSIONS OF THE INPUT-OUTPUT FRAMEWORK

More sophisticated versions of the basic input-output model exist. First it is possible to take into account a greater number of "rounds" of indirect effects - so called induced effects - by increasing the size of the matrix to include the household sector in the Leontief inverse matrix. This requires a great deal of data on household expenditure. It is also possible to breakdown final demand into certain categories of expenditure. This is done for example by Archer and Owen (1971) which looks particularly at tourist expenditure.

Another extension which is relevant to EIA is the use of input-output models which account for the environmental sector. These have been developed by Cumberland (1966), Ayres and Kneese (1969), Leontief (1970), Victor (1972) and Johnson and Bennett (1979). What these models do in their various ways is to include an interaction matrix of the environmental and economic sectors.

An example is that of Johnson and Bennett who attempt to evaluate the economic and environmental impact of a nuclear power plant in South Carolina. They use a two sector model illustrated in Figure 1.

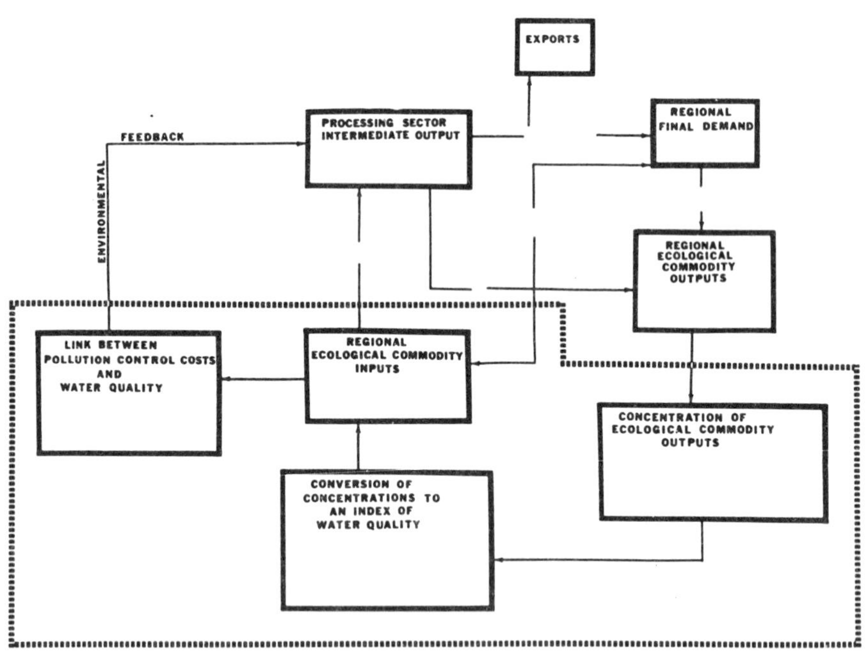

Figure 1: The Regional Environmental-Economic Impact Model Adapted from Johnson and Bennett (1979)

One sector is a standard input-output model of the economic sector as outlined above. They also have an environmental sector with flows of ecological commodity inputs and outputs as well as those associated with final demand. They also have some feedback from the environment to the economic sector via a pollution control industry for water quality.

This model is fairly sophisticated showing as it does the predicted effects on water quality as well as output, employment, income, and local government revenue over the construction period and the first four year operating period of a nuclear power plant. However as the authors point out the development of more comprehensive and complex input-output models to assess economic and environmental impacts may merely compound the errors of the simple model In many ways input-output is a crude but useful tool because of its highly restrictive assumptions about technical relationships as well as economic relationships. It may be useful for very broad analysis but when disaggregated data is needed it may break down.

The whole input-output model is to some extent unsatisfactory for a number of reasons. There are a number of technical problems.

1) It is a linear model implying that technical relationships between inputs and outputs are fixed at some average level for all firms at all levels of output. This is unreasonable when different firms have different efficiencies at the same and different levels of output.
2) It is a static framework which is unsatisfactory in a changing world not only where expectations and stock building occur (undermining also condition (1)) but also where data is collected for a past period.
3) The largest problem is one of collecting accurate and timely data. For example the latest input-output table for Scotland was constructed in 1973 - before North Sea Oil, thus seriously restricting its applicability. Moreover data collection can take some considerable time and the cost can be very high if survey data is required.

With these caveats input-output does represent a very useful tool of analysis and can be of assistance in illuminating the interdependencies within the economic sector as well as the linkage with the natural environment. With a degree of sensitivity analysis, the checking of physical predictions, and more replication of studies it could provide EIA with a firm foundation within the mainstream of economic analysis.

NOTES

1. Kosten en macro-economische gerolgen van het voorgenomen milieubeleid, Tweede Kamer, zitting 1980-81 16495 nrs. 1-2.
2. National Board of Waters (1978). Final Report for the International Bank for Reconstruction and Development of Research Project carried out in 1975-78 by the National Board for Waters, Helsinki, Finland.
3. See for example Cumberland (1966), Leontief (1970) and Victor (1972).

REFERENCES

Archer, B.H. and C.B. Owen: 1971, 'Towards a Tourist Regional Multiplier', Regional Studies 5, 289-294.

Ashcroft, B. and J.K. Swales: 1982, 'The importance of the first round in the multiplier process: the Impacts of Civil Service Dispersal', Environment and Planning A, Vol 14 (4), 429-444.

Ayres, R. and A. Kneese: 1969, 'Production, Consumption and Externalities', American Economic Review 59, 282-97.

Benedict, H.M., C.J. Miller, J.S. Smith: 1973, Assessment of Economic Impact of Air Pollutants on Vegetation in the US - 1969 and 1971, Stanford Research Institute, Menlo Park, California.

Brownrigg, M.: 1973, 'The Economic Impact of a New University, Scottish Journal of Political Economy 20, 123-29.

Cumberland, J.: 1966, 'A Regional Interindustry Model for Analysis of Development Objectives', Papers and Proceedings of the Regional Science Association 17, 65-94.

Freeman, A.M.: 1979, The Benefits of Air and Water Pollution Control: A Review and Synthesis of Recent Estimates, A report for the U.S. Council on Environmental Quality.

Greig, M.A.: 1971, 'The Regional Income and Employment Multiplier Effects of a Pulp and Paper Mill', Scottish Journal of Political Economy 18, 31-48.

Heintz, H.T., A. Hershaft, G. Horak: 1976, National Damages of Air and Water Pollution, a report submitted to the U.S. Environmental Protection Agency.

Johnson, M.H. and J.T. Bennett: 1979, 'An Input Output Model of Regional Environmental and Economic Impacts of Nuclear Power Plants', Land Economics, 55, May.

Lave, L.B. and E. Seskin: 1977, Air Pollution and Human Health, John Hopkins University Press, Baltimore.

Leontief, W.: 1970, 'Environmental Repercussion and Economic Structure: An Input Output Approach', The Review of Economic and Statistics 52, 262-71.

Lipfert, F.W.: 1978, The Association of Human Mortality with Air Pollution: Statistical Analysis by Region, Age and by Cause of Death, Ph.D. dissertation, Union Graduate School, Yellow Springs.

Liu, Ben-Chieh and E.S. Yu: 1976, Physical and Economic Damage Functions for Air Pollutants by Receptor, Environmental Protection Agency, Washington.

McGuire, A.: 1983, 'Regional Income and employment effects of a Nuclear Power Station', Scottish Journal of Political Economy Vol 30 (3) pp 264-274.

Mishan, E.J.: 1971, Cost Benefit Analysis, Unwin University Books.

O.E.C.D.: 1982, 'The Macro-economic Effects of Environmental Policies'. Draft report by the Environment Committee Group of Economic Experts.

Pearce, D.W.: 1982, The Distribution of the Costs and Benefits of Environmental Policy Report prepared for the Environment Directorate of the O.E.C.D., May.

Pearce, D.W. and C.A. Nash: 1981, The Social Appraisal of Projects, Macmillan, London.

Richardson, H.W.: 1972, Input-Output and Regional Economics, Weidenfeld and Nicholson, London.

Victor, P.: 1972, Pollution: Economy and Environment, London, George Allen and Unwin.

Waddell, T.E.: 1974, The Economic Damages of Air Pollution, Environmental Protection Agency, Washington.

AFFILIATION

A. H. Harris is a lecturer in the Department of Political Economy at the University of Aberdeen, Scotland.

Andrew I. Sors

MONITORING AND ENVIRONMENTAL IMPACT ASSESSMENT

1. INTRODUCTION

The purpose of this paper is to consider the current status of and future progress in environmental monitoring, with special reference to environmental impact assessment (EIA). In this context, a broad view of EIA is taken, including plans, development actions, projects, pollutants and products. The paper is mainly concerned with pollutant risks and impacts and specifically in relation to their effects on human health.

2. DEFINITIONS

There have been various definitions of monitoring. One of the most widely accepted was that of an inter-governmental meeting in 1971 by way of preparation for the 1972 Stockholm Conference. At the former meeting monitoring was defined as "a system of continued observation, measurement and evaluation for defined purposes". The most important fact to note is that under this definition monitoring must be carried out for "defined purposes". These are generally to be seen in the context of environmental management.

There is often some confusion as to the difference between monitoring and surveillance. In some cases, surveillance is taken to be monitoring carried out to observe trends, rather than in support of specific management objectives. In epidemiological studies, however, environmental or health surveillance takes on a much more specific meaning.

Harvey (1981) has carried out an extensive analysis of the terminology used in relation to monitoring. He has demonstrated that the terms monitoring and surveillance can mean quite different things to different workers. The most common usage appears to be a broad one, encompassing both descriptive (problem oriented) monitoring and regulatory monitoring.

B. D. Clark et al. (eds.), Perspectives on Environmental Impact Assessment, 365–380.

3. HISTORICAL BACKGROUND

As the definition above implies, environmental monitoring is not an end in itself, but an essential step in the processes of environmental management (Rockefeller Foundation 1977). It is therefore not surprising that the development of monitoring has largely followed the development of public and government concern about the environment.

There is now a widespread feeling that monitoring has failed to live up to expectations as a tool of environmental management. These expectations were probably at their highest during and immediately after the 1972 UN Conference on the Human Environment, held in Stockholm. During this period, substantial resources were devoted to the design and operation of monitoring systems. At the international level, diverse monitoring activities began to be co-ordinated and developed under the newly established Global Environmental Monitoring System (GEMS). Individual countries were also devoting substantial resources to the evaluation and further development of national monitoring activities. For example, in the U.S.A., an Office of Monitoring was established by EPA in the early seventies, and three monitoring laboratories were constructed to support the above Office. In the U.K. the Royal Commission on Environmental Pollution (1974) supported the development of a "comprehensive, unified and flexible" monitoring system. Efforts along similar lines took place in a number of other countries.

In view of the considerable resources devoted to monitoring it may be surprising that monitoring is considered not to have "lived up to expectations". Yet, more recent evaluations have largely been critical of monitoring programmes (e.g. NAS, 1977). There are many reasons for this, depending on the nature and circumstances of any particular situation. However, two of the main factors, and which are generally applicable, can be summarized as follows:

(i) Many monitoring programmes, especially the early ones, were fairly ambitious and consumed considerable resources, but were designed without clear objectives and were therefore of limited usefulness.

(ii) The scientific and technical complexity of deciding what, where, when and how to monitor has emerged gradually, but it is now clear that these questions are much more difficult to answer than originally expected.

However, there are now signs of renewed interest in

monitoring and of some progress in the design, operation and utilization of monitoring systems. This is evident at both the international (e.g. UN/GEMS, OECD, EEC) and national level.

4. THE OBJECTIVES OF MONITORING

Monitoring may be carried out in support of a wide range of management objectives. Thus, the objectives of monitoring may be classified in various ways. Munn (1980) identified the following objectives for monitoring:

(i) to determine present conditions;
(ii) to determine trends;
(iii) to understand phenomena;
(iv) to validate and/or calibrate environmental models;
(v) to make short-term predictions;
(vi) to make long-term assessments;
(vii) to optimize the utility and/or cost-effectiveness of any of the above;
(viii) to control.

5. LEVELS OF MONITORING

Monitoring systems may be operated at various levels, depending upon the nature of the problem in question and the jurisdiction of the monitoring agency concerned. These levels are generally referred to as local, regional, continental or global.

Local problems may extend for 0-100 km (e.g. urban air pollution by SO_2, NO_x, smoke).
Regional problems: 100-1,000 km (e.g. agricultural pollution, river pollution).
Continental problems: 1,000-10,000 km (e.g. pollution of the Mediterranean, the use of pesticides e.g. DDT).
Global problems: >10,000 km (e.g. buildup of CO_2 in the atmosphere, depletion of the stratospheric ozone layer).

A number of pollution problems occur on several scales. A well known example is the emission of SO_2 into the atmosphere. This causes local urban air pollution; however, dispersion from increasingly taller stacks has resulted in SO_2 transport over long distances (thousands of kms) resulting in acid rain.

Environmental impact analysis has usually been carried out for local scale (hot-spot) problems. However, it is generally believed that EIA should be carried out at progressively higher levels of decision-making; e.g. sectoral plans, policies, etc. This tendency will also result in the need for integrated or co-ordinated monitoring systems on a larger scale. In this way, monitoring for impact assessment and monitoring for national pollution control will begin to converge, and the lessons from national/regional monitoring programmes will need to be carefully evaluated in the design of monitoring systems for EIA.

6. TYPES OF MONITORING

A useful general division of environmental monitoring programmes is into those that measure <u>targets</u> which may show changes in distribution or performance and those that measure <u>factors</u> which may cause changes in the environment (Somers, 1981). <u>Factor</u> monitoring is largely concerned with measuring levels, and may be carried out:

(i) at source (emission monitoring);
(ii) at points in the environment (environmental monitoring);
(iii) at the point of exposure (exposure monitoring);
(iv) within the target (internal monitoring).

Target monitoring can occur at various levels of organization:

(i) within tissues of organism;
(ii) at the level of the individual organism;
(iii) level of species population;
(iv) sample of ecosystem;
(v) total ecosystem.

Given our poor understanding of the environment and the resulting difficulty in deciding where, what, when and how to monitor, monitoring systems have evolved in a fairly ad hoc manner.

A variety of methods have been developed, often in isolation. Each has its own uses, advantages and disadvantages, but taken together these methods provide the basis for the development of integrated monitoring systems.

The following sections contain an outline of some of the main types of factor and target monitoring methods which are applicable at least in principle, to EIA. The emphasis here is on factor monitoring. This is because, in the author's view, the specific application of the latter to EIA is particularly important. While target monitoring is also vital, it is similar in application to EIA and other methods of assessment.

7. FACTOR MONITORING

7.1 Sources of pollution

7.1.1 Emission monitoring

Sources of pollutants may be classified as point sources, area sources or mobile sources.

Point sources such as industrial emissions can be monitored fairly easily, usually by the individual discharges.

Area sources are groupings of small sources spread over cities or farmlands. Examples are domestic sources, use of fertilizers, etc. It is usually impractical to monitor each small source individually and emissions are estimated from source inventories, etc.

Mobile sources include motor vehicle emissions and possible spills or accidents during the transport of toxic substances. Here again, source monitoring is impractical as a rule, and indirect methods are required.

7.1.2 Process monitoring

In many cases it may be simpler to establish precise specifications for clean processes rather than emission limits. This approach is used by the U.K. Alkali Inspectorate in the establishment of best practicable means (b.p.m.) for the control of emissions into that atmosphere from registered works.

7.1.3 Indirect estimate of emissions

Emissions can sometimes be estimated from knowledge of production figures, etc. by using emission factors.

7.2 Monitoring in the environment

7.2.1 Regulatory monitoring

Many countries have established environmental quality standards or objectives. These limit the ambient level of the regulated pollutant in air or water. This approach has proven useful as a basis for policy decisions. However, there are a number of problems associated with it and these are briefly discussed under the section on exposure monitoring.

The monitoring of ambient quality, particularly in air, presents considerable problems of network design. It is usually preferable to monitor at locations where the pollution level is highest. However, the spatial gradients and temporal variability are usually the greatest at such locations, making it difficult to obtain representative measurements.

7.3 Trend surveillance

The measurement and statistical evaluation of trends is an important tool in the identification of pollution buildup. Surveillance takes on particular importance in the absence of statutory ambient standards or objectives. This is the case, for example, in the U.K., where there are no nationally applicable environmental quality standards or objectives. The thrust of the monitoring effort at the national level is tied up in two major surveillance networks: the National Survey of Air Pollution, and the River Pollution Survey. The former includes measurement of SO_2 and smoke levels at about 1,100 urban and 200 country sites. The latter is a periodic assessment and classification of the quality of some 39,000 km of rivers and canals in the U.K. The four classes of quality are determined by oxygen demand, suspended solids content, presence of selected toxic substances, as well as by the number of complaints by the public about pollution.

7.4 Historical monitoring

A characteristic of the environment is its variability in space and time, and this often makes it difficult to separate any new "signal" from "noise". Several processes may be at work, each with its own time-scale of variation. Perhaps the least understood and most complex example is the problem of evaluating climate change i.e. identifying change as opposed to variability. There are a number of cycles at work: seasonal, annual, sunspot, changes in magnetic field, etc. Another example is that of the natural temporal and spatial variability in stratospheric ozone; it has been estimated that if there were an effective 2 per cent depletion of ozone per annum, an additional 10 years of observations would be required before the event could be confirmed from measurements with 95 per cent confidence.

Thus, in many cases a look backwards in time is essential in evaluating the significance of present pollution levels; however, it is usually the case that environmental monitoring has not been carried out. It is then sometimes possible to use indirect methods, for example chemical analysis of tree rings, museum specimens, sediment profiles, etc. For example, analysis of snow profiles in Greenland has revealed that lead levels have increased five-fold since 1850 and one hundred-fold since 800 B.C. (Murozumi et al., 1969).

Historical monitoring may be a particularly useful tool in pre-audit for EIA, particularly if no direct monitoring measurements are available. It is also useful in identifying long-term historical trends against which more recent changes may be evaluated.

7.5 Ecological monitoring

7.5.1 Biological materials

When a pollution problem has been identified, it is often useful to obtain a synoptic picture of the scale and nature of the problem. These synoptic measurements can then indicate where more specific and precise monitoring is required. Such initial surveys may need to be carried out quickly and as cheaply as possible.

In such cases "biological materials" may be useful. For example, certain mosses have been extensively used to study the regional patterns of metal deposition from the atmosphere. It has been shown that "moss bags" made from the mosses provide quantitative and reproducable results. (Goodman et al., 1976). Moss bags are inexpensive to make, do not need electrical power, and can therefore be set up cheaply to cover a number of sites over a large area.

7.5.2 Bioaccumulator organisms

Certain organisms accumulate chemical substances and biomagnification factors of 10^5 are not uncommon. Toxic substances usually occur in the environment in such small concentrations, that precise measurements may not be possible, or only so with highly sophisticated analytical equipment. In such cases it may be more convenient and informative to measure levels in biota. An added advantage of such organisms is that they tend to reflect integrated rather than instantaneous exposure. A practical example of this is the "Mussel Watch" project. Pollutants in the marine environment are often in ppm or ppb concentrations and are difficult to measure. Molluscs (e.g. mussels) have been collected from a large number of coastal sites around the U.S.A., and the levels of selected heavy metals, pesticides, hydrocarbons, and radionuclides have been measured (Goldberg et al., 1978). These measurements provide a very useful comparative picture of pollution over a very large area.

7.6 Speculative monitoring

Many substances are emitted into the environment without anyone knowing; some may transform and transformation product(s) will remain undetected. Therefore, with the ever-increasing number and quantity of chemicals in the environment, monitoring for identification is of increasing importance. One approach which might be considered is the periodic sampling of the atmosphere for as wide a range of substances as possible, using the best and most sensitive techniques available.

7.7 Proxy monitoring

In some cases, detection of a specific pollutant in the environment raises the likelihood of certain others being present. Association may be indicated on the grounds of chemical similarity and/or geochemical affinity (e.g. Pb, Cu, Zn and Cd). In other cases, where known degradation products are found, the presence of the parent product should be suspected.

7.8 Pathway monitoring

Until recently, localized hot-spot pollution problems have been of the highest priority and often the sole concern of EIAs. In these cases, the link between emissions and the resulting exposures and effects have been determined to a reasonable degree, by using a combination of methods and experience, including ambient monitoring, epidemiological studies, etc. For example, we know roughly the relationship between the density of urban traffic and lead levels in blood of the local population. We also know the general relationship between emissions of SO_2 and ambient airborne concentrations.

However, the establishment of source-exposure relationships becomes much more difficult in cases where the pollutant reaches targets through long and numerous environmental pathways. Yet, it is just these problems which are causing increasing concern at the present time. As these pathways can be numerous, long and complex, it is not possible or practicable to monitor movement within and transfers between all compartments. It is necessary to discover the most important (or critical) pathways between sources and receptors of concern. This knowledge only becomes available through literature review and research (or descriptive) monitoring within a given model framework.

Modelling is extensively discussed in another paper in this book (see paper by Norton). However, the dose-commitment approach is particularly relevant to monitoring and warrants a mention here. This approach was developed by UNSCEAR (United Nations Scientific Committee on the Effects of Atomic Radiation) for the estimation of human exposures to radionuclides from weapons testing. It has proven successful in that field. MARC (Monitoring and Assessment Research Centre) is now applying it to a number of stable pollutants, in the first instance lead, mercury and cadmium. The method is a time-independent approach to exposure assessment and is a convenient means of comparing contributions to intake and exposure from various pathways and in expressing source-exposure relationships. The time-independence considerably reduces monitoring data requirements. Once the critical environmental pathways are known, research monitoring has done its job, and a limited surveillance system at nodal points should suffice.

7.9 Monitoring of exposure

The routes of pollutant exposure to man are ingestion, inhalation (air) and skin (dermal) contact. It is usually impractical to measure directly the total individual or population exposure to a pollutant, except in the most critical cases or where exposure is "simple". In some cases however, the total exposure may be inferred through biological monitoring (discussed in a later section). In any case, each of the above routes requires different types of measurement programmes. It is therefore usual to undertake such surveys separately and collate the results afterwards, if required.

7.9.1 Dietary surveys

Food monitoring is the basis for dietary surveys. These attempt to establish representative diets of the population. In this way it is possible to calculate the total exposure to certain toxic substances ingested through representative diets. The latter are arrived at by a mixture of survey and questionnaire techniques. In the U.K., the Government Chemist is responsible for carrying out such surveys. Selective studies on individual foodstuffs which may contain particularly high levels of certain toxins are also carried out (a good example is methylmercury in fish).

In the case of EIA, the dietary intake from locally-grown food products near the proposed development should be carefully considered.

7.9.2 Drinking water surveys

It is usually the responsibility of the water supplier to ensure that the drinking water is of adequate quality.

There is now a substantial amount of epidemiological work to investigate the possible association between the incidence of heart disease and the softness of drinking water. This may be a factor to take into account in EIAs of projects which affect water quality.

7.9.3 Exposure to pollutants in air

Exposure by inhalation, to airborne pollutants is usually estimated by monitoring of ambient airborne concentrations. Such measurements are useful, particularly if correlated with health effects by epidemiological studies. Although, apart from the problems of siting which were mentioned before, great care must be taken in interpreting ambient data. Ambient concentration-effect relationships are often derived under controlled laboratory conditions, in the work-place, etc. In such cases, measured airborne concentration is more or less equivalent or proportional to actual exposure. In a recent WHO publication (WHO, 1980), it is shown that in most practical situations the relationship between measured concentrations and actual exposures is rather complex, depending on a number of factors such as surroundings, time spent indoors/outdoors, activity undertaken, breathing pattern, etc.

7.9.4 Skin contact with chemicals

There is increasing concern about the volume of consumer chemicals in circulation. Such chemicals are part of our daily life: cosmetics, pharmaceuticals, wrappings, etc. Some of these may pose risk of adverse health effects, especially after prolonged exposure. The pattern of exposure will vary with the pattern of life habits, etc. of the individual. Prediction of exposure to chemicals arising from consumer products is an important aspect of the fast-growing legislation on toxic chemicals.

7.10 Monitoring of pollutants in targets

A valuable approach in indicating or determining human exposure is biological monitoring. This can often provide a more direct indication of total exposure than the measurement of ambient levels of pollutants in air, water, food or soil. Samples can sometimes be taken from the organ where the earliest detectable effect occurs such as cadmium in the kidney, but it is usual to use more accessible indicators of dose such as blood, urine, hair or teeth. Some biological monitoring reflects recent exposure (e.g. lead in blood) other tissues reflect integrated exposure (e.g. lead in teeth).

8. TARGET MONITORING

8.1 Monitoring of occupational exposure

Occupational data can often be valuable in assessing the risks arising from exposure to lower levels of pollution in the general population. Any extrapolation should be carried out with great care, and using expert guidance.

8.2 Pollution effects on non-human species

Surveillance of the effects of pollution on biota may occur at several levels:

(i) measurement of biochemical, physiological and behavioural changes in individual organisms;

(ii) measurement of changes in population parameters;

(iii) measurement of changes in the distribution and abundance of species;
(iv) measurement of community changes.

There are a number of examples of ecological effects surveillance being useful as early warning of hazards to man. Aldrin, Dieldrin and DDT have sometimes been found to cause reproductive failure in birds; the methylmercury problem was recognized in Sweden from observation of dying birds; increased sulphur dioxide levels have been detected by the disappearance of lichens. The last is an example of a so-called indicator species; such species may provide useful early warning if they are either sensitive to or thrive on the presence of specific pollutants.

9. THE APPLICATION OF MONITORING TO ENVIRONMENTAL IMPACT ASSESSMENT

The aim of this section is to examine briefly the applicability of the various types of monitoring discussed above to Environmental Impact Assessment. The sequence of tasks in EIA (items 1-10 underlined) have been adapted from Gilad (1979).

(1) Inventory of pollutants likely to be released into the environment as the result of the proposed action. Monitoring is of limited use in the preparation of an inventory, although previous monitoring programmes may have identified breakdown products of the primary wastes which are also to be emitted from the proposed development.

(2) Quantitative description of the health situation in the region likely to be affected by the proposed project, including morbidity statistics on diseases related to pollutants likely to be released into the environment as a result of the proposed action. Monitoring can be used in the following ways:
   (i) Definition of the region likely to be affected. Pathway models (including dispersion models) will need to be developed. The parameters required for this will be, at least partially, supplied by monitoring data.
   (ii) Pollutant related health effects. If the pollutants to be released are reasonably

well known, epidemiological data may be available, relating exposures to health effects. In most cases, however, such information is not available and the best that can be hoped for is the existence of data on occupational exposure to the substance(s) in question, which can be used to provide some indication of the hazards at lower levels of exposure.

Exposure to individual pollutants and any resulting health effects may be due to a combination of emissions from the proposed development and from other, existing sources. Therefore, a baseline survey both of the existing pollution situation and of the existing health situation are crucial. Monitoring methods are clearly important in such baseline and trend surveys.

(3) Quantitative description of the dispersive, absorptive and adsorptive mechanisms in air, water and soil.
These are largely based on research monitoring and modelling data and/or the use of results from existing monitoring programmes. Care must be taken to ensure that the specific local environmental conditions are accounted for in using data from other locations/situations.

(4) Based on point 3 above, prediction of the concentration of the various pollutants over various periods of time to which various sectors of the affected population will be exposed.
This step requires the ability to undertake exposure assessments and in particular to quantify source-exposure relationships as far as possible and required. Pathway monitoring as described above is of particular use here.

(5) Information on environmental, social and economic factors which are likely to influence the susceptibility of the affected populations to the pollutants identified above.
There are two important considerations here:

(i) There may be individuals or groups particularly susceptible to the pollutant(s): these may include the aged, the very young, pregnant women and the foetus, and people living in poverty.

(ii) There may be individuals or groups whose pollution burden will be unusually high either due to the contribution of other pollution sources and/or high exposure from the proposed new development. This group may include people with unusual dietary habits or those consuming large quantities of local, contaminated foodstuffs, or people who are occupationally exposed. Such groups may be identified by environmental monitoring or one-off survey programmes.

(6) Forecast of health effects including their intensity, duration and the time likely to elapse between exposure and the appearance of an effect.
This is mainly a question of dose-response assessment, which is generally based upon toxicological, epidemiological and clinical studies. In addition to the delay between exposure and effect, the delay between emissions and resulting exposure may also need to be considered. The dose commitment approach may be of use in this regard.

(7) Assessment of the health effects as to their importance and significance.
This is part of societal evaluation of risk or impact; monitoring does not play a part here.

(8) Means by which the health effects could be eliminated or reduced, and the associated costs, etc.
A prerequisite to a cost-benefit analysis of this kind is (i) estimation of exposure levels which do not give rise to "adverse" or unacceptable health effects; and (ii) estimation of emission levels which result in the required levels of exposure. The latter can be established by a combination of monitoring and modelling. Critical pathways monitoring will be particularly useful in the identification of individuals and populations most at risk, and in the estimation of the effects of various levels of pollution abatement.

Given the appropriate amount of information

in (i) and (ii) above, the means of reducing emissions of (and hence exposure to and effects of) pollutants can be considered and the costs of doing so estimated.

(9) Assessment of health impacts of alternatives developed as the result of the general environmental impact assessment process.
This is often a complex task in which monitoring data may be of considerable use, particularly in the following:
(i) Comparison of the extent of the region to be affected by the various alternatives;
(ii) Comparison of the distribution of risks from various alternatives.
The evaluation of alternatives should include not only the local impact of the manufacturing or other activity, but also the local, national or global risks/impacts which may arise from the products or services for which the activity is designed.

(10) Definition of the health monitoring system to be instituted and the subsequent quality control.
The post-audit may involve a considerable effort in the design, operation and utilization of monitoring and surveillance programmes. An important task here is to ensure that all such activities are carried out in support of clearly defined objectives.

10. CONCLUSIONS

Monitoring is an important tool in the process of environmental impact assessment and in any follow-up assessment and control programmes. It should be recognized that there is scarce experience in the application of monitoring in environmental impact assessment. Somewhat wider experience exists in monitoring programmes for selected pollutants on the local or national scale, but even here monitoring has been predominantly used in support of regulation. "Descriptive" monitoring, which supports the identification and estimation of risks or impacts is at a fairly early stage of development, and substantial efforts are required to ensure progress in this area.

REFERENCES

Gilad, A.: 1979, Development of Environmental Health Impact Assessment Systems, ECE Seminar on Environmental Impact Assessment, Villech, Austria, 1979.

Goldberg, E.D. et al.: 1978, 'The Mussel Watch', Environ mental Conservation 5 (2) pp 101-125. Elsevier Sequoia, Lausanne.

Goodman, G.T. et al.: 1976, 'The use of biological materials as environmental pollution gauges', In: The Evaluation of Toxicological Data for the Protection of Public Health, W.J. Hunter and J. Smeets (Eds), Pergamon Press, Oxford, U.K.

Harvey, T.E.: 1981, 'The Monitoring Paradigm of Environmental Control', IES Proceedings, Volume II, Part I, Institution of Environmental Sciences, London.

Munn, R.E.: 1980, 'The design of environmental monitoring systems'. Progress in Physical Geography 4, (4) Edward Arnold, London.

Murozumi, M., T.J. Chow and C.C. Patterson: 1979, 'Chemical concentrations of pollutant lead aerosols, terrestrial dusts and sea salt in Greenland and Antarctic snow strata'. Geochimica et Cosmochimica Acta 33, 1247-1294.

NAS: 1977, Environmental Monitoring, A report to the U.S.E.P.A. from the Study Group on Environmental Monitoring, National Academy of Sciences, Washington D.C.

The Rockefeller Foundation: 1977, International Environmental Monitoring, A Bellagio Conference, The Rockefeller Foundation, New York.

Royal Commission on Environmental Pollution: 1974, Pollution Control: Progress and Problems, Fourth Report, H.M.S.O., London.

Somers, E.: 1981, Environmental monitoring in health decision making. Environmental Monitoring and Assessment 1, 1.

World Health Organization: 1980, Analysing and Interpreting air Monitoring data, Geneva, WHO.

AFFILIATION

Dr Andrew Sors is deputy director of the Monitoring and Assessment Research Centre (MARC), Chelsea College, University of London.

PART D:

CASE STUDIES

P.L. Scupholme

THE ROLE OF ENVIRONMENTAL IMPACT ASSESSMENT WITHIN BRITISH PETROLEUM

1. INTRODUCTION

Some 72 years after William Knox D'Arcy explored for and found oil in Iran, beginning a series of events which culminated in the formation of one of the world's largest oil companies, the constraints applied to such exploration and production activities are drastically different to those experienced by D'Arcy. Not least amongst these constraints are the demands for conservation of both energy resources and the environment. For BP, the experiences of developing oil reserves in Alaska has had a lasting effect on the company's attitude to matters of environmental protection, and today the company would not embark on a similar major project without first undertaking an exhaustive environmental impact assessment.

For the last decade the demand for greater controls on industrial developments have resulted in an impressive array of environmental legislation and regulatory controls in many parts of the world. It is during this period of time that BP has invested large sums of money in searching for and producing oil and gas in the North Sea, an ecosystem characterised by a rich variety of flora and fauna. Although little is known about the way this ecosystem functions, it is clearly one of the world's most abundantly productive seas, situated in the centre of one of the most densely populated and industrialised areas of the world.

The development of BP's North Sea interests has therefore necessitated the careful monitoring of the environmental effects of each new project. The purpose of this paper is to briefly examine BP's environmental assessment procedures. This is done by reference to Forties, currently the largest producing oilfield in the North Sea. Forties was discovered in 1969 and began producing in 1975, by which time a full environmental assessment had been made. Today

B. D. Clark et al. (eds.), Perspectives on Environmental Impact Assessment, 383–389.

the company's environmental policies are largely derived from experiences gained at that time.

2. ENVIRONMENTAL POLICY

During the course of BP's North Sea venture, the company's environmental assessment procedures have steadily evolved. Today these assessment procedures are embodied within the company's environmental policy which states that "it is a primary and continuing policy to conduct the company's activities in a way which endeavours to limit the adverse effects on the physical environment in which its activities are carried out.

With regard to activities associated with the exploration and production of oil and gas, there are two essential elements to the implementation of this environmental policy.

(i) Environmental Assessments of new activities - An environmental assessment is made of any new company project at the earliest possible stage and prior to the commencement of new operations. Although there is no statutory requirement in the UK for such an assessment the planning procedures for land based activities ensure that the relevant information is made available to the appropriate authority. The scope of the assessment depends on the scale of the project and the degree of environmental sensitivity of the area in which the project is planned. The assessment would usually include a baseline study of the environmental conditions before operations commence and predictions of likely environmental changes thereafter.

(ii) Monitoring of Environmental Performance - All operating activities of the company are reviewed to ensure that the company's environmental standards and the appropriate statutory regulations are satisfied. This environmental auditing is seen as an essential element of policy implementation. In cases where necessary changes in operating procedures could significantly affect the environment, it is company policy to examine the implications of these changes and to ensure that environmental standards are maintained.

3. ENVIRONMENTAL ASSESSMENT PROCEDURES

A fundamental and often misunderstood point about BP's environmental assessment procedures is that the preparation of an environmental impact assessment (E.I.A.) document is not in itself the sole purpose of the assessment procedure. The philosophy adopted is that the assessment should be an ongoing process which at some point could lead to the publication of an E.I.A. document, but in many cases does not, depending on the scale, location and environmental sensitivity of the proposed development.

Let us firstly consider the objectives of an environmental assessment. There are two principle objectives:-

(i) to ensure that the design of the proposed project meets the company's environmental protection standards;

(ii) to ensure that the relevant statutory and regulatory controls will be satisfied.

If both objectives are satisfied then the time taken to obtain the necessary planning approvals for construction and operation will be minimised, so that management can confidentially sanction the release of funds in the knowledge that the project is unlikely to meet any expensive delays.

The way in which an assessment is carried out reflects the fact that the assessment is seen as an integral part of project management. The assessment is carried out by a group of people within the company but calls upon the specific expertise of other organisations, for example academic institutions, conservation groups etc. in an advisory capacity. The preparation of the assessment is always co-ordinated by the company's Environmental Department in close association with representatives from the Production, Engineering, Public Relations and other functions within the company. In the case of Forties and other subsequent North Sea projects such as Buchan and Magnus, this group received considerable advice from representatives of the various North Sea fishing organisations, from conservation groups such as the Nature Conservancy Council and the Royal Society for the Protection of Birds, from Government Departments and advisory bodies such as the

Department of Agriculture and Fisheries for Scotland (DAFS) and the Ministry of Agriculture, Food and Fisheries (MAFF), and from Universities, Landowner's representatives etc. Consultation with these organisations is made at both formal and informal levels and this has resulted in establishing extremely good relations to the benefit of all parties.

The timing of these consultations can be critical and yet is often difficult to optimise by the very nature of the way such large scale projects are planned and developed. It can frequently happen that the engineering design of one particular part of an offshore structure can be changed during the course of the development and this could well have a different effect on the environment to that previously envisaged. It is therefore important to continue the dialogue between all parties throughout to ensure that the overall assessment of the environmental impact of the project remains valid.

Given that one of the primary purposes of an E.I.A. is to satisfy internal management of the efficacy of the environmental protection measures within the project design, it can be seen that the wide publication of an E.I.A. document is not of paramount importance. Clearly if an E.I.A. were required by some statutory authority then there would have to be a publication, but with regard to Forties, Buchan and Magnus there was no such requirement. On the other hand there can be considerable benefit in terms of enhanced public relations by the publication of such a document, and at the same time the publication ensures that the objectivity of the assessment can be judged.

Today a typical E.I.A. document for a North Sea oilfield would normally contain a description of the existing environment, an overall description of the proposed project, an examination of drilling operations, a review of the potential ecological impacts of the project during construction and production by reference to any baseline monitoring, both chemical and biological, and a discussion on oil spill contingency response during an emergency. Finally the E.I.A. would include a list of all relevant legislation, and some comment on environmental management of the project indicating responsibilities for environmental protection etc.

4. ENVIRONMENTAL REVIEWS

As explained earlier an essential element within the company's policy on environmental protection is that operational activities should be reviewed on a regular basis. To many it may seem that such a review would be unnecessary if the environmental effects of the project were being properly monitored throughout its operations. In the case of North Sea oil production facilities such as Forties there are biological monitoring programmes to assess the effects on the seabed ecology of any discharges made from the platforms, and the chemical composition of the aqueous effluent from the platforms is monitored in accordance with statutory controls.

Nevertheless it is easy to become complacent about the environmental performance, particularly if the production activity has changed in many ways from the original design. With this in mind an environmental review of the Forties production system was carried out by BP in 1981 with the following objectives:-

(i) to assess environmental performance in the light of current and anticipated legislation and to advise management on the effectiveness of environmental measures making recommendations for improvement where appropriate;

(ii) to comment on the effectiveness and value to management of environmental assessment procedures applied during the design, construction and operation of the Forties system.

The first objective was in a sense akin to carrying out an E.I.A. of an operational project, thereby providing a sort of baseline assessment against which to judge the impact of future changes to the system. The second objective was primarily to look back at how well the original E.I.A. had benefitted the project and to comment on the extent to which the E.I.A. document had been referred to subsequently by operational management.

This environmental review of the Forties system was in fact the first time such an exercise had been carried out for an oil production system, although in the mid 1970's similar environmental audits had been carried out at two of BP's European refineries. The Forties Environmental Review was carried out by a team of four people from within BP. The review team began by compiling information regarding statutory and regulatory controls within the Forties System, consent conditions, the results of monitoring programmes, and this information was collated into a reference dossier. The team then embarked upon a period of site visits during which management and operators were closely questioned about procedures related to environmental protection. In this way it was possible to gauge the general awareness of environmental matters.

A review report was then prepared by the team and circulated to management for approval prior to publication. An agreement to publish the review report was made at the outset of the review procedure, in recognition of the fact that an unpublished review carried out by internal BP staff could be thought by some observers to be less objective than a review carried out by an independent consultant organisation.

The report made 18 specific recommendations for improvement, although in general the environmental performance of the Forties System was found to be satisfactory. It was concluded that the procedures which had been evolved during the design and construction for dealing with environmental questions were highly effective. The main principles, namely direct and continuous involvement of environmental experts, and an open approach to people and interests outside the company, are still being applied, and, it seems, with acceptable results.

The review team had difficulty in commenting in detail on the extent to which the original environmental assessment procedure had been adequate, but it did appear that in virtually all respects the effects upon the environment had been along the lines expected at the planning stage. Consents had for all practical purposes been met and various chemical and biological monitoring schemes had not brought to light any serious or unexpected environmental impact. That the environmental side of Forties was found to be in a reasonably satisfactory state, and that this position was reached on schedule without major upsets and without public hearings, was seen as strong evidence that the management policies applied during planning and construction fully met the project requirements.

An important management concern is the cost of measures for the protection of the environment. Much of the expenditure is usually to meet statutory requirements, such as air or water quality, and has to be regarded as an essential part of the project cost. In the case of Forties there has been considerable further expenditure on such things as oil spill clean up systems, and ecological and amenity protection, in line with general company policy on such matters. The results are for the most part in terms of unquantifiable but valuable assets such as good public relations and good working relationships with external organisations.

Many lessons were learnt during the Forties Environmental Review. Future reviews of other oil or gas production systems could well include at some stage the involvement of outside consultants in an attempt to be seen to provide a completely objective review.

5. CONCLUSIONS

In general terms the environmental assessment procedures within BP are regarded as an essential and valuable element of any project development with potential impacts on the environment. Regardless of whether or not it is a statutory requirement to submit a formal environmental impact assessment it is BP's experience, particularly in connection with North Sea oil development, that such procedures have enabled projects to proceed smoothly with little or no delay on environmental grounds at a time when public opinion and pressure groups have been most vocal about environmental issues. It is hoped that environmental reviews of the type described in this paper will continue to ensure effective environmental management of the company's operations and serve as a foundation for the continued assessment of the impact of operational changes to existing activities.

AFFILIATION

Dr. Peter Scupholme is Head of Environmental Services in B.P. Petroleum Development (UK) Ltd.

G. M. Dunnet

SULLOM VOE OIL TERMINAL: ITS ENVIRONMENTAL APPRAISAL

The Shetland Islands, lying about 100 miles north of the Scottish mainland, have hitherto been a remote rural community concerned primarily with fishing and crofting and cottage industries such as knitting. No major industrial enterprises have been established within the island group. Shetland has a very interesting and well documented historical past. Stone-age remains dating back to four thousand years before present are found in association with remains of Viking settlements a thousand years old. There are many Pictish brochs. Round the coastal and low lying areas of the islands are scattered crofting communities, the only town of any size being the capital, Lerwick.

With the discovery of offshore oilfields in the northern North Sea in the early 1970's, the Zetland County Council (now replaced by the Shetland Islands Council) quickly became aware that the Shetland Islands would provide the closest landfall to which oil might be brought and recognised the likely need for a crude oil exporting terminal to be sited there (see Figure 1). The prospect of a massive multi-million pound industrial complex being developed in their midst caused great concern. Accordingly they commissioned a study by independent environmental consultants to determine the most suitable site in the Shetland Islands for such a terminal. This study (Livesey & Henderson 1973) identified 6 deep-water sites into which supertankers might come for loading, and selected Sullom Voe in the north east of the main island of Shetland as the most suitable site taking account of such factors as hydrography, weather and exposure and available flat land for developments. The Zetland County Council then acquired land around Sullom Voe and under the Zetland County Council Act, 1974, became the Port and Harbour Authority and the Conservancy Authority for Shetland coastal waters. It was therefore clear that if the oil industry decided to bring oil ashore at Shetland by pipelines, the site of the terminal had been predetermined and by virtue of its planning powers the Council could ensure that no other site would be developed.

B. D. Clark et al. (eds.), Perspectives on Environmental Impact Assessment, 391–403.

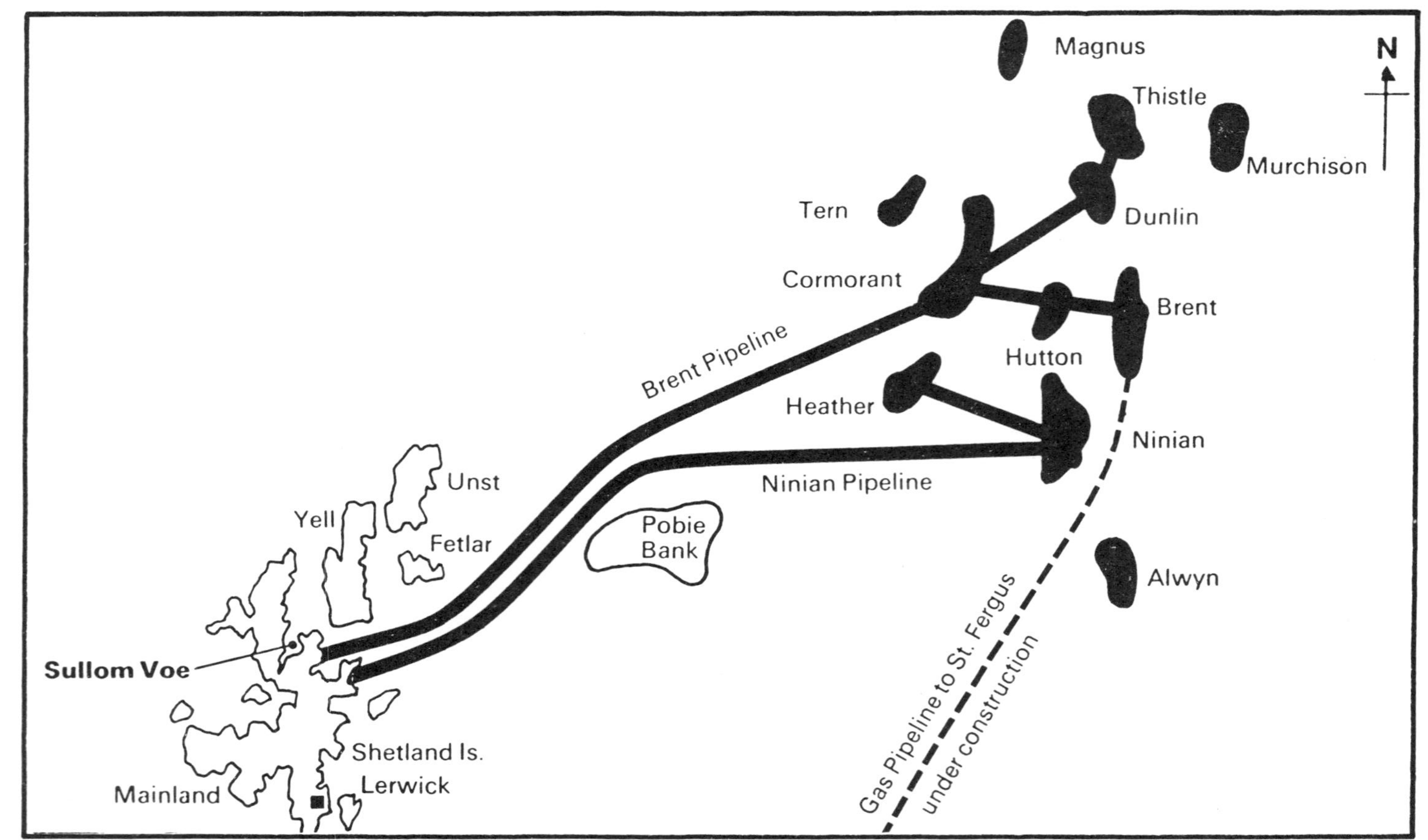

Figure 1. Location of Sullom Voe, Shetland and North Sea oilfields.

In 1974 the oil industry had decided that oil should be taken from fields in the East Shetland Basin by pipeline to Sullom Voe. At this early stage the British Petroleum Company (BP) approached the Zetland County Council with the suggestion that an environmental advisory group might be established to consider the environmental problems and implications associated with the construction and operation of the proposed terminal and to give its advice to both the industry and to the Council. At a meeting in Lerwick on 9th July, 1974 this idea was accepted and it was agreed to establish the Sullom Voe Environmental Advisory Group (SVEAG) with representation on it of both the Council and industry and a number of national environmental organisation with both scientific and management responsibilities, together with some independent university scientists. The role of SVEAG was purely advisory, and the Council had the power to grant or withhold planning consent for each stage of the development as formal plans were submitted. It was further agreed that SVEAG should have as its Chairman a member of the oil industry and that the industry would also provide the Secretariat. Later in 1975, the Sullom Voe Association Limited (SVA Ltd) was established. In it the Council and the oil industry were established as equal partners, with responsibility for the design, construction, operation and maintenance of the terminal in which the treatment, storage and loading facilities were to be used jointly by all the participating oil companies.

The major tasks of SVEAG were to consider and make recommendations of the environmental problems associated with the construction of the terminal and its jetties, the pipeline landfalls, overland pipeline routes, and also to prepare an environmental impact assessment while construction was in progress. Such an appraisal involved detailed consideration of all matters subject to planning consent and the Group was able, whenever necessary, to invite appropriate experts from outside to give specialist technical advice. Such a forum, by discussing these problems at an early stage, was able to influence the industry with regard to the environmental safeguards required for their developments. At the same time they were able to reassure the planning authority, as proper technical appraisal of the possible environmental implications of very complex processes could be carried out in their presence. (The Shetland Planning Officer was a member of SVEAG).

The general plan for the Sullom Voe Terminal was that

2 sub-sea pipelines, one from the Brent Field and operated by Shell, and the other from the Ninian Field and operated by BP, would come independently to the Shetland coastline where they would converge on Sullom Voe. Now oil from other fields is fed into these two major pipelines. At the terminal the oil would be stabilised and loaded over jetties into supertankers for export. All storage and processing facilities and jetties were to be used by all the companies with no separate facilities for each company. Owing to differences in the quality of the crude oil from the various fields this principle was not entirely adhered to with respect to the stabilisation processes. As more fields were discovered a single treatment was not appropriate for the different oils. Facilities were also provided at the terminal for the reception of dirty ballast water from tankers and for the treatment of that water before its release to the sea.

SVEAG identified a number of areas of special concern and established working groups to cover such matters as oil spill control and contingency planning; environmental baseline studies and monitoring; ornithology; fire fighting and safety; and environmental health. Each of these working groups consisted of specialists selected for their knowledge and experience from a wide variety of institutions. They carried out or commissioned detailed studies on behalf of SVEAG and reported regularly to SVEAG itself.

In preparing the Impact Assessment, SVEAG sought to describe the environmental features of Sullom Voe and its surroundings and to identify any features which were of especial value, e.g. the seabird populations. It then listed the major elements in the construction and operation of the terminal and identified possible environmental consequences. It was clear that environmental impact would be greatest in the vicinity of the terminal itself, and zones of intensity of impact were identified. This led to future consideration of components of the environment which were likely to be vulnerable to the development and which might provide early indication of changes occurring. For such purposes both baseline information and data from monitoring were essential, and appropriate projects were designed and commissioned. It is important to note, however, that this was done just before and during the early construction phase of the terminal, and that the assessment developed progressively as successive stages in the development were discussed and integrated, and eventually submitted to the Council for planning consent.

In this sense the assessment differs radically from many others which treat entire projects comprehensively at some particular time before any developments take place on the ground.

A number of different organisations, such as the Local Authority in Shetland, National Government Departments and the Nature Conservancy Council, had their own special statutory responsibilities for environmental care in Shetland, and SVEAG, in addition to commissioning its own programmes of work, served as a co-ordinator of a number of environmental projects only some of which were financed directly by the Sullom Voe Association Limited. These studies covered a wide range of environmental matters including the investigations of the physical and chemical environment, such as the oceanography and hydrography of the waters and currents in and around Sullom Voe; the chemistry of these waters and sediments: the occurrence of heavy metals in Sullom Voe and the origin, fates and concentrations of hydrocarbons derived from biogenic as well as petrogenic sources.

Another set of projects concerned the biology of Sullom Voe itself. These included:

(i) studies of the salt marshes which can readily act as traps if oil is spilt;
(ii) the biology of the soft shores, muddy and sandy environments and of the rocky shores around Sullom Voe;
(iii) studies of the plants and animals living on the inter-tidal and sub-tidal areas of the Voe; and
(iv) work on the biochemical and physiological condition of marine invertebrates.

A major programme was initiated in relation to ornithology. Shetland had already been recognised as one of the major seabird breeding areas in Britain and also had other important ornithological features such as some species of waders and divers rare as breeding birds in Britain.

All of these studies had two elements associated with them: firstly the need to provide an inventory and description of the biological populations and communities in these areas before operations began, and similarly an accurate description of the physical characteristics of the environment. Secondly the need to identify from this accumulated information an appropriate set of environmental features that could be monitored regularly in order to detect as early as possible any changes that might be caused by the construction and operation of the terminal. The scientific results of these baseline studies were published as a volume of the Proceedings of The Royal Society of Edinburgh Section B (Biological Sciences) in 1981. Some of the monitoring programmes which evolved from these studies got underway as early as 1975. These baseline and monitoring programmes are carried out by independent contractors from all over the U.K., selected by SVEAG for their known specialist competence in the particular areas.

Much of Shetland is covered with peat and the area of Calback Ness on which the terminal was to be sited on the shore of Sullom Voe had a carpet of peat sometimes 20 feet (6 m) deep. About 11 million cubic metres of peat and other morainic surface material had to be removed from the terminal site and deposited where it would pose no environmental dangers. Numerous proposals to utilise such quantities of peat commercially were considered but the enormous quantities and the rate at which it would have to be disposed precluded any solution of that kind. It was eventually decided that such peat would have to be deposited in an adjacent sea inlet called Orka Voe. In effect, much of this Voe has now been reclaimed by deposition of unwanted peat behind a series of bunds, which were built in advance of the dumping to prevent the spread of peat into the sea where it might spoil the amenity of neighbouring beaches. This decision has proved to be very satisfactory and large areas of what used to be Orka Voe will, with appropriate reseeding, fertilisation and management, constitute important new grazings. Another early problem which led to a great deal of discussion was the treatment and disposal of the oil-contaminated water that would be brought to Sullom Voe as ballast in the oil tanks of tankers. After treatment at the terminal, the water, having achieved acceptable levels of cleanliness, was to be discharged through a diffuser sited in the tideway of Yell Sound to the north of the terminal, to achieve rapid dilution and dispersal. Special monitoring programmes were designed to detect any effects that the discharge might have on the distribution and abundance of animals living on the nearby seabed, and also to detect any tainting that the commercially important shellfish might develop as a result of living close to the diffuser. So far no unacceptable changes have been detected.

SVEAG operated for 2 years and at the end of that time published, "Oil Terminal at Sullom Voe: Environmental Impact Assessment". This was issued by SVEAG at a Public Seminar which they held at Sullom Voe on 9th June 1976. The document was produced for public consumption and not as a technical document, since the technical information was to be published elsewhere.

SVEAG came to an end in the summer of 1976. It was already recognised that certain changes in the composition and structure of the Group and its terms of reference were required. Some of these could not easily have been foreseen, but it was felt in particular that the Group should have an independent Chairman and Secretariat and not appear to be run by the oil industry. There were other problems: for example, the question as to whether the storage for crude oil should be in under-ground caverns or above-ground tankage. These alternatives had quite different costs associated with them as well as quite different environmental implications. SVEAG found it difficult to favour one rather than the other on environmental grounds, and were accused by the Council of being influenced by economic considerations. The Council also considered the Group to be biased in favour of the industry. These and other considerations led to the withdrawal of the Council from SVEAG in May 1976, just before the public seminar at which the Environmental Impact Assessment was presented. SVEAG therefore foundered on political difficulties. It should, however, be said that by this time the major part of the development was proceeding with very careful attention being paid to environmental problems, and without a costly and delaying public inquiry into the development. SVEAG was disbanded but it had been an interesting, unconventional and imaginative approach to dealing with the very complex environmental issues involved in a development of this kind and magnitude, sited in the relatively remote and non-industrial Shetland Islands.

For the next year or so there was no forum for environmental discussions, but none-the-less many of the environmental projects which had been established by SVEAG were continued by the contractors. The gap left by the dissolution of SVEAG emphasised the desirability, and indeed the need, for such a forum for discussion and advice during the further development, commissioning and operation of the terminal. Discussions took place between the industry and the Council and in the summer of 1977 a new group, The Shetland Oil Terminal Environmental Advisory Group (SOTEAG) was formally established by the Sullom Voe Association Limited with the following terms of reference:
"SOTEAG shall examine and advise on the environmental

implications of the Terminal at Sullom Voe during all stages of construction, site rehabilitation, commissioning and operation. The area for review shall be restricted to within the zone for which the Shetland Islands Council is responsible for environmental matters and will include the activities of tankers trading to Sullom Voe".

There are some fundamental differences between SOTEAG and SVEAG. SOTEAG has as an independent Chairman, a professor from the University of Aberdeen, and a Secretariat established in the University. SOTEAG retains equal representation of the Shetland Islands Council and the oil industry, but has a much better representation of national organisations and local interests than did SVEAG. As with SVEAG, members give their services to SOTEAG without cost, though SVA Ltd meets the expenses of its meetings, administration and environmental programmes. Advice is given formally to SVA Ltd by a reporting system designed to ensure that interpretations and conclusions from monitoring activities, as well as advice on specific matters are fed back to both the Council and industry. (Figure 2)

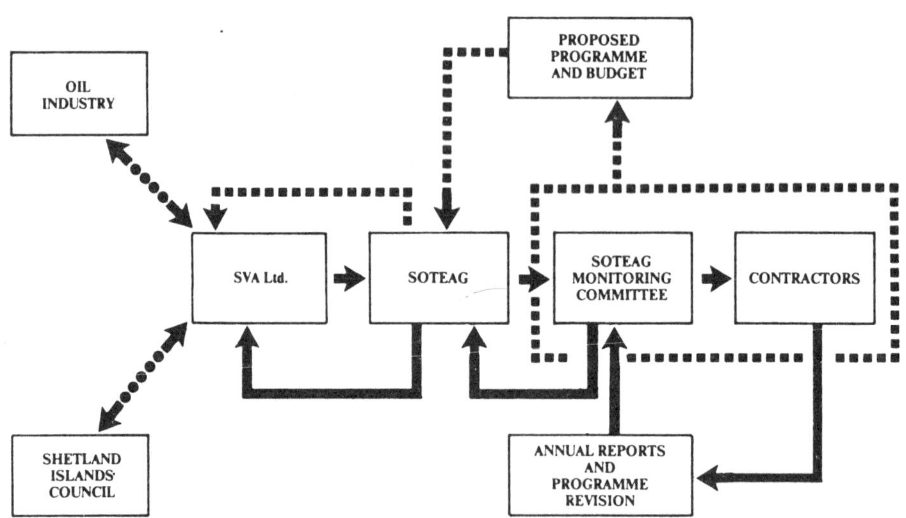

Figure 2. SVA/SOTEAG Monitoring activities flowchart.

SOTEAG has 3 major sets of activities. It continues to give advice to both the industry and the Shetland Islands Council on matters relating to new developments and planning applications. Whenever appropriate research is commissioned to obtain any necessary information and specialised consultants are called in to advise. Secondly, SOTEAG has a major input into the environmental aspects of oilspill contingency planning. For example various sensitive environmental areas in Sullom Voe and its approaches have been identified and the Group made recommendations relating to the appropriate treatment of spilt oil in these areas. Environmental advice has also been given to other technical advisory committees who are responsible for the design and implementation of systems containing and recovering spilt oil. Such advice includes the identification of sacrificial beaches to which spilt oil may be taken for recovery, and questions of access to these beaches; the kinds of treatment that are acceptable for different areas of the shore and of the Voe and Yell Sound - including spraying with dispersants, mechanical and hand-cleaning of beaches, and doing nothing. In this context, the need to make arrangements for safeguarding sheep which graze on seaweeds on the foreshores in winter, was not adequately recognised and only after the "Esso Bernicia" spill of fuel oil killed and damaged sheep in January, 1979, was a system of fencing foreshores introduced.

Thirdly, SOTEAG is responsible for environmental monitoring. The early baseline studies commissioned by SVEAG gave rise to the identification of appropriate environmental characteristics which are suitable for monitoring in order to detect changes in the environment brought about by the operations of the terminal. A number of such studies are now commissioned and from time to time others are added from baseline investigations which arise as new problems emerge during the operations of the terminal. A list of current monitoring activities is given below:

SOTEAG MONITORING PROJECTS AND RELEVANT STUDIES UNDERTAKEN BY OTHER ORGANISATIONS:

Commissioned by SOTEAG:

Chemical monitoring:

Hydrocarbons	Scottish Marine Biological Association/Newcastle University
Heavy metals	Scottish Marine Biological Association

Biological monitoring:

Macrobenthos	Oil Pollution Research Unit
Rocky Shores	Oil Pollution Research Unit
Soft Shores	Dundee University
Salt Marshes	Imperial College

Atmospheric monitoring:

Lichens	British Museum (Natural History)/ British Petroleum

Ornithological monitoring:	Aberdeen University

Jointly funded SOTEAG monitoring and baseline studies:

Otters:	Natural Environment Research Council, Institute of Terrestrial Ecology

Ornithology:

Red-Throated Diver Survey	Royal Society for the Protection of Birds
Black Guillemot Survey	Edward Grey Institute for Field Ornithology, University of Oxford
Biological Effects Monitoring	Natural Environment Research Council, Institute for Marine Environmental Research

Terminal operator:

Statutory monitoring at effluent outfall diffuser:	Heriot-Watt University

Department of Agriculture and Fisheries for Scotland:

Hydrography
Fisheries effects
Macrobenthos production
Ballast water discharge
Hydrocarbons

Natural Environment Research Council (N.E.R.C.):

Monitoring of Shetland seal stock:	Sea Mammal Research Unit

As with SVEAG independent expert contractors are commissioned to undertake this work. The current cost of environmental monitoring and baseline studies is of the order of £150,000 per year. SOTEAG has set up a specialist committee, its Monitoring Committee, to ensure that well designed, credible monitoring programmes are devised, and has recently commissioned an external audit of the programme with a view to reviewing the long-term commitment.

To date our monitoring programmes show, that apart from effects known to be due to the "Esso Bernicia" spill, almost no detectable environmental change has occurred. Annual studies of the benthos and of fore-shore communities of plants and animals show a range of variation considered to be well within naturally-occurring fluctuations. Although breeding seabird populations in Shetland generally continue to increase, some changes in the numbers and distribution of wintering seaducks and divers have been described, only some of which may be associated with operations at the terminal. As expected traces of petrogenic hydrocarbons are now being detected in the sediments of Sullom Voe, at very low, and acceptable levels, but at least proving that the monitoring is being effective.

A number of points important to the concept of environmental impact assessment have emerged during the course of the Sullom Voe developments. The first is that since the initial plan for the terminal and its operational processes was established there have been major changes. The biggest of these has come from an external source, from the Inter-Governmental Maritime Consultative Organisation (IMCO), in relation to the operations of tankers. In the early 1970's most big tankers carried ballast water in their oil tanks and therefore on arrival at an oil terminal had to discharge the oil-contaminated water into tanks onshore. In these tanks the oil and water separated and the water was further cleaned in a series of filters and other devices before being discharged into the sea. At Sullom Voe, this effluent is discharged into the relatively fast currents of Yell Sound, and must not exceed an average maximum concentration of 15 ppm. IMCO decided that in the 1980's tankers must carry segregated ballast: i.e. their ballast water would be carried in tanks dedicated for that purpose and which would never carry oil. Such ballast water was deemed to be "clean" and the practice is that it is discharged directly into Sullom Voe over the side of the ship and not passed into the effluent treatment plant. So now thousands of tons of low salinity water, taken on at a previous port and containing various contaminants from these locations, are being discharged directly into Sullom Voe. Such a procedure was not envisaged at the time of the original impact assessment. Clearly it is a very good reason for a continuing environmental advisory group, rather than a "one-off" environmental impact assessment at the beginning of a development. SOTEAG now has to consider what the environmental impact of this segregated ballast water is likely to be, and this is complicated by the variety of systems for the discharge of the water from different tankers. This practice may also affect existing monitoring programmes, and modifications, or even new monitoring, may be necessary to detect environmental changes arising from this new operational practice.

Secondly when the impact assessment was produced by SVEAG, contingency plans were devised to deal with spillages of crude oil. This decision was based on experience at other crude oil loading terminals where there were no bunkering facilities for tankers. There has only been one major oilspill at Sullom Voe and that was a spill of fuel oil, not crude oil, and it occurred in mid-winter. At such low temperatures fuel oil is very difficult to handle and it was found that the equipment that had been provided for the retention and recovery of crude oil was not at all appropriate for fuel oil. Once again the predictions and precautions in the original plan were irrelevant and had to be changed.

Thirdly, as the major developments at the terminal itself got underway, satellite developments were proposed. SOTEAG has been asked to consider the environmental impacts of an LPG plant for the local distribution of liquid petroleum gases within Shetland. This was proposed for a neighbouring site to the terminal, and it had a new set of environmental implications, and also influenced the monitoring programmes established around the terminal. So far this proposal has not gone ahead but it indicates the clear need for continuing environmental surveillance associated with large developments of this kind.

REFERENCES

Livesey & Henderson Reports: 1973, Master Development Plan and Report related to Oil Industry Requirements, 6 Volumes, Zetland County Council.

Proceedings of the Royal Society of Edinburgh: 1981, The

Marine environment of Sullom Voe and the implication of oil developments, Section B, Volume 80, Latimer Trend & Company Ltd., Plymouth.
Sullom Voe Environmental Advisory Group (SVEAG): 1976, Oil Terminal at Sullom Voe Environmental Impact Assessment, Sandwick, Shetland. Thurleprint.

AFFILIATION

Professor George Dunnet is head of the Department of Zoology at the University of Aberdeen, Scotland.

F. E. Dean

SOME ENVIRONMENTAL ASPECTS OF GAS DEVELOPMENTS IN THE UK

1. INTRODUCTION

British Gas operates one of the most complex natural gas transmission systems in the world. It has had to design, construct and commission this within a very short time scale and within the very complex town and country planning legislation, without any special privileges and at a time when the public is taking a greater interest in environmental matters than ever before. In order to do this it has not only to use the most modern engineering techniques, but also to demonstrate its care and concern for the environment.

British Gas has always set very high standards for the reliability and the safety of its installations and these do not result in cheap solutions. Reassessments are made regularly to ascertain if any relaxation can be made but never at the cost of putting people at risk.

The national need for energy is often in conflict with the desire for the conservation of the environment and is thus sometimes at variance with agricultural requirements. In an endeavour to solve these problems British Gas has devoted much time and effort in the past years. This has resulted in the development of environmental impact analysis techniques to define the problems, followed by the implementation of solutions to them. This has necessitated the expenditure of considerable sums of money but the return resulting from the reduction in the overall planning and construction time covers this several times over.

2. THE UK GAS INDUSTRY TODAY

The Gas industry came into public ownership in May 1949 and since then has passed through several marked changes. For example, the number of plants actually making gas has fallen from nearly two thousand to, incredibly, two. The source of the gas was, for nearly a hundred and fifty years, coal heated at atmospheric pressure in closed retorts. In the

B. D. Clark et al. (eds.), Perspectives on Environmental Impact Assessment, 405–427.

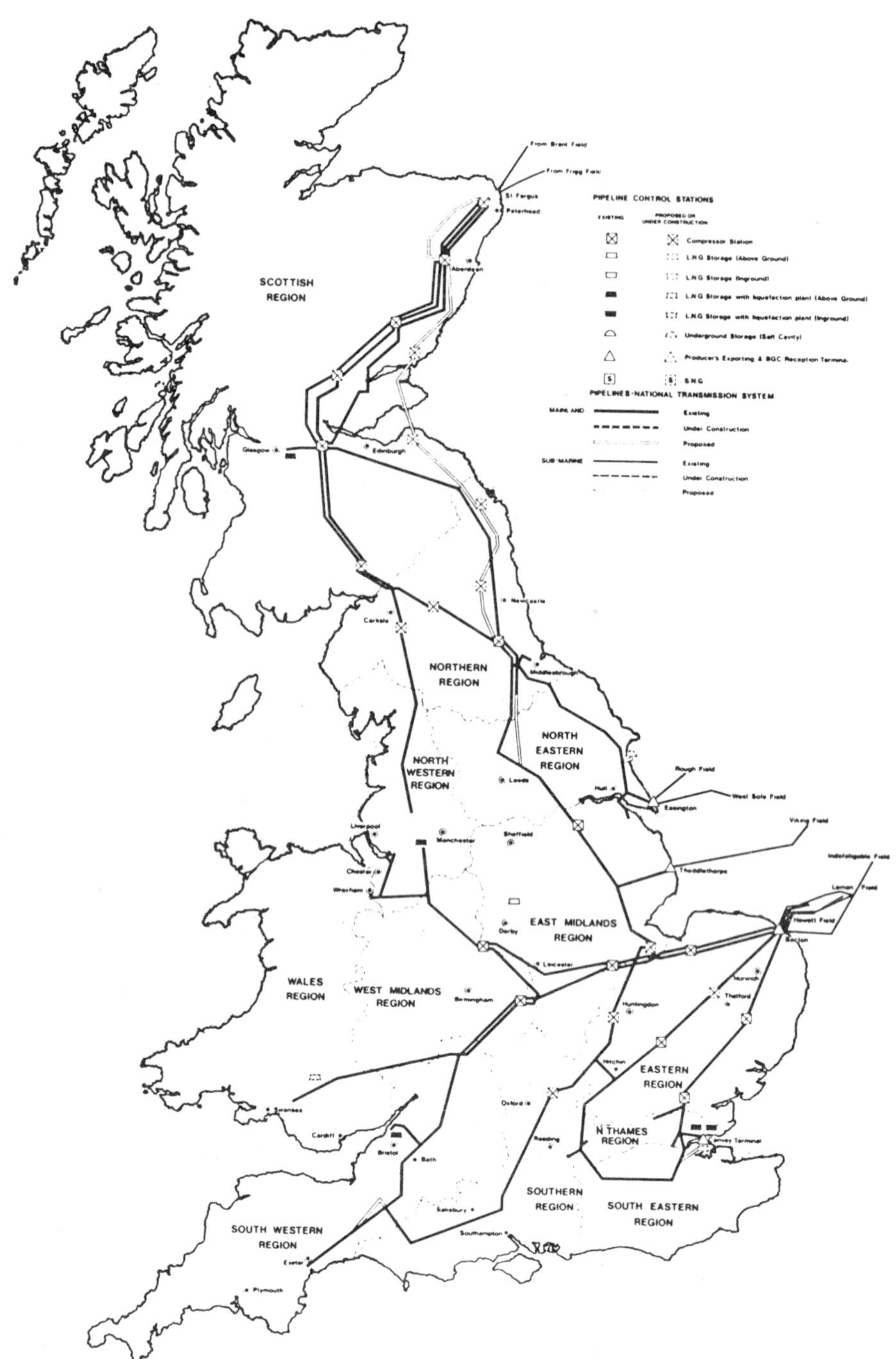

Figure 1 National Gas Transmission System April 1980.

late 1950's to early 1960's, light oils became the prime feedstock producing gas at pressures of 1.05 x 10^5 - 1.75 x 10^5 kg/m². Finally, in the late 1960's, natural gas from the continental shelf fields was brought ashore and now forms nearly 100% of the gas used. Originally the gas was made in works situated within cities and towns, often with no connection between adjacent systems. The industry now has a nationwide high pressure transmission system taking gas from the coastal terminals inland to the centres of demand.

The function of this system (Figure 1) is to receive the gas from the offshore fields, purify it, control its flow and finally to deliver it into smaller regional networks which in turn supply local distribution systems. To do this, the national grid comprises some 5600 km of large diameter (90 cm - 106 cm) high pressure (7.03 x 10^5 kg/m²: 70 bar) pipeline connecting the five (plus one under construction) terminals to some 200 "offtake" stations where it passes under controlled conditions into the regional systems. To assist the gas flows there are 14 (with more planned) large compressor stations and several very large storage installations for seasonal and security requirements. To give some idea of the scale of these projects, a coastal reception station typically occupies some 80 - 120 ha or more, a compressor station on a site of 8 - 12 ha whilst a storage installation can consist of four tanks of 20,000 tonnes capacity of liquefied gas, each being 50 m in height and 50 m in diameter.

All of the above has been constructed within the last fifteen years and almost all of it on "greenfield" sites in relatively remote rural areas in which industrial developments of any kind are conspicuous by their absence. When it is realised that the gas industry has no special powers of its own - except for underground pipes - and has to work within the normal town and country planning legislation, the magnitude of the problems that might have arisen can be visualised.

3. THE STATUS OF THE GAS INDUSTRY IN THE UK PLANNING SYSTEM

3.1 The Gas Acts and other legislation

The Gas Industry was nationalised by the Gas Act 1948 and the appointed vesting date was the 1st May 1949. Since then the Gas Act 1965 and the Gas Act 1972 has established the Gas industry in its present form of the British Gas Corporation and the Gas Act 1980 dealt with certain aspects of the Corporation's obligation to supply gas to larger consumers.

The Gas Act 1972, section 2(1) places upon the Corporation the duty "to develop and maintain an efficient, co-ordinated and economical system of gas supply for Great Britain, and to satisfy, so far as it is economical to do so, all reasonable demands for gas in Great Britain". The Act goes on to give the Corporation various powers to enable it to comply with this duty and these include the power to acquire and hold land and subject to the approval of the Secretary of State for Energy, the power to acquire land by compulsory process. Schedule 2, Part 1, para 1 of the Gas Act 1972 states "The Secretary of State may authorise the Corporation to purchase compulsorily any land in Great Britain which they acquire for or in connection with the exercise and performance of their functions under any enactment".

In carrying out large new developments in greenfield locations the Corporation has to comply with the requirements of the Town and Country Planning Acts. The only exemptions the Corporation has from these requirements are contained in the Town and Country Planning General Development Order and in particular, Class XII - Development under local or private Acts or Orders and Class XVIII (D) Development by Gas Undertakings. Underground pipelines do not require planning permission.

The Corporation has to comply with the Control of Pollution Act and similar legislative requirements in addition to the Health and Safety at Work etc. Act. The British Gas Corporation is under a special obligation with respect to its development programme for by virtue of the Countryside Act 1968, section II, it is one of the public bodies required "in the exercise of their functions relating to land ... (to) have regard to the desirability of conserving the natural beauty and amenity of the countryside".

Environmental Impact Analyses have been undertaken as part of the preparation of planning applications for submission to local planning authorities. They have been useful in this context. It is now being proposed that such analyses become a mandatory part of the process.

3.2 Environmental and Land Use Restraints

There is also a framework of international policies within which the gas industry has to operate. This is epitomised in the United Nations Declaration issued at the end of the Conference on Human Environment held in Stockholm in 1972. This proclaims that:

(i) "Man is both creature and moulder of his environment ... a stage has been reached when ... man has acquired the power to transform his environment ... on an unprecedented scale. Both aspects of man's environment, the natural and the man-made, are essential to his well-being ... even the right to life itself".

(ii) "The protection and improvement of the human environment is a major issue which affects the well-being of peoples and economic development throughout the world; it is the earnest desire of peoples of the whole world and the duty of Government".

The UK is one of the most densely populated areas in the world and the pressures on land use are very great. In spite of this, large stretches of land are given protection, to a greater or lesser degree, from housing or industrial development. These areas include National and Country Parks, Nature Reserves, Areas of Outstanding Natural Beauty, as well as "green belts" around towns and cities. In another context, there are large numbers of Sites of Special Scientific Interest or of Archaeological Interest which also have statutory protection. In total, these protected areas amount to some 40 per cent of the total land use. Another 10 per cent comprises the built up areas in towns and much of the remainder is high quality agricultural land, whose value to the nation is very high. Land above 300 metres, representing a significant proportion of the total, poses difficult technical problems as well as being usually of high scenic value.

4. GAS INDUSTRY ENGINEERING AND TECHNICAL REQUIREMENTS

4.1 System Design Parameters

All the significant supplies of natural gas for the UK come from under the seas surrounding the coast of Britain and therefore there is a need for gas reception terminals to be situated on or very near to the coastline. The pipelines from the gas fields usually take the shortest practicable route to the nearest landfall for economic reasons, and this requirement has dictated, within fairly close limits, the geographical location of the four existing terminals. A fifth, for the Irish Sea Gas, required a landfall within a fairly restricted area of the NW England coastline, an area already extensively developed for industrial, residential, recreational and amenity purposes. The planning authority in this case required an assessment along the lines suggested in Department of Environment Research Report 13, to be carried out before they granted planning permission.

British Gas has the responsibility of transmitting the gas from these coastal terminals to the centres of demand, this being done by a combination of transmission pipelines and compressor stations. Duplication of pipelines results in a reduction in the number of compressor stations and postpones their construction until later in the programme. It is possible to calculate the optimum solution to the problem of relating number and diameter of pipelines to the number and spacing of compressor stations. This spacing must take into account the throughput rates that can be achieved by modern compression equipment, the pressure lift that can be obtained at each station and the largest diameter of transmission pipelines that is available to meet the strict engineering standards required. Until recently this latter dimension had been 90 cm in diameter, but techniques now proven by British Gas, have allowed this to be increased to 106 cm, thus deferring the need for intermediate compressor stations still further in the future.

All the compressor stations and the storage installations mentioned require certain minimum areas of land and have to be on or close to the routes of the pipelines.

4.2 Engineering Standards

British Gas has always set extremely high standards for the safety and reliability of all of its installations. These standards are based on the best British and American current standards and codes of practice as minima. The standards with an environmental content such as those used for the design, proximity distances from residential and industrial areas and other safety aspects, material requirements and other matters, are included in specifications drawn up by the following organisations; the American National Standards

Institution (ANSI), National Fire Protection Association (NFPA), British Standards Institution (BS), American Petroleum Insitute (API) and the Institute of Petroleum (IP). In addition to these, British Gas has a number of its own standards which are mandatory for use on its own projects. These are in many cases based on existing British or American standards but they often specify even high standards for such items as valves, high pressure and cryogenic pipe work.

4.3 Transmission Main Restraints

In conveying gas from the coastal reception terminal to centres of demand, the high pressure pipelines have to pass through open country, semi-rural areas and may come close to more densely populated regions for short distances. The safety requirements under these widely differing conditions are set out in the Institution of Gas Engineers Relevant Code of Practice. Basically this divides the country into three main zones - rural, suburban and city centre - and pipelines into groups according to the working pressure. The code then relates proximity distances from houses, schools and factories, etc. to pressure, diameter and type of country. Quite apart from the "normal" problems of pipe routing and laying, other major factors arise when crossing rivers, estuaries, mountains and peat bog areas, all of which involve not only the use of the most advanced engineering techniques, but also those applying to ecological and environmental considerations.

4.4 Technical Restraints

4.4.1 Site location

The selection of the shortest route for a high pressure large diameter gas transmission pipeline, so that it may pass as far as possible through open country, will usually have to take precedence over the choice of sites for compressor stations and peak shaving facilities, since the cost of pipelaying is extremely high. Suitable sites have to be looked for which are as close as possible to the proposed pipeline route and, if this requirement cannot be met, then severe cost penalties will be incurred for the increased length of the connecting lines. This problem is discussed in greater depth later.

4.4.2 Site area

The actual area required for each site depends on a number of factors, apart from the actual number of compressor units that may be required or the number of storage tanks proposed. These are:

(i) The inter-relationship of items of plant. The engineering requirements of an installation often dictate that various items of plant are situated in close proximity to each other to avoid excessive lengths of inter-connecting pipework, electrical cables, etc. If visual supervision is required from a control room, then it must be located in a particular position on the site. The layout of roads must be such that traffic does not have to pass under pipe bridges and cable racks and must give unrestricted access at all times to emergency vehicles. The size of and the internal layout of buildings is often determined by the need for various items of equipment and rooms to be situated adjacent to each other.

(ii) The safety distances on sites. These are laid down in the codes which have been previously mentioned. This includes requirements for the distance of control rooms from working plant, siting of electrical equipment in relation to hazardous areas, the siting and bunding of tanks relative to each other and to properties outside the site boundaries.

(iii) Security requirements. The security requirements on any particular installation will vary depending on the importance of any particular site and the nature of the operations carried out, but generally require certain minimum distances between the security fence and the nearest items of plant. These requirements can increase significantly the total acreage required, in some cases by as much as 100 per cent.

(iv) Interconnecting pipe systems. Certain minimum space requirements are needed for interconnecting pipe systems between items of plant on site. For example, provision has to be made for the expansion and contraction of pipework, by installing expansion loops to reduce stress levels to an acceptable limit. These items are also space consuming.

(v) Height of equipment and buildings. Equipment on sites is generally less acceptable to planning authorities in direct proportion to its height. Unfortunately, many engineering structures, particularly tanks, have an optimum relationship between height, cross sectional area and cost. As the tank height is reduced and the diameter correspondingly increased, the cost for a given volume of storage increases quite markedly. There is also an associated penalty relating to land take as the proximity distance required from the centre of the tank to the site boundary is also a function of its diameter.

5. ENVIRONMENTAL CONSIDERATIONS

5.1 Technical Considerations

In this section, the Gas Industry's own technical requirements are not considered as it is axiomatic that any site chosen must meet the specifications laid down by the Gas Industry. Although it may be possible to relax this, there may well be financial penalties involved in departing too far from the original parameters. These extra costs have to be equated with any environmental penalties associated with the optimum specifications. It is when these penalties cannot be assigned monetary values that the difficult decisions have to be made.

Large installations in remote areas, inevitably cause a reaction amongst the other public utilities. The effect on water supplies and drainage of the area must be considered at the very earliest stage. Many installations have no option but to be sited within the catchment areas of reservoirs and obviously, could have a deleterious influence on them. Construction work can soon alter the surface water run off, both in quantity and quality, and the appropriate remedial action will have to be known in advance. Attention has to be focussed on even the minor processes used on the plant. For example the use of salts of chromic acid as a herbicide in cooling water systems is unacceptable where the overflow would pass into a local watercourse and thence into a fishing stream. Power supplies to other users at the extremities of distribution systems could be endangered by the installation of large electric motors if suitable starting procedures were not drawn up and implemented and appropriate reinforcement provided. Communications can prove to be a very emotive point, particularly when VHF or microwave radio masts are involved. By definition, the "line of sight" routing of the radio links means that the aerials have to be on hilltops where their visual effect is out of all proportion to their size. They can be the cause of much ill will being directed at the industry and their introduction is usually fiercely opposed. The lack of planning permission for one mast can cause lengthy delays to the commissioning of the whole system with consequent operational and financial penalties.

The subject of the potential pollution of the air or water supply is always in the mind of the general public. Fortunately, the gas obtained from the North Sea is of extremely high quality, usually dry and sulphur free and requiring the minimum of treatment. Provided that leakage and spillage control is adopted, there is no real pollution problem. The problem of odour is one that is always present in an industry dealing with hydrocarbons and which is under a statutory obligation to sell a gas with a characteristic smell! It is extremely important that adequate care is taken in the design of the odorant storage and dosing system to ensure that no nuisance is caused to nearby residents. To this end, sophisticated and effective systems have been built and operated. The discharge to the atmosphere of even minute quantities of odorant vapour can cause an immediate and hostile reaction and one which will remain in the public memory for a long time. A similar problem arises with the venting of odorised gas and the design of stacks has to take account of the local topography and climatology to ensure that the discharged gases cannot return to ground level under any possible atmospheric conditions.

The plant construction period can be one of much environmental controversy and relations can become strained between the developer and local authorities and residents if care is not taken. A very simple example is the disposal of surplus spoil. The quantities which will have to be handled, and the point and method of disposal must be defined at the planning stage, and it is not now permitted in the UK to tip spoil without first obtaining the requisite planning and other statutory consents. A variant on this theme concerns the discharge of large volumes of sea water after use for leaching out salt caverns. It can only be pumped back into the sea under carefully controlled conditions relating to salinity, suspended matter and trace elements, all of which could have a deleterious effect on fish populations and subsequently the local fishing industry.

5.2 Social Considerations

In many instances, the social implications of development can be as important as the technical. The Gas industry's new installations are not labour intensive, requiring only a small number of mostly specialist staff. This feature can be a mixed blessing if the site is in an area of high unemployment and occupies a significant proportion of the land available for industrial use. At the other end of the scale, social problems arise during the construction phase. The importation of a large labour force into an area can cause housing difficulties, overload the social services and have problems concerning the education of the children of the workforce. In some cases, where the contractors are international in character, a language problem has emerged and specialist school teachers have had to be moved into village schools to cope with the situation. All of these factors have to be considered at an early stage and the appropriate action taken. The road, rail and air transport facilities all have an influence and any unplanned-for demands can create serious operational difficulties and be the source of much bad feeling towards the developer. The undisciplined use of the public road system by contractors' vehicles is a very common example of this. On the other hand, oil and gas developments in a region can bring new business to a railway line leading to rejuvenation rather than closure.

Pollution by noise is a major social problem in our industrialised society and compressor stations, for example, could be a classic example using, as they do, aircraft type gas turbines which have an unfortunate emotive connotation. Fortunately there is no weight problem with land based machines, and adequate silencing, although expensive, is perfectly practicable. The policy of British Gas in this matter is to do what is necessary to ensure that, no nuisance will be caused to the nearest residents to a site. In practice this means that the night-time ambient noise level shall not be increased by more than 5 to 10dBA at the nearest house.

5.3 Ecological Considerations

Finally, but by no means the least important, there is the effect on the flora and fauna of the site itself and any such implications that there might be over a much wider area. The use of a lake or marshland as a staging post or feeding ground for migrant bird life can have a major effect on the proposal to site an installation in its vicinity. Factors such as these can only be determined by a detailed investigation by qualified ecologists over a long period of time. It is fortunate in this context that there is available a vast fund of knowledge and expertise in the voluntary organisations of this country which devote their interests to the study of wildlife. The Royal Society for the Protection of Birds, with its quarter of a million members, is probably the largest, but the total number of such bodies is very large and the Gas Industry is very grateful for the co-operation it receives from them as well as from the official bodies, such as the Nature Conservancy Council, the Institute of Terrestrial Ecology and the Countryside Commission. In some cases, the very size of the site required and the limited above ground facilities that appear on them do have ecological advantages. Some of our installations have been sited adjacent to nature reserves and they exist in perfect harmony. In at least one instance, a nature reserve is being formed on land which forms part of a major installation. This proposal was included as one of the conditions relating to the issue of the planning consent and has been welcomed, not only by British Gas, but also by those organisations responsible for the protection of wildlife as this new reserve will take the pressure of public interest off another important nature reserve in the locality.

5.4 Visual Considerations

The non-visual considerations which have been referred to will apply in varying degrees to different sites, but the problems inherent in the search for acceptable quality of visual appearance arise for all practical purposes, in the assessment of each and every site. In Great Britain it is almost inevitably the case that any possible site will be visible, from housing, roads or railways, from surrounding high ground etc. The introduction of new artefacts into existing landscapes which, because of the nature of the national transmission system, are of great visual impact, presents some very sensitive issues and problems.

5.4.1 Visual assessment techniques

One area of study which has received a good deal of attention is concerned with techniques to achieve the accurate graphic representation of buildings and plant in a landscape. The traditional artist's impression is simply not good enough and there is a need for a relatively quick way of achieving accurate representations of a development on a particular site as seen from a number and variety of viewpoints.

Several computer programmes have been investigated and studied. In the case of a compressor station at Moffat, in Scotland, the photo-montage system, programmed by the Computer Aided Design Centre of Cambridge, has been utilised. The method is based upon a photograph of the site taken from a selected viewpoint and data is fed into the computer to cover such factors as the co-ordination of a viewpoint, the height of the camera, the focal length of the lens and the enlargement factor. Selected objects on or near to the site which appear in the photograph are accurately surveyed for location, level and height, and the resulting data is programmed. In the case of sites without significant reference, points such as open moorland, it is necessary to create one by erecting a tower of scaffolding or flying a balloon. Data concerning the location co-ordinates of buildings, their levels and heights are then added to the programmed information and the print-out plotter produces an overlay to the same size as the original photograph, which sets the proposed buildings accurately in the view.

Not only can print-outs be rapidly obtained for any number of viewpoints, but it is also a simple matter to assess accurately the effects of design amendments such as raising or lowering the base levels or location of the buildings. The method is particularly useful in the study of the screening effects of relatively minor topographical variations or the value of existing and proposed tree planting. A similar technique can be used to prepare intervisibility studies, i.e. maps showing areas from which installations will or will not be seen. By use of suitable computer programmes, it is possible to obtain these rapidly and to include the effect of changing site layouts, heights of plant and the effect of existing or proposed off site landscaping. The same technique is also of value for locating sites for radio masts which have to be within line of sight of each other.

5.4.2 The use of colour

There is a school of thought that contends that if a fence, an item of plant, or even a large tank is painted green it will somehow merge or be absorbed into is natural background or landscape setting. The appreciation of the infinite variety of greens in nature and the effects of differing light conditions and the seasonal variations quickly highlight the superficiality of such an approach. If camouflage is seriously to be attempted the art is very sophisticated and the chances of achieving a satisfactory effect are low. By contrast the positive use of colour - and sometimes strong colour - provides the designer with an important and valuable tool with which to fashion a new, interesting and acceptable feature in the landscape. The potential for such an approach must grow out of a really studied appreciation of the site and area in question; there are no standard answers which would satisfy both the texture and richness of a lush Cotswold landscape, on the one hand, and, say, the watery diffusion of an estuarial panorama, on the other.

The differentiation and articulation of the various component parts of an installation by an imaginative colour scheme can affect the resulting visual scale and transform an otherwise complicated composition or even confused jumble of buildings, plant and pipework into a visually clearer and more organised picture. On one site an accent of bright red may well be an interesting incident when seen from a passing train, on another the use of recessive blues and muted greys will be appropriate to a Scottish moor. This use of colour as a positive architectural and landscape element is widely acknowledged as a successful feature of much of the British Gas Corporation's recent work. It has been based on the fact that often the new installation will be visible as an additional element in a landscape; the problem is to design it to be a visually acceptable and interesting addition, in sympathy and harmony with its setting, and not an apologetic, half-hearted statement wishing it were invisible when that can never be.

5.4.3 The use of landscaping

It has previously been pointed out that visual quality can

never be regarded as something to be applied to an otherwise predetermined solution, like icing on a cake.

Landscape - both existing and new - is a factor in the visual equation which has to be integrated into the design solution in the earliest conceptual stages. During the initial analyses of the characteristics of alternative sites the location of say, a tree belt which might form a visual background to items of plant when seen from an important viewpoint, together with its condition and life span, must be studied and evaluated in the way that other conditions, such as drainage, bearing pressures, means of access, and the like, are assessed. New landscape must not be conceived in irrelevant terms concerned with the rose-beds outside of the office building, but in terms of ground formation, modelling and tree planting which is in scale with the new installation and its existing natural environment. Its function is to act as an additional element to assist in the visual composition of the new artefact in its landscape setting. Although new planting can often be used as a screen to ground level areas such as storage compounds or car parks it is, of course, quite futile to think that new planting can effectively screen a cluster of 150 ft high storage tanks.

Another important feature of new planting, derives from the fact that the life of the trees will well exceed that of the installation. Planning schemes must be so devised that a logical philosophy will be apparent even when the equipment and buildings have been removed. In this context, British industry must accept its long term patronage role in relation to landscape in the way that the great landowners of the past invested for the benefit of the future generations.

6. THE PROBLEM OF CONFLICTING INTERESTS

It follows from what has been said above that the national need for energy supplies - and also other industrial projects - is often in conflict with the equally strong national desire for the conservation of the natural environment, the preservation of national and local amenities and the needs for agricultural production. This statement applies not only to the general concepts but also to the more specific relation of the design of individual projects and its effect on the local area. Examples of this conflict of interests are:

(i) The "anywhere but here syndrome". Everyone accepts the advantages of the modern industrial society but nobody wants the necessary development in their own back yard.
(ii) The conservation movement and the need to increase industrial production and thus increased energy usage.
(iii) The "alternative or renewable energy source" projects versus the effects they would have on the environment.
(iv) The conflict of land use requirements, housing and industry versus agriculture, recreation or nature conservation.
(v) Engineering, technical or economic practicability to be weighed against environmental and amenity interests.

During the past 15 years or so, British Gas has devoted much time and effort to provide satisfactory solutions to these problems. It was realised at a very early stage that before they can be resolved, the problems have to be identified and that this can be done by the use of what is now called Environmental Impact Assessment and Analysis. This technique although extremely useful, must never be regarded as an end in itself.

6.1 The Functions of Environmental Impact Statements and Analyses

Mr George Dobry, Q.C., in his report of 1976 gave a good definition of an environmental impact statement. He stated that such a study should be "the orderly assessment of the overall impact of a planning or existing development on the environment in terms of both physical and socio-economic effects". Such studies must include secondary effects of development as well as the more obvious primary ones. For example, not all effects of industrial development are necessarily harmful to local ecosystems; the elimination of public access or the cessation of farming has been shown to encourage the return of wildlife, as has been demonstrated at the oil refineries on Teeside and the gas terminal at Bacton.

A major and basic component of any impact statement must be a detailed description of the existing environment to provide a baseline from which changes can be measured or predicted. This should also identify those features which are most sensitive so that they can be analysed in depth. A second feature should be a detailed description of the project, indicating the reasons for the project, the nature of the process, materials used and resources which will be required. This section will itemise those features most likely to affect the overall environment, the magnitude of these effects and the methods to reduce or eliminate them. The final section, which is sometimes produced as a separate report, is the analysis which uses scientific methods of evaluation and should be carried out by multi-disciplinary teams, experienced in both natural and social sciences.

Although some parameters are common to all sites, such as land use, and can be reasonably subjected to numerical and economic analysis, others, such as visual appearance, are unique to one and cannot wholly be dealt with in an objective manner and any assessment must include an element of subjective and qualitative judgment.

6.2 The Content of an Environmental Impact Statement

In general terms, the content of an Environmental Impact Statement has been concisely expressed by Mr J Calvert in a paper to the Institute of Civil Engineers in 1975. He posed it in the form of five questions:

(i) Why should the project be built and what alternative options have been evaluated?

(ii) When should the project be constructed and what long term planning has been carried out to verify this decision?

(iii) Where should the project be constructed and what alternatives were considered?

(iv) What are the effects of construction and operation upon the national, social and economic environment?

(v) How should the project be developed in order to ensure that the environment is reasonably protected?

In the case of the British Gas Industry, answers to the first two are usually provided by the overall Corporate and Economic Planning functions of the industry and form an essential part of the national Capital Development Programme which covers a rolling period of five years.

The third question is much more difficult to answer. The general location of a gas industry development is determined by the location of offshore gas fields and the technical requirements of the national transmission system as outlined above. These, however, only give a broad indication of areas in which suitable sites have to be found. The exact location which must, of course, be suitable from an engineering point of view, is often determined by environmental considerations and cannot be defined until the work required for answers to the fourth and fifth questions has been carried out. To obtain these, very detailed studies have to be carried out in a wide range of subjects. Although not all of these subjects may be appropriate in every instance, the questions have to be asked and the answers very carefully considered because values can change fundamentally from site to site. For example, Marram and Sedge grasses are a common feature of dune systems and contribute to their stability. If these dunes are essential for flood protection purposes, as in the Netherlands, then the grasses are a vital feature of the ecosystem and their protection is imperative. In other cases, they may be merely decorative and their significance is less. Some questions are more difficult to answer than others and require the expenditure of a great deal of time and effort. Some, it may be said, are incapable of being answered at all and must remain a matter of opinion. A visual impact study alone can produce a report as long as the rest of the analyses put together, whilst the monitoring programme required for atmospheric or water pollution assessment may cover a period of many years, starting long before construction commences and continuing during the life of the installation.

7. SITE IDENTIFICATION AND EVALUATION

It will have become clear from the information given above, that the selection of a site for a major industrial project is a rather more complex operation than the mere sticking of pins in a map or choosing the apparently cheapest site which is what the average objector seems to assume. It starts in the office with desk studies covering many aspects, of which only two will be considered here.

7.1 Geographic Considerations

In the preparation for any site selection process or for any environmental impact statement, no matter how simple, the restraints imposed by geography, both natural and man made, must be ascertained at the outset of the exercise. The first step in so doing is to prepare a series of sieve maps, on which are included all known constraints on development. These include existing and future land usage, designation and degree of protection, centres of population and, as far as possible, all the geomorphological information which may affect site location. This latter information would include such items as the nature of the coastline, sea bed conditions, the existence of peat bogs or marsh land and similar features. From these sieve maps, it is then possible to prepare a master map which designates "go" and "no go" areas and enables attention to be focussed on the former. If such a survey reveals no possible sites, as is likely in environmentally sensitive areas, then it is necessary to go back to the original site specification and decide what items, either technical or environmental, can be relaxed in order that a satisfactory solution is obtained.

Usually however, the search reveals more than one site which appears to be acceptable and then more detailed investigations have to be made. One by one the alternatives are rejected until one site appears as the optimum and for which planning consent application can confidently be made. The advantage of this procedure is that much of the argument that normally arises at this stage or even the public inquiry stage, if that is reached, can be eliminated by the publication of the studies of the alternatives that have been considered.

7.2 The Cost Benefit Equation

One aspect of the alternative sites or solution problem which must be examined is the major one of extra costs. To obtain value for money it is necessary to adopt the least cost solution. This may be thought to be stating the obvious, but a real difficulty lies in defining the true least cost when reviewing the project as a whole and taking national as well as gas industry factors into account. The true cost of any project can be divided between four cost centres, some of which are easy to evaluate and control, and some of which are more subjective and open to varying interpretation. They are:-

(a) Technical and engineering requirements;
(b) Environmental requirements and effects;
(c) Aesthetic considerations;
(d) The effect a current project may have on those in the future.

A least cost solution could possibly be prepared for

each of these in isolation, but the sum of these need not necessarily be the true least cost for the project since the requirements for each may often be mutually incompatible. The main difficulties arise when identifying all the costs that have to be included and then measuring these in terms which are meaningful. Cost centre (a) for example uses techniques which are well established and form an essential part of normal project management. The questions to be resolved here are those relating to levels of expenditure and standards to be adopted as explained above. These decisions, however, will affect the other cost centres which have to take into account the more subjective and often emotional factors that occur in our modern society. These can only be resolved by the use of the techniques mentioned previously. Then comes the extremely difficult task of using the results of these analyses to the best advantage by the extremely delicate balancing of costs and benefits. The situation is not hopeless, however, and the decisions which have been made were not arbitrary since it has been found possible to quantify, at least to some extent, the major imponderables and put fairly firm financial figures in the equation.

7.3 Cost Contours

When a number of alternative sites are available a detailed assessment has to be made of each one. Not the least of these assessments is the cost penalty involved in moving from the optimum site as determined by technical considerations. A good example of this type of problem is that of a compressor station. By definition such a station has to be on or adjacent to the pipeline at or near a predetermined point. If it is separated from the line, then connections have to be laid to and from the site with consequent increase in costs. These increases are not just those due simply to the cost of the lines, but must also include an increase in running costs. The extra length of the line required increases the total distance between the stations and there is therefore a larger drop in the gas pressure which, in turn, increases the fuel consumption of the unit. A similar result would occur if the station were merely moved from the optimum point. The varying nature of the sites themselves can affect costs considerably, for example on a hillside as against in the valley bottom, or on boggy moorland, or rocky outcrops rather than good quality agriculture land. By taking all these factors into account it is possible to prepare for each area of search a cost contour map showing the financial penalties applying to the various locations. This enables not only British Gas, but also the planning authority to make a realistic assessment of the merit or otherwise of specific sites. Such a cost contour map is shown in Figure 2.

8. OBSERVATIONS AND CONCLUSIONS

One of the main objectives of this paper has been to try and answer the question of whether the British planning system is working reasonably well in relation to those projects initiated by British Gas and whether any improvements might be made particularly in view of the proposed EEC Directive on Environmental Assessment of Industrial Projects.

British Gas has had an Environmental Planning Department for about 15 years and its experience and policy has been as follows. The Corporation does carry out environmental impact assessment on all its major above ground installations and uses it as a fundamental part of its decision making process. Also it believes in a controlled public participation procedure and issues as much information as is required to enable planning authorities to make a balanced decision on applications before them. Since 1967 the Corporation has carried out over 70 major projects, and within the last twelve months has received no fewer than eleven planning consents. The average time taken to obtain all of these consents over the fifteen year period and including the two public inquiries at Bacton and Hirwaun is under 26 weeks.

There have been two variations of the proposed procedure which have been adopted with some success by Dorset County Council and in Highland Regional Council. In both cases the Planning Committee held what was verbally a "mini public inquiry" at which the proponent and the objectors presented their cases to the committee meeting in public after which they considered the application either there and then or at a later meeting. All the necessary documentation had been prepared by the proponent and circulated in advance. It seems that this procedure has much to commend it and prevents the often unproductive confrontation that takes place in the more formal atmosphere of a Public Inquiry.

Finally, to sum up, the UK planning system has much to

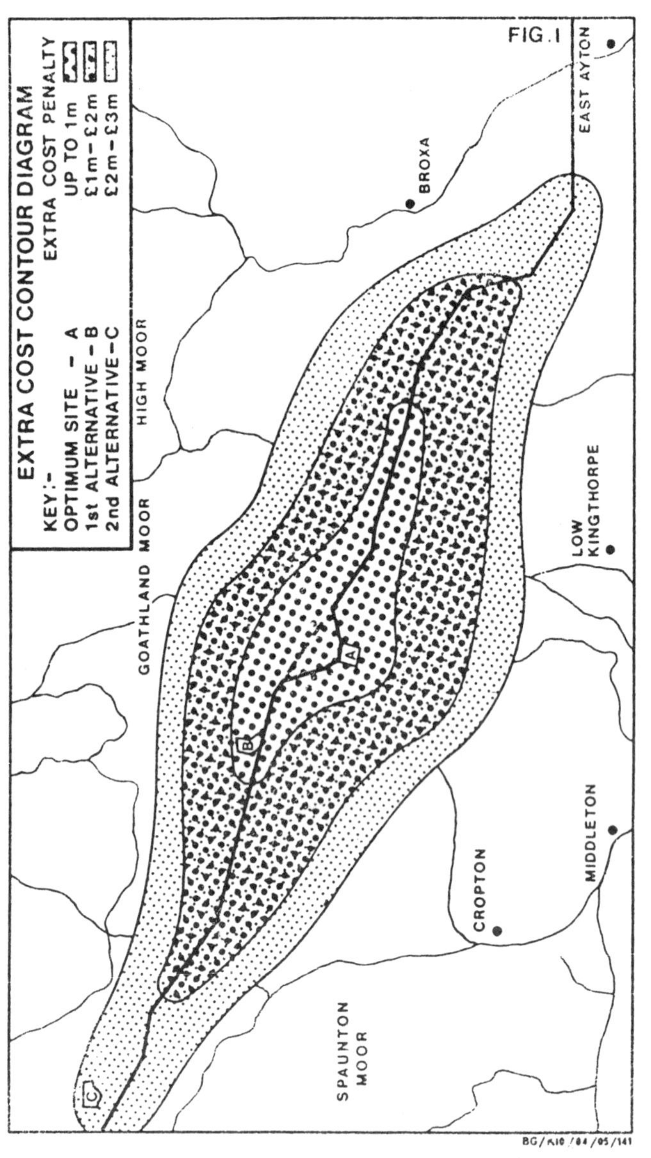

Figure 2 Extra Cost Contour Diagram

commend it, can work reasonably well and quickly and many of the delays often attributed to it are frequently caused by other influences. It is doubtful if the more formal approach to environmental assessment contained in the proposed Directive would in fact produce better solutions in less time than that introduced by existing UK legislation.

ACKNOWLEDGEMENT

The opinions contained in the paper are those of the author and may not represent official British Gas policy. The author wishes to thank the Corporation for permission to present this paper and also colleagues for help in its preparation, particularly Mr A Ward of the Construction Department, British Gas and Mr G Graham, Senior Partner of Architects Design Group.

REFERENCES

Calvert, J.: 1975, Informal discussion on Environmental Impact Analysis. Institute of Civil Engineers, 21st July 1975.

Catlow, J. & C.G. Thirlwell: 1976, Environmental Impact Analysis, Department of the Environment Research Report 11, London.

Dobry, G.: 1975, Review of the Development Control System, H.M.S.O.

AFFILIATION

F.E. Dean is now a consultant, Francis E. Dean and Associates and was formerly Chief Environmental Planning Officer at the British Gas Corporation.

Peter Hills

ENVIRONMENTAL ASSESSMENT OF COAL EXTRACTION PROJECTS: THE VALE OF BELVOIR EXPERIENCE

1. INTRODUCTION

The emergence and subsequent development of environmental assessment procedures and techniques in the United Kingdom have been closely associated with energy sector projects. The development of the UK's North Sea gas and oil resources from the late 1960s onwards created a significant demand for various types of onshore facility. Many of these facilities were complex and large-scale developments located on greenfield sites, often in areas of considerable natural beauty and a relatively high degree of environmental sensitivity. Given these factors it is not surprising that much of the experience of environmental assessment work in the UK that has been gained over the past ten years has been related to oil and gas development projects, a situation that has been extensively reviewed in the literature (see, for example, Clark et al., 1980).

The UK has, in the early 1980s, a comprehensive energy resource base. Clearly, oil and gas have been major growth industries over the past fifteen years but it is also the case that the future prospects for the UK coal industry are predicated on a substantial increase in production and use over the next twenty five years. Such an increase in coal production and use, which it should be emphasised is by no means a certainty, would necessitate a number of major new coal extraction projects. Given the nature of coal industry activities and the type and scale of environmental impacts that coal extraction can generate, it is perhaps inevitable that the industry's operations are receiving much greater scrutiny, both internally and externally. Some form of environmental assessment is now a standard component of National Coal Board (NCB) feasibility studies for new underground mining projects.

This paper deals with one such assessment that was carried out some five years ago in connection with the NCB's largest ever proposed new mining project. This project

B. D. Clark et al. (eds.), Perspectives on Environmental Impact Assessment, 429–450.

NOTTINGHAM
A46
Normanton
Bingham
A52
Bottesford
Sedgebrook
Barrowby
GRANTHAM
Granby
Cropwell Bishop
Cotgrave
NOTTINGHAMSHIRE
Plungar
Belvoir
VALE OF BELVOIR
Harby
Stathern
Knipton
A607
Hose
A606
Long Clawson
LEICESTERSHIRE
Croxton Kerrial
Saltby
LINCOLNSHIRE
A1
Scalford
Ab Kettleby
Coston
Saxby
Asfordby
MELTON MOWBRAY
A46
County Boundary
Coalfield Area
Proposed Mining Sites
Railways
Disused Railways
Major Roads
Scale 0 1 2 3 4 5 6
Kilometres
VALE OF BELVOIR
MINING PROPOSALS

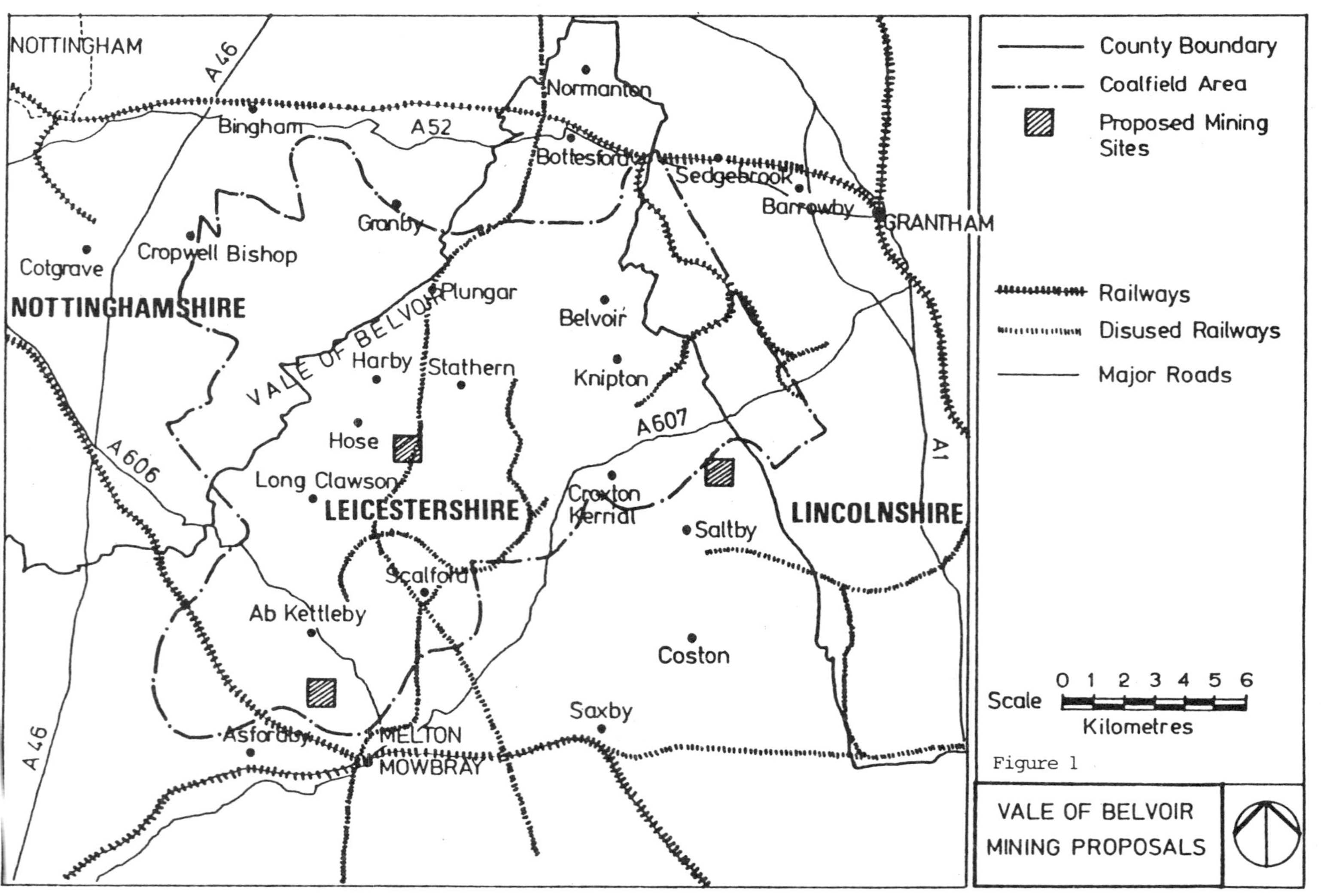

Figure 1

would involve the exploitation of the North East Leicestershire, or Vale of Belvoir, coalfield, located approximately 25 km to the south-east of Nottingham, in the East Midlands region of England. The status of this particular proposed development has yet to be finally resolved for reasons that will be explained in a subsequent section. This means that it is not yet possible to provide a comprehensive evaluation of the role that environmental assessment has played in the planning process. The paper will therefore focus on the procedures and techniques employed in the assessment exercise and will subject these to a critical evaluation.

2. THE BELVOIR COALFIELD DEVELOPMENT PROPOSALS

Prior to discussing the analytical procedures used in the impact assessment of the proposed project, it is useful to provide some limited background information on the nature and scale of the development. Detailed exploration of the coalfield commenced in 1974. By 1977 it had become clear that the Belvoir coalfield represented a very large energy resource. It is reputed to be the largest coal find in Western Europe, with recoverable reserves in excess of 500 m tonnes, underlying an area of over 230 km^2 of mainly agricultural land in north east Leicestershire, south Nottinghamshire and west Lincolnshire (Figure 1). The NCB commissioned two feasibility studies on the development of the field and these were published in 1976 and 1977. In this paper the discussion will centre on the assessment that was undertaken as a part of the second feasibility study (Leonard and Partners, 1977).

The NCB submitted planning applications for the scheme in August 1978. The Board were seeking permission to construct three new deep mines and related works at Hose, in the Vale of Belvoir, at Asfordby, near Melton Mowbray, and at Saltby, near Grantham. When fully on-stream, these mines would produce 3 m, 2.2 m and 2 m tonnes of coal respectively, per year, yielding an annual total of 7.2 m tonnes of saleable product. The NCB's original proposals provided for the mines at Hose and Asfordby to be constructed simultaneously, with work on the Saltby mine commencing four years later. The total construction period for the three mines and related works such as coal preparation plants, workshops and stores, would be 13 years. Land take for construction purposes would amount to a total of 100 ha. for the three sites.

A major problem associated with the coal deposits in

the area is the amount of waste material in the coal seams. Approximately 30% of the total material extracted would consist of waste rock. A total of 2.72 m tonnes of such material would be produced each year and would have to be separated from the saleable product and then disposed of in an appropriate manner. The NCB's preferred disposal option for the waste would follow standard UK practice of surface tipping on sites adjacent to the coal mines. Tips would be progressively restored to agricultural use and would also be used to screen mine sites to reduce visual and noise impacts. The land required for tipping under the NCB's proposals would amount to almost 700 ha. in total, although the NCB always maintained that less than 10% of this total would be out of agricultural at any one time.

The average labour force required during the construction phase of the project was estimated at 850 persons. The permanent mine workforce would be built up progressively, commencing in the third year of construction with just 50 men at Hose and Asfordby. By the 13th year of the construction period, as the mines were all fully on-stream, this workforce would have grown to 3800 men, together with 300 managerial and clerical staff. Data relating to the key impacts of each proposed mine site are presented in Table 1. The background to and nature of the NCB's proposals are described in more detail in Herrington and Hamley (1978), HMSO (1982) and Lewis (1978).

Characteristics	Asfordby Mine	Hose Mine	Saltby Mine	Totals
Annual Average Output of Saleable Coal	2.2 m tonnes	3 m tonnes	2.0 m tonnes	7.2 m tonnes
Annual Output of Waste	0.55 m "	1.5 m "	0.67 m tonnes	2.72 m tonnes
Total Mine Site Area	176 ha	347 ha	300 ha	823 ha
Waste Tip Area (50 years Production)	114 ha	272 ha	230 ha	616 ha
Total Mine Workforce	1100	1550	1150	3800

Table 1: Belvoir Prospect: Main Characteristics of Proposed Mine Developments.

These, then, were the main features of the proposals that formed the basis of the NCB's planning applications submitted in 1978. In January 1979 the Secretary of State for the Environment 'called in' the applications to determine them himself following a public inquiry. This public inquiry commenced in October 1979 and ended, after 84 days of formal proceedings, in early May 1980. The inspector responsible for chairing the inquiry submitted his report to the minister in late 1980. At the end of March 1982, the minister announced that he had decided to refuse the NCB planning permission to develop the coalfield on the basis of the proposals examined at the inquiry. His decision was based on a number of significant considerations but one of the most important was undoubtedly the scale of the environmental impact of local tipping of waste from the mines. In mid-1982 the NCB, government departments and local planning authorities entered into discussions concerning the nature and scale of revised planning applications for the coalfield. There is a strong possibility that the NCB may seek approval for a rather less ambitious scheme involving the construction of only one new coal mine, to be located at Asfordby.

3. THE BELVOIR IMPACT ASSESSMENT

As noted in the previous section, a feasibility study, dealing with both the underground and surface aspects of the development of the coalfield, was published in 1977. The recommendations presented in this study - the 700 page 'Belvoir Prospect' - formed the basis of the NCB's feasibility study were to formulate development proposals that would:

(a) be economically and operationally efficient;
(b) produce a safe and secure working environment;
(c) be the most acceptable in environmental terms.

It was necessary therefore for possible development strategies to be evaluated in relation to:

(i) mining factors;
(ii) surface operational factors;
(iii) environmental factors.

The consultants' task was essentially to identify "where viable mines could be sited in an acceptable manner". The degree of acceptability of alternative sites would therefore depend on reconciling underground and surface operational factors with environmental considerations. To achieve this requirement the consultants adopted a three stage process:

(i) A sieving exercise to identify suitable locations

within the overall study area for the establishment of mines or spoil tips.

(ii) An impact analysis of alternative mine and/or tip complexes to select those causing the least overall impact and as an aid in the design of selected alternatives.

(iii) A comparative exercise bringing together surface and underground factors to identify an acceptable means of exploiting the reserves.

3.1 Constraints on the Assessment

Before describing in detail certain aspects of the methodology of the assessment exercise, reference should be made to two important constraints which strongly influenced the scope and form of the analysis employed in the feasibility study.

Firstly, certain types of impact were specifically excluded from the consultants' brief for the study. Impacts falling into this category included manpower requirements, housing and social impacts of the proposed project, infrastructure requirements, subsidence, various economic impacts and aspects of the waste disposal problem. These impacts were excluded for various reasons. Some, like housing and infrastructure, were regarded as matters requiring consultation between the NCB and local planning authorities and this liaison did in fact take place through the mechanism of a joint working party, which met regularly over the two year period prior to the public inquiry. Certain other impacts, including subsidence, had still to be investigated in detail by NCB specialists at the time the environmental assessment was being prepared.

The second major constraint on the assessment exercise reflected the influence of underground mining considerations on the mine site selection process. Detailed exploration and proving of the coal seams had indicated that it would be operationally desirable to exploit the coalfield from at least three points. This meant, in effect, that there would have to be a mine in each of the three zones of the field; south, central and east. This produced some difficulties in the assessment as it limited the range of viable mine site locations that could be evaluated in the exercise.

The consultants' assessment was, therefore, limited in scope from the very outset and, not surprisingly, it could never be regarded as a comprehensive appraisal of large-scale and complex development proposals. Mining factors constrained the form of the environmental assessment at a very early stage and were to have a very powerful influence on the outcome of the exercise.

3.2 Methodology of the Assessment

The first stage of the assessment involved a sieving exercise. This has been described in some detail elsewhere (Lewis, 1978; Williams, Hills and Cope, 1978) and will only be briefly summarised here.

A range of restraints considered to indicate 'more preferred' or 'less preferred' areas were plotted on a 0.5 x 0.5 km map grid. Three categories of restraints were considered; operational, terrestrial and activity factors. Operational restraints included road and rail access and service availability or conflict with major services. Terrestrial factors included ground slopes greater than 1 in 25, significant wildlife and vegetation habitats to be avoided, areas of high agricultural land value and difficult land (marshy areas, streams) to be avoided for engineering reasons. Activity factors related to the proximity of settlements - buffer zones of 1 km for mines and 0.5 km for spoil tips were used in the analysis - to operations, areas of high landscape value and statutory designation (conservation areas, Sites of Special Scientific Interest). Areas of potential conflict between mining and the existing environment could be readily identified by superimposition of the various restraints on a set of maps.

Composite restraint maps were used to examine the possibility of siting mines at locations which would be 'ideal' from the viewpoint of underground factors, especially the distribution of the coal reserves. The three site constraint entered the analysis at this stage. It was found however that the ideal locations in each of the three coalfield zones conflicted badly with the mapped restraints. The consultants argued that the ideal mining locations would therefore be unacceptable in terms of surface environmental factors.

The sieving exercise provided a generalised framework for a preliminary assessment of possible environmental conflicts. Subsequent analysis was geared to a more detailed examination of the impact of specific mine site options and development strategies. Operational factors permitted the identification of 9 mine site options. These were based on

Table 2: Method of calculation of the visual impacts of pithead facilities, on the existing resident population.

Assumptions - the 3 km visual range of the top of a 62.5 m winding tower or top of a 30 m coal preparation plant was chosen to represent the worst case for visual exposure.
- maximum area within 3 km zone is 2825 ha.
- maximum population in study area may be achieved with a mine in the vicinity of Asfordby
- potentially exposed poulation at Asfordby 13750 but assumed maximum set at 15000

Calculation - impact is expressed as a % calculated on basis of:
area within visual range of tower (calculated by site intrusion analysis)(At)/ total area within 3 km zone (A) + settlement population within visual range of tower (Pt) / maximum possible population in study area within 3 km zone (P)

Thus,

$$\left(\frac{At}{A} \times 5\right) + \left(\frac{Pt}{P} \times 5\right) = 10 \text{ maximum} \times 10 = \text{\% visual impact}$$

Example:

Asfordby $\left(\frac{2018.8}{2825.0} \times 5\right) + \left(\frac{13753}{15000} \times 5\right) =$

$$3.57 + 4.58 = 8.15 = 81.5\%$$

Source: Leonard and Partners (1977)

various types of mine configuration (total mines, where all required activities are carried out and satellite mines, which have limited operational capabilities), coal preparation facilities and tipping arrangements in various combinations. From these 9 options, 6 development strategies were identified, each strategy involving one mine in each zone to meet the previously mentioned constraint. Mine site options and development strategies were subject to environmental assessment.

3.3 <u>Selection of Impacts</u>

The NCB's consultants argued that they selected a range of 'significant' impacts on which to assess the environmental implications of the proposed development. Mine site options were subject to an assessment which incorporated only four factors:

(i) visual impact;
(ii) noise impacts;
(iii) landscape loss;
(iv) agricultural loss.

Development strategies were assessed using two additional impact categories:

(v) road capacity;
(vi) traffic intrusion.

Each of these impact categories will not be discussed and the method of calculation of levels of impact will be illustrated.

(i) <u>Visual Impact</u> This form of impact was analysed with regard to three types of effect:

- (a) exposure of the resident population to proposed mines;
- (b) exposure of the landscape and built fabric;
- (c) exposure of motorists and vehicle passengers.

All categories of visual impact were expressed in % terms in relation to the worst (hypothetical) case obtaining in the study area. Thus, the higher the calculated % value the higher the level of adverse impact. The method of calculation of the visual impact of pithead facilities (winding towers or coal preparation plants) is presented in Table 2.

(ii) <u>Noise Impact</u> Noise impacts for both the construction and production phase of the proposed development were calculated. These impacts were estimated in relation to the exposure of the residential population. In the case of the construction phase calculations related to the daytime worst

Table 3: Method of calculation of the loss of landscape quality due to the adverse visual impact of mine buildings.

Assumptions
- the 3 km zone of visual impact was used
- maximum affected area within zone could be 2825 ha.
- for worst case calculation it is assumed that all landscape within the potentially affected 3 km area is of the highest quality and therefore weighted x3
- Class 1 landscape x3
  Class 2 landscape x2
  Class 3 landscape x1

Calculation - example of Hose mine site

Area of Class		
	1 landscape	850 ha.
	2	737 ha.
	3	325 ha.

expressed as % of worst case:

$$\frac{(850 \times 3) + (737 \times 2) + (325 \times 1)}{2825 \times 3}$$

$$\frac{2550 + 1474 + 325}{8475} = \frac{4349}{8475} = 51\% \text{ impact}$$

Source: Leonard and Partners (1977)

case and the level of impact was expressed as a % derived from:

$$\frac{\text{Total population within 50 dB(A) contour}}{\text{Total maximum population achievable within 3 km zone}}$$

Production phase noise impacts, also expressed in percentage terms, were calculated on the basis of exposure to noise emanating from both the mine site and associated rail facilities.

(iii) Landscape Impact This measure was used to assess the visual impact of pithead facilities on the quality of the landscape. Three categories of existing landscape value were used and the amount of land falling into each of these categories within the 3 km zone surrounding possible mine sites was calculated. Weighting values of 3, 2 and 1 were applied to these categories of landscape value in declining order of quality. As in the previous impact measures, a worst case % assessment was used. An example of the method of calculation is presented in Table 3. The use of this particular measure is somewhat dubious in view of the fact that it closely resembles the measure used to calculate the visual impact of mine buildings on the built fabric and associated landscape mentioned above. The landscape impact measure does utilise three categories of landscape quality (compared with just two of the same classes used in the earlier measure) but does not incorporate components relating to areas of conservation and special scientific interest. Nonetheless, the two measures of impact (built fabric and associated landscape and landscape) are certainly highly correlated and appear to be measuring much the same thing. In both cases of visual impact it must be emphasised that the assessment deals only with the predicted effects of mine buildings and fails completely to take into account the much more dramatic and controversial issue of mine spoil tips and their impact on the visual quality of the environment.

(iv) Agricultural Impact The objective of this component of the overall assessment was to indicate the cost implications of utilising alternative sites for mine construction and waste tipping. The mode of assessment reflects an attempt to take into account differences between different sites in terms of their suitability for restoration. Although the consultants' report presents an extensive and detailed costing exercise the actual basis of the derived figures is somewhat unclear. What the summary data for agricultural impact attempt to show is the loss of agricultural productivity due the presence of the coal mines. There are various problems with this particular analysis. For example, it is assumed that mining occurs only over a fifty year period, when even at the time of preparation of the environmental assessment it was known that the proposed mines could have a life expectancy in excess of 75 years. It is also assumed that the productivity levels of restored waste tips will match that of land prior to tipping. In order to present their data in the form of a % of the worst case a quite arbitrary figure is used to set the worst possible case level of 100%. Such is the complexity and lack of clarity in the explanation of this part of the overall assessment that it is extremely difficult to establish just what is being measured and how it is being measured.

(v) Road Capacity/Usage Limitation Impact This component of the assessment is an attempt to consider possible problems on the transport network associated with mining strategies rather than individual mines in isolation from one another. Traffic intensities or overloads were measured in terms of flows above the theoretical capacity of the network or above the flow projection for 1995. Both goods vehicles and private vehicles were considered in the analysis. The major problem with this aspect of the assessment would seem to arise in assignment process, when vehicle movements are allocated to specific links in the network. The heavy goods vehicle component associated with the mine locations was, by the consultants' own admission, allocated in a very simple fashion - it was assumed 'that 50% of the traffic would go to the nearest main town and that the remainder would gravitate towards the south'. Private vehicle movements to and from the mine sites were assigned to the network by a standard transportation gravity model. The inherent weakness of this part of the overall assignment procedure was its dependence on various assumptions regarding the distribution of miners' housing within the study area. It was noted above that housing provision was specifically excluded from the consultants' brief and yet for the purposes of assessing both types of transport-related impact it was necessary to derive some hypothetical spatial allocations for housing. These allocations necessarily involved estimates of total housing demand as well as location of future housing provision for the workforce. Clearly, if the project were to have been approved and the housing provision did not match the spatial distribution hypothesised in the consultants' impact assessment then the projected levels of both types of transport impact are meaningless.

(vi) Traffic Intrusion Impacts This represented an attempt to estimate the impact on local settlements in terms of noise and other hazards associated with road vehicles. The assessment related to both the construction and production phases of proposed development. As in the case of the previously mention categories of impact, the intensity of adverse environmental effects was presented in % terms. The assessment procedure for traffic intrusion involved the identification of settlements that would be likely to experience an increase in total traffic of at least 25% due to the presence of a mine and also those settlements experiencing an increase of 50% or more in heavy goods vehicle movements associated with either the construction or production phases of the development. Comparisons between the different strategies were made on the basis of the numbers of people in existing settlements that would be exposed to increased traffic flows. The results of the analysis were presented for the production phase of the project and for the construction and production phases combined. The latter results involved the weighting of impacts of the two phases. Production phase impacts were deemed to account for 90% of the overall impact and construction phase effects for just 10%. The assumptions underlying this weighting are far from clear and it could be argued that the initial impacts of a dramatic increase in traffic movements, especially heavy goods vehicles, should be weighted more heavily.

3.4 Option and Strategy Assessments

The impact assessment described above yielded a series of quantitative measures of the intensity of certain types of adverse environmental effect for various mine site options. These data were initially presented to assess the relative merit of the different options. At this stage of the analysis, surface cost factors were also introduced into assessment and so comparisons reflect both environmental and cost implications of the proposals. The mine options and strategies compared in the assessment are listed in Table 4. Tables 5 and 6 illustrate the way in which the option and strategy assessments were presented in the consultants' report. An important point to emphasis about these analyses is that only one possible mine site option was regarded as

Table 4 Belvoir Mine Options & Strategies

POSSIBLE MINE SITES	
ASFORDBY	south
HOSE LANGAR COTGRAVE	central
SALTBY STENWITH NORMANTON	east

OPTIONS	Option No.		
2M Tonne Mines	1	ASFORDBY	total mine
	8	SALTBY	total mine
	9	STENWITH NORMANTON	overland conveyor to coal preparation plant
3M Tonne Mines			
	2	HOSE	total mine
	3	LANGAR	total mine
	4	LANGAR HOSE	total mine satellite mine
	5	LANGAR HOSE	mineral outlet service mine
	6	HOSE LANGAR	total mine overland conveyor to coal preparation plant
	7	COTGRAVE HOSE	mineral outlet service mine

STRATEGIES				
Strategy	Option Nos.	South Mine	Central Mine	East Mine
A	1,3,8	ASFORDBY	LANGAR	SALTBY
B	1,2,8	"	HOSE	"
C	1,6,8	"	HOSE LANGAR	"
D	1,5,8	"	LANGAR HOSE	"
E	1,4,8	"	LANGAR HOSE	"
F	1,7,8	"	COTGRAVE HOSE	"

Source: Williams, Hills & Cope (1978)

Table 5 Performance of Belvoir Mine Options

These factors are summarised below by relating the score for each factor as Above Average (+), Average (0), Below Average (-), for the group as a whole.

Surface Consideration	2M tonne mines Option 1	8	9	2	3	3M tonne mines Option 4	5	6	7
Visual	-	+	0	+	+	0	0	0	-
Noise	-	0	+	-	0	0	0	0	+
Agriculture	0	+	-	-	+	+	+	+	-
Landscape Loss	+	0	-	+	+	-	-	0	0
Cost	0	+	-	+	+	-	-	0	0
Total +	1	3	1	3	4	1	1	1	1
0	2	2	1	0	1	2	2	4	2
-	2	0	3	2	0	2	2	0	2

2M tonne mines		3M tonne mines	
top	+	1st & 2nd	+
middle	0	3rd & 4th	0
bottom	-	5th & 6th	-

By summarising the factors in this manner it is assumed that all the factors are of equal weight and importance. This is obviously not so, but the table can be used as an indicator.

Accepting this limitation for this purpose, the following order of merit can be deduced:

South - Option 1 - no choice

Central - Option 2 & 3 appear to be better than the other options. Option 3 being on surface grounds slightly better than Option 2. The next best would be Option 6.

East - Option 8 appears significantly better than Option 9.

Source: Leonard & Partners (1977)

Table 6 Performance of Belvoir Mine Strategies

Strategy	A	B	C	D	E	F
Environmental Impact						
Visual	+	+	0	0	0	–
Noise *note: equal 2nd	*0	–	*0	*0	*0	+
Agricultural loss	+	–	+	+	+	–
Landscape loss	+	+	0	–	–	0
Transportation Impact						
Road capacity/usage *Note: Approx equal 2nd	+	*0	*0	*0	–	–
Traffic intrusion *note: equal 1st	–	*+	*+	*+	–	*+
Surface cost	+	+	0	–	–	0
Total +	5	4	2	2	1	2
0	1	1	5	3	2	2
–	1	2	0	2	4	3

Source: Leonard & Partners (1977)

feasible in the southern zone of the coalfield (at Asfordby). It also became clear that only one feasible option really existed in the eastern zone, where the site at Saltby appeared to consistently outperform the alternative Stenwith/Normanton option but particularly in cost terms. Thus, the major contribution that the analysis could make related to the selection of a possible mine site in the central zone of the coalfield.

By the time the assessment reached the strategy evaluation stage the mine sites in the southern and eastern zones of the study area had already been determined and the mines at Asfordby and Saltby consequently figured in each of the 6 assessed strategies. The strategy assessment revealed that:

(i) all strategies has a considerable impact on the environment;

(ii) no individual strategy was clearly superior on all surface factors (environment and cost);

(iii) no single strategy had consistently better results in the environmental assessment.

In order to further explore the performance of the various strategies the consultants then decided to carry out a weighting exercise in which the relative significance of the different types of environmental impact was hypothesised and built into the analysis. Where impacts had been assessed in population terms (i.e. in relation to the size of the potentially affected population), weights on a scale of 0-100 were applied (0 = imperceptible, 100 = totally unacceptable). Agricultural impacts were assigned 15% of the total weighting and surface cost factors 25%. The weighting values eventually arrived at after a series of calculations were as follows:

Visual	17
Noise	12
Landscape	17
Road capacity	7
Traffic intrusion	7
Agriculture	15
Total environment	75
Surface cost	25
Total	100

The previously derived scores were converted to a common scale length and were then multiplied by the weighting factors to obtain new values which were then summed for each strategy. It transpired that the strategy performance was relatively insensitive to the weighting process and the consultants concluded that there was no significant alteration in the rankings with the weighted values.

3.5 Strategy Selection

The only unresolved question at the strategy selection stage of the assessment was the location of the proposed mine in the central zone of the coalfield. Three preferred strategies were identified on the basis of, firstly, surface considerations, and secondly, underground mining considerations. In terms of surface factors, the preferred alternative was strategy A (see Table 4). This strategy, however, was evidently not viable as a mining proposition, which makes it rather tempting to ask why it was included in the first place! The second ranked surface strategy was B, which was the first ranked underground solution. Strategy C was a 'poor third' as a surface solution but was ranked equal to B on underground criteria. The strategy selected by the consultants was in fact strategy B and this formed the basis of the NCB's planning applications. The rationale for this was basically that even though strategy A was the preferred surface solution it would take 18 months longer to achieve full production with a total mine at Langar and a satellite mine at Hose rather than a total mine located at Hose. Costs would also be significantly greater (approximately 30%). As a satellite mine would be needed at Hose anyway it would not have been possible to avoid any development at that sensitive location and it was really a question of whether this development should be a total or satellite mine. The consultants opted for the cheaper Hose total mine solution embodied in strategy B.

3.6 Weaknesses of the Assessment

The Belvoir assessment is distinctive, if not unusual, in the context of UK experience in the field of EIA. It was the first assessment of a major new deep mine project in the UK. It was based on a highly quantitative approach to the assessment task and few, if any other, UK studies have tackled EIA in this way. Although it may have been a novel approach to assessment it also served to reveal many of the weaknesses associated with an over-reliance on the quantitative presentation of impact measures. It did, however, have other significant deficiencies. Overall, the weaknesses of the exercise may be summarised as follows:

(1) Lack of comprehensiveness - this may in part be attributed to the nature of consultants' brief but it is still doubtful that a major project of the size and complexity of the Belvoir coalfield scheme can be effectively assessed using only 6 types of 'significant' impact.

(2) A failure to present clear and concise baseline studies of the existing environmental conditions.

(3) Weak conceptualisation of the impact categories employed in the study as, for example, in the apparent duplication of measurement in the case of visual impacts on the quality of the landscape discussed previously.

(4) Arbitrary selection of worst case situations, all of which were in any case hypothetical and provided no direct comparison with existing conditions - it is doubtful whether a % of a hypothetical worst case has very much meaning at all.

(5) Lack of clarity in the explanations provided for the method of calculation of various categories of impact - it is frequently far from clear how the results of parts of the analysis are actually derived.

(6) Poor presentation methods - aspects of the environmental assessment are dealt with in three separate parts of the consultants' report thus making it difficult to relate tabular presentations with data presented in map and diagramatic form.

(7) Related to the above point, there is considerable repetition in the consultants' report with the same material relating to the environmental assessment appearing several times - although the descriptions of impact measures and methods of calculation do not always convey the same meaning.

(8) Poor quality of the written report - it is too long, lacking in clarity and at times contradictory and inaccurate in terms of expression and there are various errors in the data used in the calculations.

(9) A failure to clarify the relationship between the environmental assessment of surface factors and underground mining and surface cost factors - the rationale for the trade offs between environmental

considerations and mining and cost factors is far from clear.

(10) A failure to adequately account for the application of such an elaborate procedure when it would appear the range of feasible options for mine site locations was so severely constrained.

These are just some of the criticisms that may be directed at this particular assessment and they are by no means exhaustive. It would not be overstating the case to suggest that one impression to be gained from examining the assessment is that it represents a rather elaborate (and costly) justificatory exercise, where decisions in principle had already been made. There is little evidence to suggest that the assessment made a major contribution to the design of the project and it is certainly ironic that one of the most significant impacts excluded from the assessment proved to be the single most important factor in the minister's decision to refuse the NCB permission to develop the coalfield. The impact concerned was waste disposal from the mines and its impact on the quality of the visual environment.

If the Belvoir assessment has made any contribution at all to the development of EIA work in the UK then it has probably been in rather more of a negative than positive sense. Quantitative approaches to EIA seem to have had a poor reception in general. Clearly, measurement of baseline conditions and the prediction of potential levels of impact are important dimensions of the overall assessment task. However, it is imperative that the conceptual and methodological bases of such work are presented in an explicit manner, and within a framework that is logically consistent. There are obviously additional problems involved when indices are weighted and the rationale of weighting procedures tends to be poorly explained.

The lessons that can be drawn from the Belvoir experience are both methodological and procedural/administrative. On the methodological side the experience suggests that a totally quantitative approach to EIA is extremely demanding of the analysts involved and may not be an appropriate mechanism for revealing the nature and extent of possible adverse changes in the existing environment. In fact, quantification may obscure rather than illuminate such changes. Despite the many pages of calculations, maps, diagrams and text, the Belvoir impact assessment certainly does not present a clear and concise appraisal of the environmental changes that would have resulted from the project had it been approved in its original form. On the procedural/administrative side, it is a good illustration of the perhaps unnecessary difficulties and analytical shortcomings that can occur in a situation where intending developers are not provided with quite explicit guidance on what is expected in an assessment and where they, or their consultants, work in isolation from environmental planners and other professionals in local planning agencies. It is certainly an indictment of any kind of regulatory planning system that a major public project on the scale and of the complexity and cost (£800 m in 1977) of the Belvoir coalfield should be determined in the absence of a comprehensive and rigorous environmental assessment.

Editorial Note: Following the Secretary of State for the Environment's refusal to grant planning permission for the original plan, the NCB submitted revised proposals in June 1982. In October 1982 Leicestershire County Council voted to accept the revised proposals, agreeing to the principal mine being at Asfordby, the least environmentally sensitive of the earlier three-pit plan. There has not been another public inquiry despite the proposals being a departure from the Country Structure Plan. A Government backed Working Party set up in mid 1982 to look at the issue of waste disposal reported in October 1983, recommending three sites as worthy of serious consideration for remote disposal of colliery spoil. The three sites (one being a limestone quarry and the other two, worked-out brick pits) are argued as being technically feasible and environmentally worthwhile, providing the best surface disposal solution on environmental and agricultural grounds. (Remote Disposal Working Party, 1983).

REFERENCES

Clark, B.D., R. Bisset and P. Wathern: 1980, Environmental Impact Assessment: An Annotated Bibliography, London, Mansells.

Herington, J. and W. Hamley: 1978, 'The price of coal in Belvoir', Geographical Magazine, 50, pp. 307-314.

HMSO: 1982, Report of the Vale of Belvoir Coalfield Public Inquiry, London, HMSO.

Leonard and Partners: 1977, The Belvoir Prospect, Surface Works Report to the National Coal Board (South Nottinghamshire/South Midlands Areas).

Lewis, J.J.: 1978, 'Progress in the development of the North East Leicestershire Coalfield', Colliery Guardian, 226, (8), pp. 501-521.
Remote Disposal Working Party: 1983, Final Report, Department of the Environment, London.
Williams, K., P. Hills and D. Cope: 1978, 'EIA and the Vale of Belvoir', Built Environment, 4, (2), pp. 142-151.

AFFILIATION

Dr Peter Hills was formerly a lecturer at the Institute of Planning Studies, University of Nottingham, England and is now a senior lecturer at the Centre of Urban Studies and Urban Planning, University of Hong Kong.

Peter Nelson

ASSESSMENT OF WATER RESOURCE DEVELOPMENTS

1. INTRODUCTION

This paper reviews some of the lessons learnt from a recent study into the environmental effects of water storage schemes in North West England. It indicates the important types of impacts which occur and those impacts which should be considered in Environmental Impact Assessment (EIA). Particular emphasis is placed on organization and methods because the author considers that it is the way in which expertise in EIA was used in the case study, rather than the extent of the knowledge itself, which is likely to be of greatest interest and application in other countries.

The North West Water Authority (NWWA) Regional Water Resource Studies provides the case study, being one of the first EIAs of its kind to be conducted in England (EIA procedures have been more widely used in Scotland as a result of the studies required for North Sea Oil Development). It has been described as a pioneering approach (Anon, 1979) and welcomed by many professional and private organizations. Although large by UK standards, the size and scale of this project would be dwarfed by many existing and proposed water storage schemes in other countries. The magnitude of individual environmental impacts would also be small, though comparable in type with those arising from similar development elsewhere in the world. It is argued, however, that the study provides many examples of the benefits of adopting the EIA method of approach in project design and evaluation, and that the broad principles can be applied equally successfully in developed and developing countries alike.

2. BACKGROUND TO WATER RESOURCE PLANNING IN BRITAIN

In order to put the NWWA study in perspective, a few introductory remarks are appropriate about the way in which water resources are administered in the United Kingdom. Since 1974, responsibility for the planning of water resource development in Britain has rested largely with the Regional

B. D. Clark et al. (eds.), Perspectives on Environmental Impact Assessment, 451–478.

Water Authorities of which there are nine in England, one in Wales and five in Scotland.

Each regional water authority is required to prepare plans showing how it will administer, develop and improve water resources (covering the whole hydrological cycle) to meet future requirements. These plans are submitted to the Secretary of State for the Environment and the Minister of Agriculture, Fisheries & Food for formal approval. The topics covered are:

- The conservation, augmentation, distribution and proper use of water resources and provision of water supplies;
- Sewerage and treatment, and disposal of sewage and other effluents;
- Restoration and maintenance of the wholesomeness of rivers and other inland waters;
- The use of inland water for recreation;
- Enhancement and preservation of amenities in connection with inland waters;
- Use of inland water for navigation;
- Land drainage and fisheries management in inland and coastal waters.

Although regional water authorities have considerable autonomy in preparing plans for water resource development, any new proposals for development can only proceed following the confirmation of a Water Order, or the enactment of a private Act of Parliament, together with the successful completion of standard planning procedures. In Britain, any major proposal for development by either private or public bodies is subject to careful scrutiny guided by planning legislation, and it is normal for such proposals to be considered at a public inquiry. As a result, environmental issues are receiving increasing attention in the formative stages of project design.

3. THE NORTH WEST REGION AND ITS EXISTING WATER SUPPLY

North West England is a very diverse region in both a physical and economic sense. North of the River Ribble much of the land is mountainous and subjected to high rainfall (1500-2000 mm per annum). Soils are often poorer than in the south, there is also a high proportion of moorland and hill country in which livestock farming and forestry are the predominant land uses. Fewer than one million people live in this northern zone, much of which is designated either as National Parks or Areas of Outstanding Natural Beauty. South of the Ribble, the lowland plains are more fertile, although exploitation of coal and other natural reserves during the industrial revolution has led to the growth of major conurbations around Liverpool and Manchester where more than 6 million people live.

This southern zone creates the major demand for both domestic and industrial uses, and the annual consumption in the region is around 2,400 megalitres per day (Ml/d). Present supplies are obtained largely from inland catchments feeding reservoirs in the north and east of the region, although some is also imported from the Welsh Mountains. A small amount, approximately 10% of the total, is pumped from underground aquifers in the central belt. Scope for any significant expansion of output from these underground sources is, however, restricted by saltwater incursion of the coastal aquifers, siltation of boreholes, and penetration of nitrates and tip leachates. The major source of future water supplies for the region is, therefore considered by the Water Authority to lie in further development of the northern river catchments (Oldfield, 1978).

4. EVENTS LEADING UP TO THE NWWA STUDY

Following its formulation in 1974, the NWWA began detailed studies covering both the supply of, and demand for, water within its region. The authority decided to investigate alternative sources which would be capable of meeting different levels of projected demand should the need arise, recognizing that forecasts of demand for all natural resources are subject to major uncertainties. More than 30 schemes proposed by its predecessor authorities were considered, each with a capacity to provide by itself an increase of between 10 to 30 percent in the overall supplies of the region (that is a yield of between 300 and 900 Ml/d). From these 30 schemes a shortlist of 4 was selected, each of which had different characteristics (Figure 1).

Three of the options had in common, the storage of water in an inland reservoir, and release of supplies to augment natural low flows in "regulated" rivers which would convey the water towards the main centres of demand in the south of the region. Here the similarity ended, since 1 scheme involved increasing the use of an existing reservoir by pumped storage at Haweswater in the English Lake District; another required the construction of a new reservoir in the

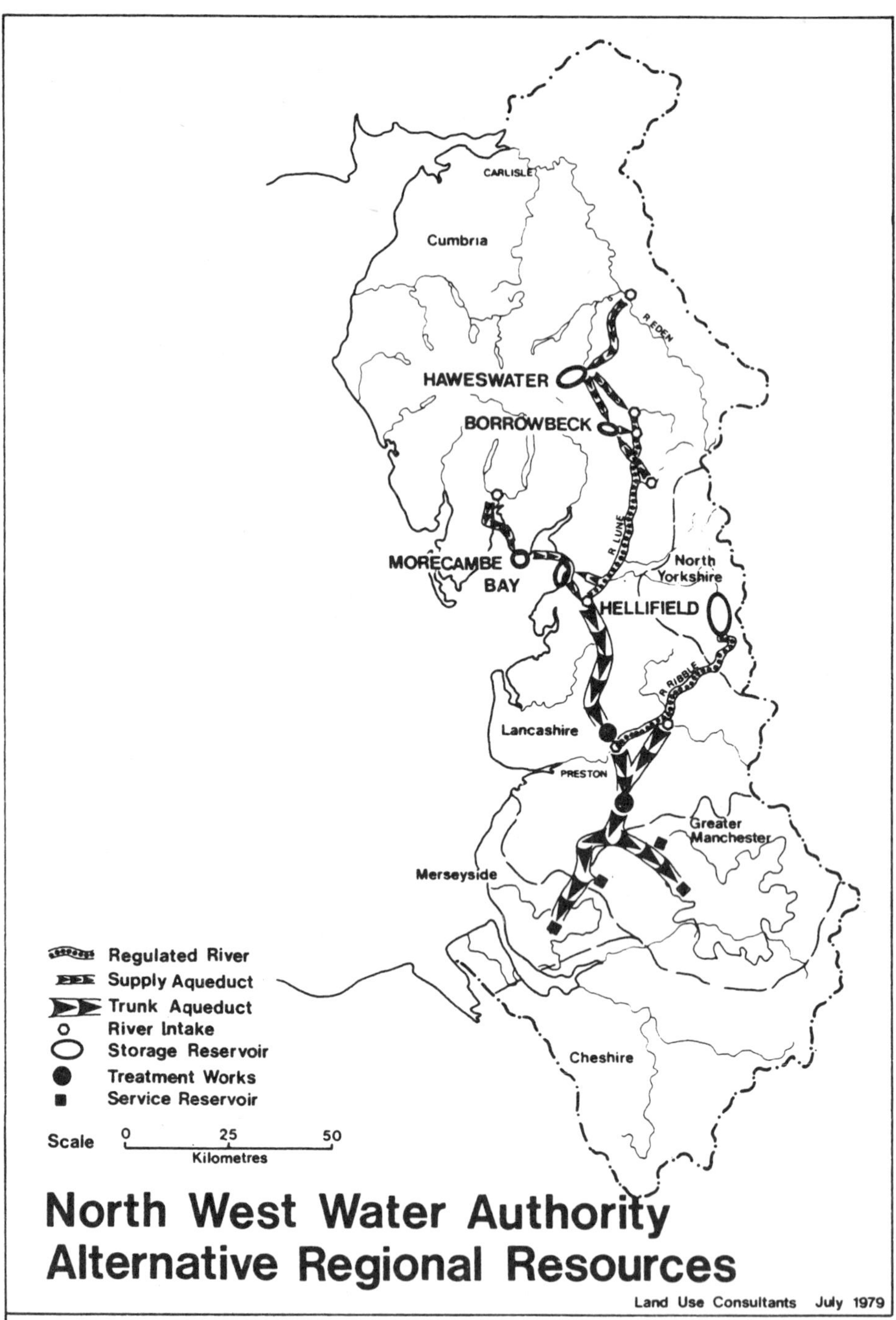

Figure 1: North West Water Authority Alternative Regional Resources

upland valley of the Borrowbeck, whose natural water supply would also be supplemented by pumping; and the third involved flooding a flat low-lying agricultural floodplain on the headwaters of the River Ribble at Hellifield. The fourth alternative differed radically from the others in proposing the creation of "bunded reservoirs" in the intertidal mud-flats of Morecambe Bay, or estuarial barrages across the mouths of rivers flowing into the Bay.

Having drawn up this shortlist of alternatives, NWWA decided to carry out further studies to determine the engineering feasibility and economics of developing and operating the schemes, and to assess their environmental consequences. In order to supplement its own engineering and economic expertise, 3 firms of Consulting Engineers (Messrs. Babtie Shaw & Morton; Binnie & Partners; and Rofe Kennard & Lapworth) were commissioned to examine different reservoir sites and to advise on technical matters such as storage capacity, types of dam, sources of refill water, location and outline design of pumping stations, and the construction and routeing of aqueducts. These engineering studies proceeded in parallel with the EIA.

NWWA recognized that there were no obvious precedents to follow in deciding that the environmental consequences of each alternative should be assessed at this early stage in the design process. The advice of the Chief Planning Officers for the 7 Structure Planning Authorities (SPAs) in the region was sought, and after discussion it was agreed that a joint study should be undertaken by NWWA and the SPAs, with planning and landscape consultants being appointed to devise the programme and co-ordinate the work. Submissions from leading environmental practices were invited by NWWA, and Land Use Consultants (LUC) was appointed to the task, early in 1977.

5. MANAGEMENT OF THE STUDIES

From the outset, it was recognized that a clear division of work and responsibilities would be needed to ensure satisfactory co-ordination of the work of NWWA's internal staff, the 3 firms of Consulting Engineers, the Environmental Consultants, the staff of the Planning Authorities, other specialist bodies and selected government departments. An outline of the organization framework which was created is given in Figure 2 and a brief description follows.

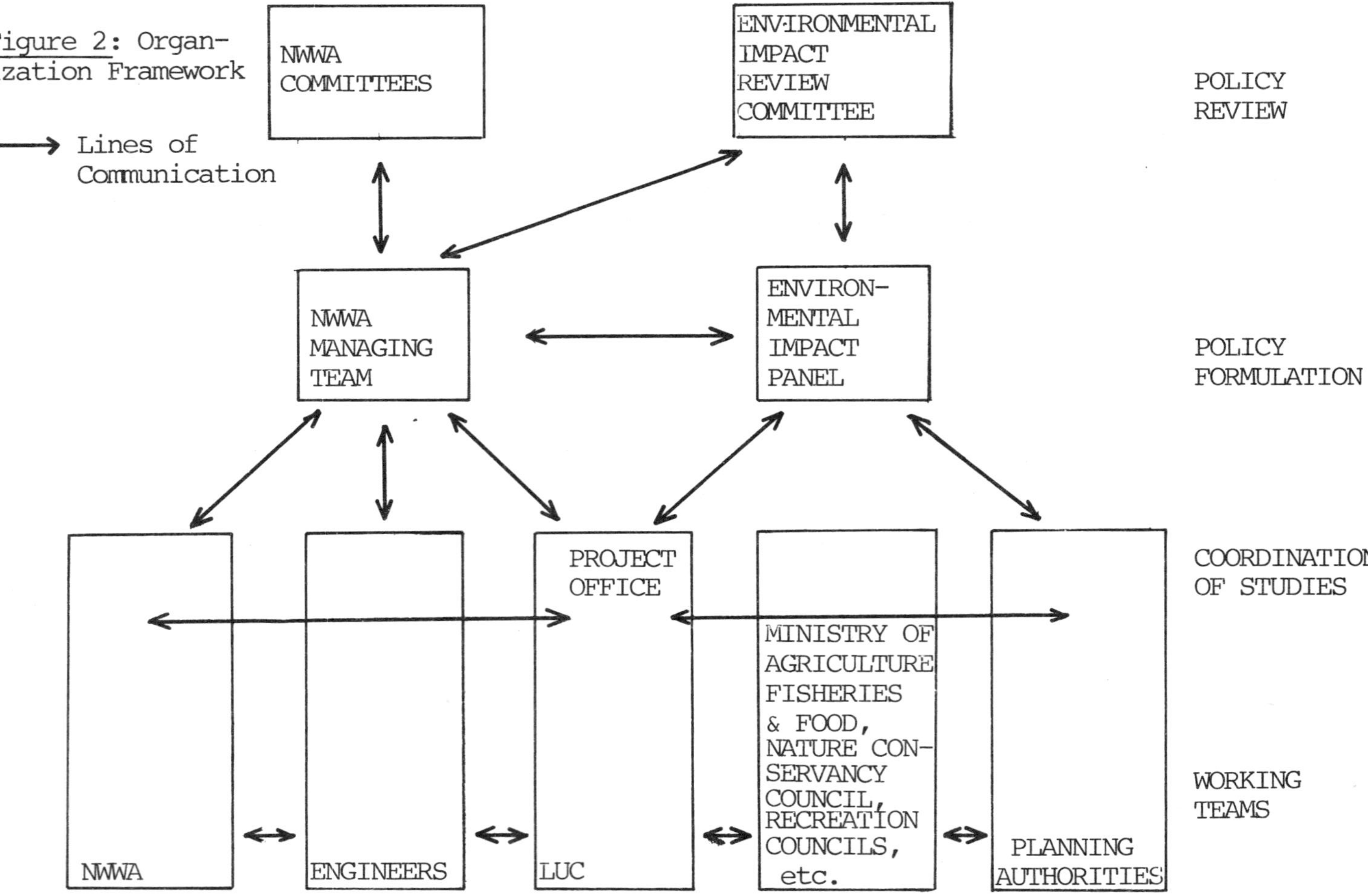

Figure 2: Organization Framework

When the studies started, NWWA emphasized that a "preferred" scheme had not been selected and that an entirely independent and impartial review of the engineering, economic, and environmental advantages and disadvantages of the different options was required. For this reason, a clear distinction was drawn between the professional conduct of the technical studies by qualified staff and consultants, and the decision-making role of elected Members of Water and Planning Authority Committees. The policy-makers were kept advised of the purpose and progress of the studies through a group of their own members known as the Environmental Impact Review Committee (EIRC), but all decisions concerning the direction of the studies were made by Chief Officers, who met at 6 weekly intervals as the Environmental Impact Panel (EIP). Their instructions were relayed to senior members of staff in each of the different organizations who acted as Co-ordinators and met at monthly intervals. Only when the studies had been completed and published did the decision-makers begin to form their own opinions on the respective merits of the alternatives.

This review of the study findings was accompanied by a 6 month programme of consultation in which local people and interested groups were invited to make their views known through a select list of official bodies including the SPAs, the District and Parish Councils. Altogether, 64 organizations commented on the findings of the studies. Their opinions were summarized in a further report for the benefit of the Water Authority Members, with whom the responsibility of deciding the future strategy of water resource development in the region rested.

Early in 1980, the NWWA decided to reject 1 of the 4 alternatives because of the excessive adverse environmental impacts which would result from its development, in spite of the fact that it had emerged as the cheapest of the options. This was the Hellifield reservoir scheme which would have flooded prime agricultural land and caused major social disruption. In itself, this decision provides strong support for the use of EIA techniques, since it demonstrates that it is possible to provide sufficient information about matters of subjective judgement which cannot be quantified in economic terms, to allow the decision-maker to form an opinion of their relative worth. There are, however, many examples which can be produced from the studies themselves where the use of EIA actually helped to reduce the potential costs of scheme variants. This was achieved by identifying

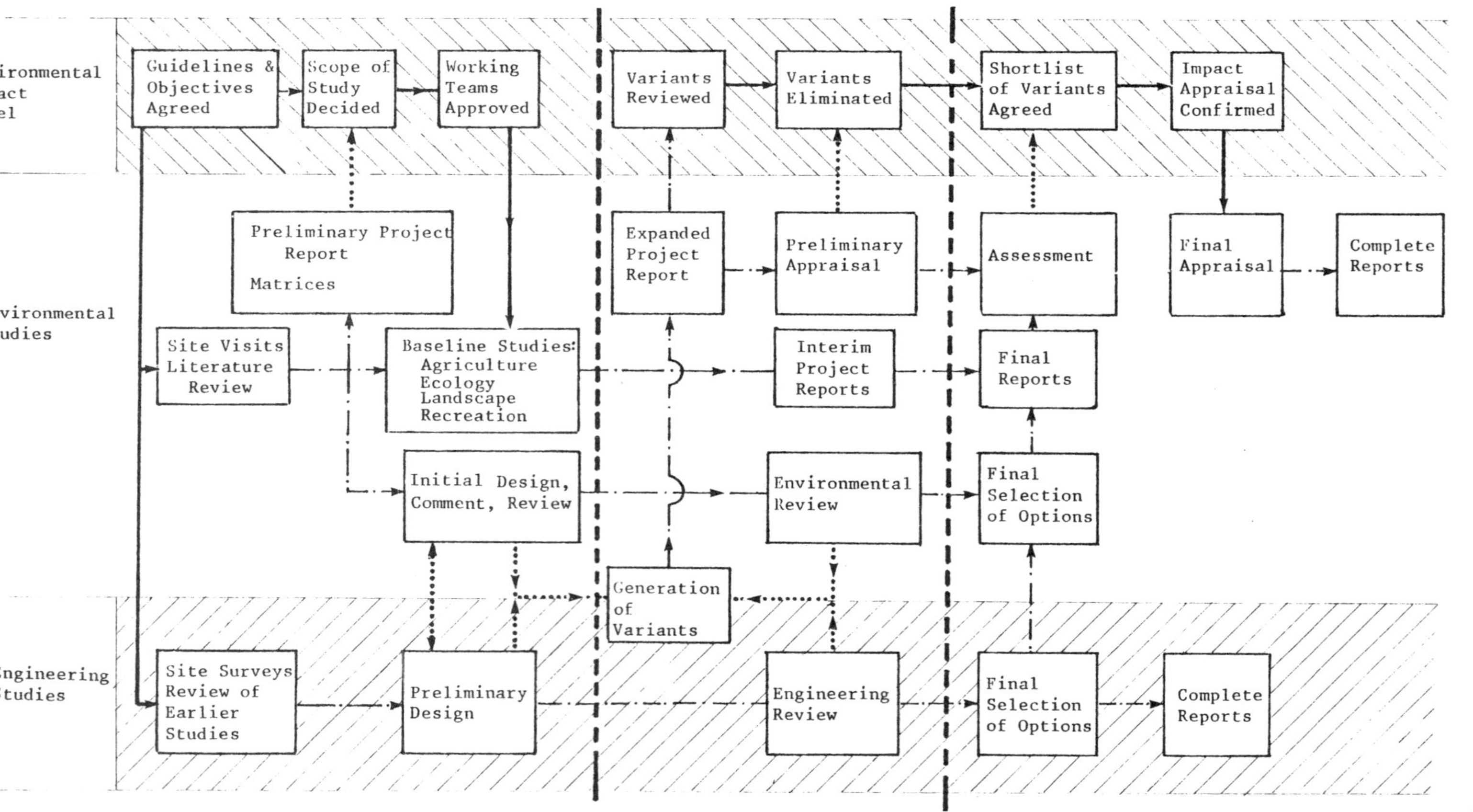

Figure 3: Network Diagram of NWWA Regional Water Resources Studies

design solutions which would harmonize rather than conflict with the natural environment.

A final decision has not yet been taken on the choice of scheme which would be promoted if the need arises. NWWA has stated, however, that the high engineering and environmental costs of developing a storage reservoir in Morecambe Bay makes this an unattractive proposition, and future plans are likely to concentrate on either the Haweswater or Borrowbeck schemes to regulate the River Lune.

6. STUDY METHODS

Having described the broad concept and approach of the Regional Water Resources Studies, it is appropriate to examine the ways in which the two parallel programmes of the engineering and environmental teams were dovetailed together to provide a full EIA. The process is illustrated in Figure 3.

For convenience, the engineering studies are described first, although it should be noted that the environmental appraisal began 3 months earlier. Preliminary engineering work included the review of earlier desk studies and commissioning of borehole site investigations at 2 of the 4 sites. It was followed by a wide ranging exploration of different reservoir sites, alternative dam designs, water intakes, aqueduct routes, and the effects which construction would have on existing services and facilities. Each different component was described as a variant of 1 of the 4 basic schemes. The 53 variants generated by this process, were, ultimately, reviewed in varying degrees of detail according to their technical and economic feasibility in the final engineering reports, which were published at the end of the joint studies (Babtie Shore & Morton 1978, Binnie & Partners 1978a, Binnie & Partners 1978b, NWWA 1978).

In parallel with the preliminary engineering work, baseline studies were conducted by LUC, the staff of the Planning Authorities, and other specialists, into the environmental character of the potential reservoir sites including their land-uses, local economy, ecology, landscape character, social welfare and historic and cultural traditions. The findings of this work were presented by LUC in a preliminary Project Report, which was circulated to all study teams. Joint discussions followed, in which potential interactions between the engineering proposals and the existing environmental character of land which might be developed, were identified and recorded on matrices. These documents provided supporting data for a series of internal study conferences, at which the merits and demerits of successive pairs of variants were compared, and those which were judged to be the least satisfactory on both engineering and environmental grounds were discarded. The direction of the environmental studies was reviewed continuously by the EIP, which also acted as the executive body in ratifying decisions on the selection and elimination of variants, as shown in the top row of Figure 3.

This process of refining the design of the variants by considering their engineering, economic and environmental characteristics, and then eliminating those which were the least suitable, was repeated 3 times, over a period of 9 months. Finally, when the process of reduction had reached its practical limits, a full description of the unavoidable impacts of the remaining 14 variants, was prepared as an Environmental Impact Statement (EIS) and the details were published in a report entitled the Environmental Appraisal (Land Use Consultants, 1978a) supported by technical appendices (Land Use Consultants, 1978b, 1979).

The method of EIA described here was developed from the PADC model produced by the unit led by Brian Clark of Aberdeen University (Clark et al. 1976). In essence the essential stages can be summarized as:

(i) Examining existing planning policies relevant to the proposal;
(ii) Surveying the environmental characteristics of the areas potentially affected by each scheme variant;
(iii) Analysing the characteristics of the engineering proposals;
(iv) Defining the probable interactions between the engineering proposals and the environmental character of the affected area, and assessing the relative significance of particular impacts;
(v) Examining the scope for reducing or modifying the effects of adverse environmental impacts by altering the engineering proposals;
(vi) Making a final assessment of those environmental impacts which were inherent and unavoidable.

7. AREAS OF ENVIRONMENTAL IMPACT ASSOCIATED WITH WATER STORAGE

For the reasons given in the Introduction, this part of the paper is not intended to provide an exhaustive review of potential impacts which can arise in water storage schemes, but rather an illustration of the ways in which particular issues of concern identified in the NWWA study have influenced the outcome of the joint engineering and environmental investigations.

Reference has already been made to the matrices which were produced at an early stage in the study, by comparing development proposals and site characteristics. Figure 4 shows a typical example, with site characteristics listed across the column and descriptions of the individual components of development against the rows. Major impacts are recorded by the notation 2, minor impacts by the notation 1, and areas of possible but uncertain impact with a question mark. Matrices of this type were produced for each of the 4 schemes. Interesting variations in the extent and scale of individual impacts were found between them. To begin with, probable areas of impact were defined simply from prior knowledge of the circumstances in which adverse and beneficial changes in the environment had been brought about by similar water storage projects. As such, the main value of each matrix was to provide a convenient checklist of potential impacts. Subsequent studies of individual impacts were written up in the form of reports since the matrices were found to be unsatisfactory as a means of conveying complex technical information. A detailed schedule of potential impacts associated with water storage schemes is provided in Appendix A.

The basic components of any water storage scheme include; available sources of supply; the means of storage; and ways of transporting water between the source, storage area, and the consumer. In some instances direct impoundment of the source river, through the construction of a dam, may meet the first two requirements, as in the case of the Hellifield reservoir on the River Ribble. There are also many examples throughout the world of irrigation and hydroelectric power schemes where water consumption occurs immediately below the dam site and the river channel, and gravity can be relied upon for water transportation at all stages. In the case of the NWWA studies, however, 3 of the 4 schemes involved pumping water from 1, or more sources into an

ENVIRONMENTAL IMPACT STUDY : NWWA PRELIMINARY ENVIRONMENTAL MATRIX

Operational Phase Only

HAWESWATER - EDEN + LUNE Family of Variants

300 Ml/d)
600 Ml/d) Schemes
800 Ml/d)

Development Characteristics / Site Characteristics	*Environment*	Geomorphology	Climate	River regime	Ground water	River/estuary ecology	Terrestrial ecology	Archaeology/historic interest	Landscape	*Land Use*	Agriculture	Mineral reserves	Settlements	Public utilities/services	Transport & communications	Recreation/tourism	*Health, Welfare Community*	Community structure	Local economy	Safety, welfare, convenience	Perceived environment	Interested parties.
Intake																						
Source water		–	–	1	–	2	–	–	–		–	–	–	–	–	?		–	–	–	–	?
Intake works		–	–	1	–	1	–	?	?		–	–	–	–	1	?		–	–	–	1	?
Preliminary treatment works		–	–	–	1	–	–	2	2		?	–	–	–	–	?		–	–	–	2	?
Pumps from prelim. treatment to storage reservoir		–	–	–	–	–	–	?	2		1	–	?	1	–	–		–	–	–	?	?
Aqueducts to storage reservoir		1	–	–	?	–	?	?	1		1	?	?	–	?	?		–	–	–	?	?
Reservoir																						
Reservoir inlet arrangements		–	–	–	–	–	2	?	2		–	–	–	–	?	?		–	–	–	?	?
Reservoir		?	1	–	–	–	2	2	2		2	–	2	–	?	2		–	–	?	2	2
Reservoir draw-off/ discharge arrangements		–	–	2	–	2	?	?	?		–	–	–	–	?	?		–	–	–	?	2
Pumps to full treatment works		–	–	–	–	–	–	–	–		–	–	–	–	–	–		–	–	–	–	–
Aqueduct to full treatment works		–	–	–	–	–	–	–	–		–	–	–	–	–	–		–	–	–	–	–
Regulated river		–	–	2	?	?	–	–	1		?	–	1	–	–	?		–	–	–	?	2
Treatment																						
Abstraction from regulated river																						
Reserve storage		(Not completed in this example)																				
Full treatment works																						
Supply																						
Pumps to supply																						
Aqueducts to supply																						
Delivery arrangements																						

Figure 4: Example of Scheme Matrix

indirect storage reservoir. It was, therefore, necessary to consider the potential environmental impacts of abstracting water from different catchments, both in terms of its effect on the source rivers, and through the biological consequences of mixing water with different physical and chemical properties in the storage reservoir, with possible indirect effects on subsequent treatment processes. The uncertainties surrounding the extent of potential ecological changes which might arise from inter-river transfers of water, was one of the major areas of environmental concern identified during the course of the studies. It played an important role in eliminating certain variants of the Borrowbeck proposal, as illustrated in the following example.

Plans to dam the Borrowbeck, a tributary of the River Lune, would provide a reliable yield of up to 200 ml/d from the direct catchment. In order to increase the yield, however, it would be necessary to pump in additional water. The studies revealed that these could be drawn first from the Lune itself at Low Borrowbridge and then either from another tributary of the Lune, the River Rawthey, or from an entirely different river system, the River Eden, as shown in Figure 5.

Each of these variants was examined in detail, and it was concluded that the construction of an intake at Low Borrowbridge (1) would be the most logical second stage in the development of the Borrowbeck Scheme providing an increase in yield of 100 ml/d. The choice between the two variants for the third stage of development was, however, shown to be balanced in economic terms even though the engineering requirements were very different. Water taken from the River Eden (2) would be pumped through a pipe and tunnel aqueduct (for which three alternative routes were examined) and discharged to the upper headworks of the Lune, to be abstracted 15 kilometres downstream at Low Borrowbridge and pumped against a head of 60 m into the reservoir. This variant would provide a yield of 425 ml/d. On the other hand, water taken from the River Rawthey or its confluence with the Lune, could be pumped direct by a longer pipeline to the reservoir. The Rawthey has a more variable regime than the Eden and a lower average flow. As a result the yield of this variant would be 395 ml/d. On engineering and economic grounds there was little to choose between the 2 variants, but the EIA confirmed that the Rawthey option (3) would be the least damaging, with the result that the Eden variant was eliminated. The reasons supporting this

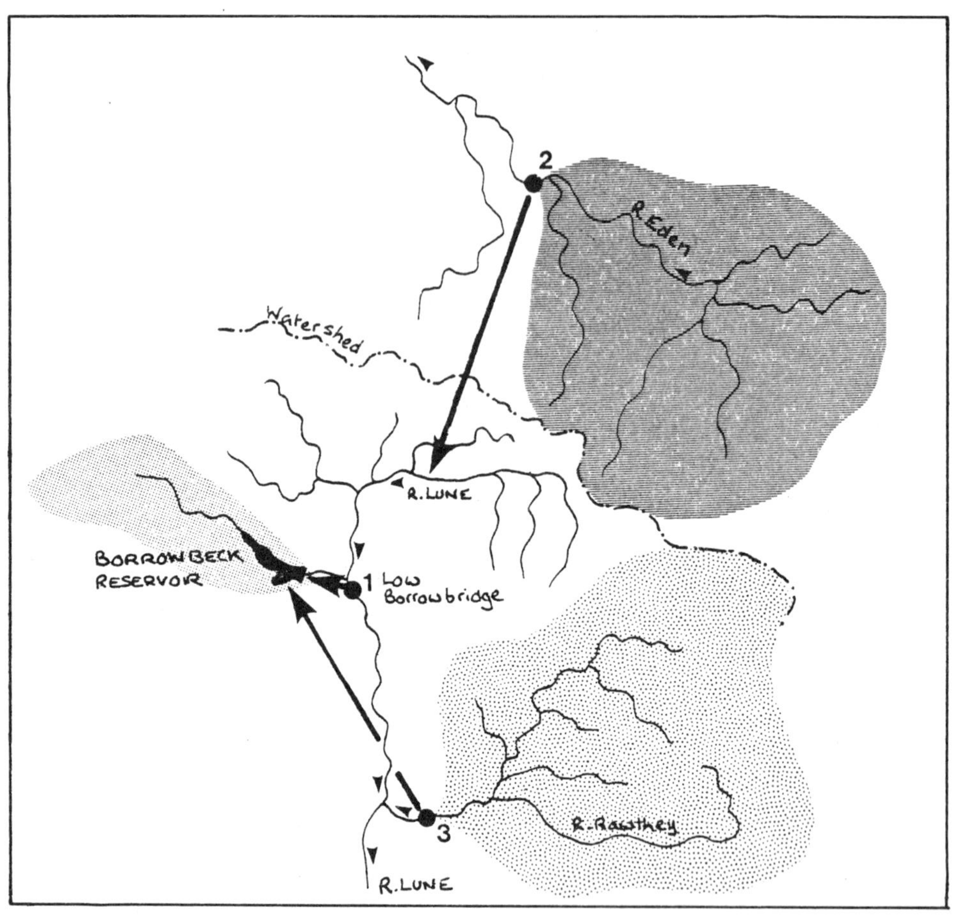

Figure 5: Borrowbeck Catchments

decision are outlined below.

The major disadvantage of the Eden variant lay in the fact that this river has a much higher nutrient content than the River Lune, as a result of the run-off of manure and fertilizers from rich agricultural land. Release of quantities of this enriched water to the purer headwaters of the River Lune could have stimulated the growth of algae, and seriously impeded the homing instincts of salmon and other game fish which are though to rely largely on the ability to detect minute differences in the chemical properties of water.

Construction of the aqueducts between the Eden and the Lune were also shown to have potentially damaging effects upon recognized Sites of Special Scientific Interests, whose ecological value stemmed from the rare plant communities growing on water-logged calcareous soils which could be disturbed. Finally, the effects of discharging increased quantities of water from the Eden to the Lune at times of low natural flow would have altered the present regime of the latter river sufficiently to require the construction of weirs, and stone walls along the river bank to prevent scour and erosion. This would have adverse ecological effects and be visually intrusive.

By comparison, the aqueduct from the River Rawthey to Borrowbeck could be constructed partly in tunnel and partly as a pipeline laid along the bed of a derelict railway. None of the problems of mixing waters of different quality would arise and no special measures would be required to protect the river banks. The Rawthey variant (3) was eventually confirmed as the least damaging solution to the development of the Borrowbeck, but certain adverse environmental impacts still remained. For example, concern was expressed by the biologists that if too much water was pumped from the Rawthey instead of the Lune at Low Borrowbridge, the migration patterns of salmon and sea-trout returning to spawn in the Rawthey could be upset by the change in the relative volume of water in the two rivers. This fact was, therefore, described in the final EIS, so that it could be taken into account in assessing the relative merits of the Borrowbeck scheme by comparison with the other 3.

The effects of changes in water chemistry and biology brought about by the mixing of waters from different sources, and through prolonged storage in reservoirs, emerged as an important environmental consideration in each of the other 3 reservoir schemes. For example, in the case of plans to pump water from the River Eden into the existing oligotrophic (low nutrient) reservoir at Haweswater, it was concluded that the new supplies of enriched water which would be provided by certain variants, could trigger off explosive growths of blue-green algae and the diatom, Asterionella. This can be sufficiently prolific to clog the filter screens of water treatment and greatly increase the complexity and cost of providing potable water.

Similar problems were identified in the case of the Hellifield reservoir on the River Ribble. Here it was concluded that the large expanse and shallow depth of water would provide an ideal "culture vessel" for the growth of both algae and emergent vegetation. This rapid growth of vegetation coupled with the lowering of water levels in summer and the exposure of large areas of mudflats, could result in the decay of organic material and the formation of ideal breeding conditions for black fly and other biting midges.

In the case of the fourth alternative, namely bunded reservoir construction on the inter-tidal mudflats of Morecambe Bay, the only real cause for concern over water quality arose from possible sources of contamination by salt-water and the introduction of Salmonella organisms by sea birds feeding on local refuse tips. In both cases, however, it was concluded that adequate control and treatment could be provided to meet the most stringent safety standards in excess of those laid down by the World Health Organization.

In the English context, the irritation and occasional annoyance of biting midges, and the odour of rotting vegetation caused by the formation of a shallow reservoir would be regarded as a major public nuisance and a loss of local amenity. There are, of course, many examples particularly within the tropics, where ecological changes brought about by flooding land, as part of a water storage scheme, has had major implications for public health through the spread of Schistosomiasis (Bilharzia) and other debilitating diseases spread by waterborne vectors. A number of retrospective studies have been conducted into the problems stemming from the creation of large man-made lakes following experience gained in projects, such as the damming of the Volta Lake. A notable example of such a post mortem analysis, concerns the Nam Pong Project in North East Thailand (Ludwig 1979). Relatively few full scale Environmental Impact Studies of the sort described in this paper have, however, been conducted in advance of the proposed development.

Risks to public health and welfare from water storage were not singled out as a significant issue in the Environmental Appraisal. Nevertheless, the degree of disruption which would be caused by certain of the variants during the construction period, and the psychological effects upon local communities of creating new reservoirs were described as important considerations at Haweswater, Hellifield and Morecambe Bay. The final study report (the Environmental Appraisal) noted that, although no more than 7 houses would be flooded, even in the most extensive scheme at Hellifield, the livelihoods of several hundred people could be affected directly through loss of employment. The existence of a large reservoir would also sever lines of communication and disrupt the social patterns and traditions of the isolated villages. Construction on this reservoir would have necessitated the diversion of a 3 km length of main trunk road and railway, which in itself would increase the level of disturbance.

In addition to adverse environmental impacts likely to be generated during the construction phase, major areas of concern were seen to result from disruption of existing land uses, the inundation of sites of high nature conservation interest, loss of existing recreational resources, and permanent changes in the appearance of areas of high landscape quality. Whenever possible, attempts were made to quantify the physical extent of such changes and to provide yardsticks whereby the effects of each of the four sites could be measured. In the case of agricultural land-loss, detailed surveys were made by the Ministry of Agriculture, Fisheries & Food, to establish the number of farms with land potentially affected, their size, type of management and farming practice, number of stock and annual production. When it came to assessing the extent of the disruption and damage which would be caused by reservoir construction, however, no absolute figures could be given. This was due to the dispersed nature of many of the farm holdings with land occupying not only the valley floor which would be flooded, but also the valley sides and hilltops. Although, in many cases, only part of the holdings would be affected, the land in question was often the best quality and used for fattening livestock or producing winter feed. This meant that reservoir construction could threaten the economic viability of some farms even though substantial areas of land remained above top water level.

Detailed surveys were undertaken by ecologists to

establish the types and relative distribution of plants and animals likely to be destroyed by flooding, and their rarity and conservation value was assessed. Although the areas of particular habitats likely to be affected could be quantified, the significance which was attached to these findings was acknowledged to be largely subjective. Similar conclusions were reached in relation to assessments of landscape quality, although in both cases considerable emphasis was placed on the national importance of some of the reservoir sites as confirmed by planning policies for the Lake District and Yorkshire Dales National Park, and the designation of Morecambe Bay as a National Nature Reserve and Grade 1 site of international important for wildlife conservation.

Not all the consequences of reservoir development were viewed as potentially damaging, and opportunities for recreational use were identified at 3 of the 4 sites. Two schemes in particular, Morecame Bay and Hellifield, were shown to have great potential as regional recreational centres for water sports with considerable opportunities for promoting tourism. At the same time the secondary and tertiary impacts of recreation and tourism on remaining land uses and the local economy of surrounding areas were noted. Particular emphasis was placed on the need for caution in assessing the costs and benefits of developing what would amount to largely seasonal industries. For example, it was estimated that while the equivalent of up to 200 full-time jobs would be lost in agriculture in Hellifield, only some 40 jobs would be generated by recreational use of the reservoir, many of which would be part-time and only available during the summer months. Pleasure boating and fishing would increase the level of demand for hotel accommodation and other tourist facilities, but these would also be seasonal in nature. As a result, the structural stability of the local economy would be weakened by comparison with the present position, where agriculture both supplies and places demands on other sectors of the economy throughout the year.

The possibility of creating viable commercial fisheries at both the Hellifield and Morecambe Bay sites was considered; neither Haweswater not Borrowbeck emerged as suitable due to their physical characteristics, high altitude and relative infertility. It was thought that Hellifield could support a good salmon and trout fishery, since the Ribble is already an important game fishing river. The reservoir, however, would also provide an ideal habitat for coarse fish like pike and perch which are not regarded as edible species.

The pike is particularly aggressive, feeding on other young fish, and intensive management would be needed to control coarse fish populations.

8. USE OF ENVIRONMENTAL DATA

The examples quoted here represent only a small proportion of the environmental issues examined during the course of the studies. They illustrate, however, two basic types of impact. On the one hand there are those which are inherent in the development proposal and cannot be avoided, such as the loss of agricultural land through inundation; on the other there are impacts whose adverse effects may be ameliorated or mitigated in some way by careful design, as in the case of the choice of water intake and aqueduct routes for the Borrowbeck scheme.

Throughout the joint studies, close liaison was maintained between the engineering and environmental teams. Preliminary reports and drawings were circulated for discussion and wherever possible, refinements were made to individual variants in order to exclude impacts falling into the second category.

At the end of the design process, a full statement of unavoidable impacts was prepared. It was recognized that not all impacts would be of equal importance in influencing the selection of a preferred scheme, so their relative significance was indicated by dividing them into categories according to the response obtained to the question:
Will a given impact be either:
(i) Adverse (A) or Beneficial (B)?
(ii) Of Strategic (S) or Local (L) importance?
(iii) Of Long-term (Lt) or Short-term (St) duration?
(iv) Irreversible (IR) or Reversible(R)?
An example of the table produced for each scheme variant using the code letters is given in Figure 6, which illustrates Hellifield scheme variant HEL/300. A final summary of the environmental appraisal for each of the 14 variants is illustrated in Figure 7. A certain ambiguity arises in this classification of impacts, since not all the criteria are mutually exclusive. The following descriptions were provided to expand the definitions.

(i) Adverse and Beneficial impacts are to some extent, self-explanatory, but occasions arise where an impact may have adverse effects on one section of the community or a particular inherent group and beneficial effects on

HELLIFIELD SCHEME VARIANT — HEL/300

Area and Description of Impact		Nature and Extent of Impact	Significance
PHYSICAL ENVIRONMENT			
ECOLOGY			
Statutory/Notified Sites affected by proposals	Yorkshire Naturalists' Trust 'Green Form' site	Occupies reservoir basin - primary interest migratory birds	L,A,B,R,St
Ecological Zones: Conservation Status³ and %age of total Survey area affected	High	0.6%	L,A,IR,Lt
	Medium	7.7%	
	Low	91.7%	
Effect on Reservoir and River Ecology	Existing Fishery	Serious effect on migratory fish (Ribble Salmon and Trout)	S,A,R,Lt
	Potential Fishery	Potentially high - with management as game fishery	S,B,R,Lt
	Algal Growth Potential	Potentially high - local nuisance	L,A,R,St
	New Habitats	Improved habitat for migratory birds	L,B,R,St
LANDSCAPE			
Impact on existing landscape caused by:-	Dam	Not visible from public vantage points	L,
	Reservoir	Fundamental change in character of valley	S,A/B,IR,Lt
	Other Works	Major impact through road/rail diversions	S,A,IR,Lt
Scope for restoring/modifying landscape formed by:-	Dam	Limited	–
	Reservoir	Considerable, tree planting - vegetation of drawdown zone	–
	Other Works	Some potential for landscaping road/rail diversions	–
LAND USE AND ECONOMY			
AGRICULTURE			
Total area of agricultural land lost		1,050 ha	S,A,IR,Lt
Best quality land as %age of affected area		90%	
No. of farm holdings affected		50	
No. of farmhouses/buildings affected		Two farmsteads and 58 buildings destroyed	S,A,R,Lt
Annual production from affected farm holdings	Milk 9,085,000 litres	Severe and disproportionately high loss of agricultural production on most affected holdings	S,A,IR,Lt
	Beef 1,428,000 kg		
	Mutton/Lamb 287,000 kg		
	Wool 20,800 kg		
RECREATION			
Effect on existing recreational activity and levels of use	Fishing	Loss of valuable game and coarse fishing	L,R
	Outdoor Pursuits	Loss of Gildersleets Centre	L,R
		(Both of the above activities could be replaced following reservoir development)	
Recreation development potential and levels of use	Water Area	725 ha (assumed allocation for recreational use)	S,A/B,R,Lt
	Land Area	60 ha	
	Average Summer Use	2,960 people 1,130 cars	
	Peak Summer Use	11,600 people 4,330 cars	
TRANSPORT/COMMUNICATIONS AND UTILITIES			
Diversion/Relocation of major routes and communications	A.65 Trunk Road	4.5 km bypass of Settle and Giggleswick	S,A,IR,St
	Mainline and Branch Railways	4.3 km diversion	S,A,IR,St
	Minor Roads	8.0 km diversion on reservoir's west shore	L,A,IR,St
	Sewerage and Public Utilities	Relocation of sewers and works, gas and electricity	L,A,IR,St
LOCAL ECONOMY			
Jobs Lost		30 - 60	S,A,IR
Temporary Jobs Provided		say 600 Man/years	L,B,R,St
Permanent Jobs Provided		15 - 20	S,B,R,Lt
LOCAL COMMUNITY			
Communities affected by construction:-	Reservoir Works	Hellifield, Long Preston, Giggleswick, Rathmell, Settle	S,A,IR,St
	Aqueduct Works	None	
	Diversionary Works	Rathmell, Wigglesworth, Long Preston	S,A,R,St
	Construction Traffic	Settle, Giggleswick	
Communities affected by operation:-	Reservoir Works, etc.	None	
	Recreation Activity	Long Preston, Wigglesworth, Rathmell	S,A/B,R,Lt
	Recreation Traffic	Settle, Giggleswick	
HISTORIC AND CULTURAL			
Historic or Cultural Sites affected:-	Listed Buildings	3 buildings	L,A,IR,St
	Historic Sites/Buildings	3 un-notified sites	L,A,IR,Lt
	Archaeological Sites		

Notes: 1 Vegetation types affected
High = Deciduous Woodland
Medium = Semi-natural Grassland and Mixed Woodland
Low = Hay Fields, Improved and Permanent Pasture

Key: L = Local; S = Strategic; A = Adverse; B = Beneficial; St = Short Term; Lt = Long Term; R = Reversible; IR = Irreversible

Figure 6: Summary of Environmental Impact for the Hellifield Scheme Variant - HEL/300

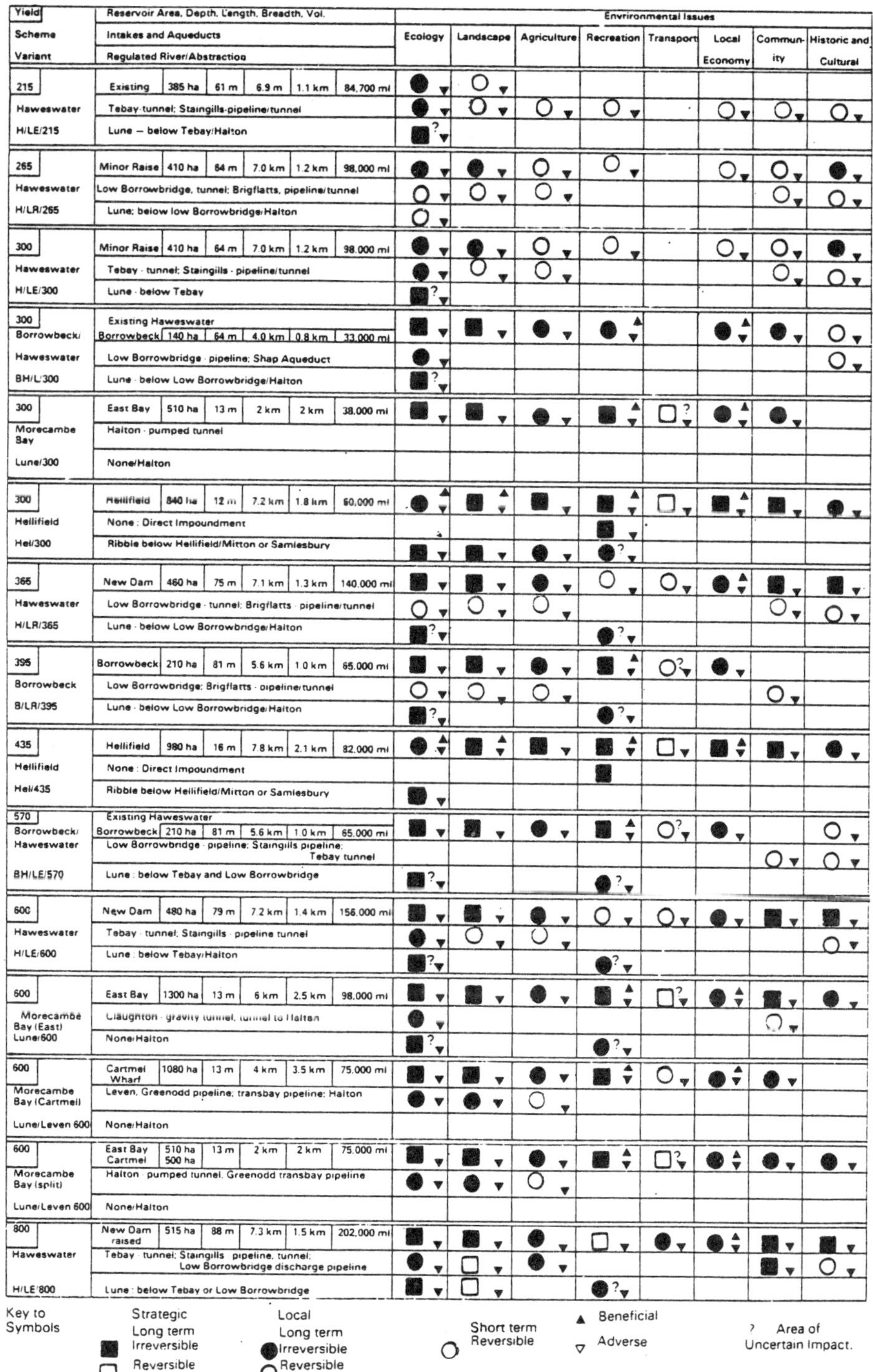

Figure 7: Summary of Main Environmental Issues Affecting all Scheme Variants

another. For example, construction of a reservoir may destroy agricultural land and put farmers out of business, but at the same time provide a new fishery and recreational opportunities on which new forms of economic activity could be developed.

(ii) Strategic impacts were defined as those where significant changes in the existing environmental character of the area could be expected to occur; and where concern about such changes would figure prominently at the level of a regional planning strategy. The word Local was used to describe impacts where some change in the environment would occur in discrete areas, but where the effects would not be of significance in the overall planning framework.

(iii) Long-term and Short-term impacts did not have any specific timescales attached to them, since the significance of the duration would vary from one type of impact to another. For example, destruction of woodland would cause an immediate short-term change in vegetation cover but its real significance would result from the succession of ecological changes resulting over many decades. In the same way the physical displacement of people from a reservoir site might be accomplished in a short space of time but the psychological and socio-economic effects could be very long lasting.

(iv) Reversible impacts were defined as those which were capable of being overcome, either by modifying the nature of the engineering works, or through the natural process of re-adjustment which would follow initial development. Irreversible impacts were those that could not be overcome by any means.

In itself, this "short-hand" description of impacts may appear simplistic, but where the method is applied conscientiously to every potential impact identified in the initial matrix, it can play a valuable role in prompting the designer to examine the ramifications of particular proposals.

Another strong argument for using a non-numerical code for assessing the significance of individual impacts is related to the ultimate use to which the information would be put. The stated objectives of the study were to present both the engineering and environmental findings in a form in which they would be readily intelligible to elected decision-makers, who were not necessarily experts in either discipline. Under these circumstances, it was decided that no attempt should be made to develop a system for weighting the significance of the different forms of environmental impact. It was felt that in the final analysis, a largely subjective and political judgement would be required in determining the degree of importance which should be attached to environmental gains and losses in the disparate areas of ecology, agriculture, recreation, landscape and local community welfare. At the same time it was clearly desirable that this element of judgement should be based on the best available information and this is where the type of evaluation called for in an EIA is of greatest value. Every effort was made to write the final published report in a concise and factual style, in language which everyone would understand, and the issues and choices posed by the 4 alternatives were clearly identified with maps, photographs, diagrams and sketches.

9. DISCUSSION ON THE BENEFIT OF THE EIA PROCESS

The joint engineering and environmental studies took 14 months to complete. Given this relatively short period of time for a major strategic study, little opportunity existed for original research. As a result, predictions of the extent and severity of potential impacts often had to be made on the basis of limited data using established principles of scientific and planning theory for guidance. The NWWA's Director of Resource Planning, who led the studies, pointed out that it was more important that his Authority should know of the existence of any major areas of uncertainty which might prejudice the successful promotion of a particular scheme, than that the studies should be held up until all the answers had been obtained through research. The studies had been planned sufficiently far in advance of the need for development to allow such matters to be investigated once the strategic decisions had been taken.

In many parts of the world, hard facts may not exist on matters as diverse as the performance of river regimes, the ecological characteristics of water catchments, and the social and economic conditions of populations in remote areas. In such circumstances there may be an understandable reluctance to explore the environmental consequences of water development proposals which might appear to conflict with economic benefits of water supply, hydro-electric power, flood control, irrigation, and navigation. This is particularly the case where the process of EIA is presented as a fully comprehensive and rigid scientific review of every conceivable impact, as in the case of many examples from the USA, which could lead to delay and even litigation. Nevertheless, as many studies have shown, the economic and social disbenefits of schemes which have not been planned with due regard for the environment, can cancel out or largely discredit the economic advantages which justify their promotion. There is, therefore, a clear argument for undertaking the type of environmental impact study which has been described in this paper from the outset of the design process.

It is sometimes suggested, and the review has been strongly supported by some participants during the WHO/PADC Aberdeen Course, that EIA is already adequately covered in major engineering feasibility studies and that there is no need for a separate process. It is certainly true that there have been some excellent investigations of this sort in different countries, and in fact one of the schemes discussed here (the proposal for reservoir storage in Morecambe Bay) was subjected to a most detailed technical, economic, and environmental review by NWWA's predecessor authorities and the former Water Resources Board in 1970. This earlier study, however, concentrated almost entirely upon the environmental conditions which would follow the development and made no attempt to assess what resources of value would be lost during its construction.

The author would argue that unless feasibility studies are given sufficiently wide terms of reference to explore both the adverse and beneficial consequences of a given development proposals and to look at alternatives, there will always be the temptation to concentrate effort and attention on those characteristics which will show the project in its most favourable light. As a result, the justification for taking particular courses of action or introducing modifications to the design become obscured. The decision-maker is ultimately confronted with a complex package of proposals which must either be approved in their entirety or rejected out of hand. By this stage in the development programme, the full design cost will have been incurred, and time for considering viable alternatives may well have run out, affecting the decision in favour of proceeding with the development, even if doubts have been raised about its adverse environmental consequences.

By comparison, EIAs of the sort typified by the NWWA study are designed to assist the decision-maker in making the best choice of alternatives for future development. They provide an opportunity to integrate technical, economic and environmental parameters from the outset of the design, and to demonstrate the full range of consequences which would stem from the development. Provided they are undertaken sufficiently early in the planning process, they afford opportunities for the views of local residents, interested groups, and other official bodies to be taken into account, and provide a vehicle for informing those people most likely to be affected by the development.

The use of EIA will not remove the need to take controversial and difficult decisions about major development proposals, which may affect the livelihoods of hundreds of people; nor will it ensure that no adverse environmental impacts will occur. It should, however, provide politicians and administrators with a vehicle for informing those people most likely to be affected by the development, about the steps which will be taken to protect their interests, or to compensate them, and establish an authoritative point of reference and baseline against which both adverse and beneficial environmental changes can be monitored. In this way, EIA should both add to the sum of knowledge about the way in which man's activities affect his environment and lead to improvements in the standards of design and construction of major engineering projects.

APPENDIX A

CHECK LIST FOR WATER STORAGE PROJECTS

The following list is not intended to be fully comprehensive; nor is it suggested that every topic should be investigated to the same level of detail in every case. It is provided only as a guide to indicate the range of issues which may be relevant to the assessment of major water projects.

1. Grounds for Development
   a) Statement of need.
   b) Availability of resources.
   c) Alternative courses of action considered (where appropriate).

2. Description of Proposal
   a) Location, area.
   b) Purpose - water supply; power generation; irrigation; commercial fisheries; flood control; navigation; recreation & tourism.

c) Hydrology - catchment areas; rainfall; run-off; river regimes; groundwater movements.
d) Geology - bearing capacity of strata under dam site, fault lines and potential seepage planes seismic stability.
e) Type and general form of structures (e.g. weirs, intakes, pipelines, treatment works, canals).
f) Construction Materials - type, source, methods of transportation.
g) Energy requirements.
h) Waste disposal.
i) Plant and Equipment.
j) Labour requirements.
k) Housing needs.
l) Timescale for construction.
m) Finance.

3. <u>Existing Legislation & Policies affecting the Project</u>
a) Relationship to National Regional or Local Development Plans.
b) Land ownership.
c) Riparian rights.
d) Navigation rights.

4. <u>Description of the Existing Environment</u>
a) Climate.
b) Landform.
c) Topography.
d) Landscape.
e) Land-uses.
f) Ecology.
g) Transport, communications, and public utilities.
h) Population density, distribution, age and sex structure.
i) Health and social welfare.
j) Settlements.
k) Employment and local economy.
l) Cultural traditions.
m) History, archaeology.

Environmental Impacts which have been shown to be of special significance in different water development studies include the following:

(1) Water abstraction and delivery to storage
   (i) Effects of reducing flows at source on river ecology and fisheries;
   (ii) Damage and disturbance through pipeline/canal construction to existing land-uses and ecology.

(2) Water storage
   (i) Adverse changes arising from:
   - flooding of agricultural land
   - loss of timber crops
   - disruption of hunting-fishing regimes
   - loss of terrestrial habitats
   - fluctuations in water level (drawdown)
   - increased range of aquatic habitats supporting waterborne vectors of disease with attendant health risks
   - eutrophication and increased water pollution
   - requirements for population resettlement.
   (ii) Beneficial changes resulting from improved:
   - fishery resources
   - communications
   - water supplies
   - power supplies
   - recreational and tourist potential.

(3) Release and use of stored water
   (i) Adverse effects arising from:
   - increases in discharge and flow rates below the dam
   - bank erosion
   - channel scouring
   - loss of spawning grounds for fish
   - alterations to rates of siltation and deposition.
   (ii) Beneficial changes resulting from improved:
   - irrigation
   - flood control
   - power generation.

REFERENCES

Anon.: 1979, 'Pioneering Study is Praised', Planning 320, June.

Babtie, Shaw & Morton assisted by Rofe, Kennard and Lapworth: 1978, Storage at Inland Sites, Engineering Report,

Hellifield, Borrowbeck and Haweswater. July.

Binnie & Partners: 1978a, Storage in Morecambe Bay, Engineering Report, July 1978.

Binnie & Partners: 1978b, River Intakes, Engineering Report, July 1978.

Clark, B.D., K. Chapman, R. Bisset, P. Wathern: 1976, Assessment of Major Industrial Applications: A Manual, Project Appraisal for Development Control Team, (Aberdeen University) Department of the Environment, Research Paper 13.

Land Use Consultants: 1978a, Environmental Appraisal, NWWA Regional Water Resource Studies, November 1978.

Land Use Consultants: 1978b, Technical Report, NWWA Regional Water Resource Studies, November 1978.

Land Use Consultants: 1979, Report on Consultations, NWWA Regional Water Resource Studies, November 1979.

Ludwig, Harvey F.: 1979, Post Mortem Environmental Analysis of Typical Multi-purpose Dam/Reservoir in South East Asia, Presented to WHO Regional Conference on EIA, New Delhi, October 1979.

North West Water Authority: 1978, Aqueducts and Treatment Plants, Engineering Report, September 1978.

Oldfield, J.B.: 1978, Environmental Impact Analysis, Practical Experience, Paper presented to the Symposium on Engineering and Environment, Institution of Water Engineers & Scientists, London.

AFFILIATION

Dr Peter Nelson is an environmental consultant with Land Use Consultants at their Branch Office in Glasgow, Scotland.

Brian S. Duffield and Susan E. Walker

THE ASSESSMENT OF TOURISM IMPACTS

1. INTRODUCTION

Environmental Impact Assessments (EIAs) are concerned with the impact of large scale developments on the environment, and developments associated with tourist activity can be at a scale qualifying for such consideration. Yet, despite the growing number of studies charting the impact of tourism developments, few EIAs have been undertaken, integrating the effects of tourism on the physical, economic and social facets of the environment. The methodology developed for a full EIA can easily be applied to tourism developments (see, for example, Rajotte, 1978). Studies which have been undertaken have been predominantly retrospective with little anticipatory or prescriptive research. Nevertheless, the research which has been carried out provides a basis both for identifying the diverse impacts of tourism on the environment, and for assessing the methods and techniques used to isolate these impacts.

2. THE NATURE OF TOURISM AND TOURISM IMPACTS

In assessing the impact of tourism developments one is not only faced with the tangible effects of major developments, situated in specific locations, but also with the need to consider the more spatially diffuse, but no less considerable, impact of tourism as an economic and social activity. Indeed, even those staying in tourist centres or resorts exert a considerable influence on the surrounding locality. The impacts of tourism developments, therefore, are not confined solely to the structural changes associated with such developments but are also related to what Doxey (1975) calls the dimensional changes, i.e. those impacts that occur as a result of ever increasing numbers of tourists within an area. Such a distinction recognises the dynamic nature of tourism as an agent of change.

Moreover, tourism is not one industry but an amalgam of many and encompasses developments as disparate as the

B. D. Clark et al. (eds.), Perspectives on Environmental Impact Assessment, 479–516.

modest "pension" or bed and breakfast establishment to the integrated resort incorporating purpose-built accommodation and facilities for tourists. As such, developments in a tourist area may be to a greater or lesser degree 'planned'. The major cities of the western world have been attracting visitors for several centuries, whilst on the other hand, there are now examples of the 'instant' resort resulting from a selection process made by computer, against pre-selected parameters, from a range of alternative sites (Bosselman, 1978).

In attempting to accommodate the variable nature of tourist activity, numerous classes of tourism developments, types of tourism, tourists and tourism impacts have been developed. Butler (1980), for example, has recently argued for the concept of a 'cycle of evolution' related to tourist areas, involving the following stages:

(i) Exploration
(ii) Involvement
(iii) Development
(iv) Consolidation
(v) Stagnation
(vi) Decline or rejuvenation.

Cohen (1972), on the other hand attempts to categorise the tourist along a continuum of novelty and familiarity. He identifies four 'tourist roles':

i) the organised mass tourist;
ii) the individual mass tourist;
iii) the explorer;
iv) the drifter.

More importantly, Williams (1979) argues that not only does the scale and character of tourism vary over time and space, but it is important to appreciate the linkages and inter-relationships between the stages of development of tourism, the characteristics of tourists and tourist activities, and the nature of the accompanying impacts. Drawing from the categorisations of others, Williams has constructed a conceptual framework to incorporate these interrelationships (Figure 1).

A further problem of identifying the impacts of particular forms of tourist development is that the effects are often represented in isolation without being placed in the spatial and environmental context of the area concerned. Communities associated with tourism developments are very diverse, not only in their historic and cultural development and geographical settings, but also in their ability to accommodate change and in the nature of the economic forces that affect them. Tourism-related investment induces a very distinctive form of economic activity. Employment of large numbers of residents (and incomers) in the tourist industry creates new working relationships and new forms of social stratification, and transforms social insititutions and accompanying beliefs, attitudes and values, as well as causing a physical impact upon the environment.

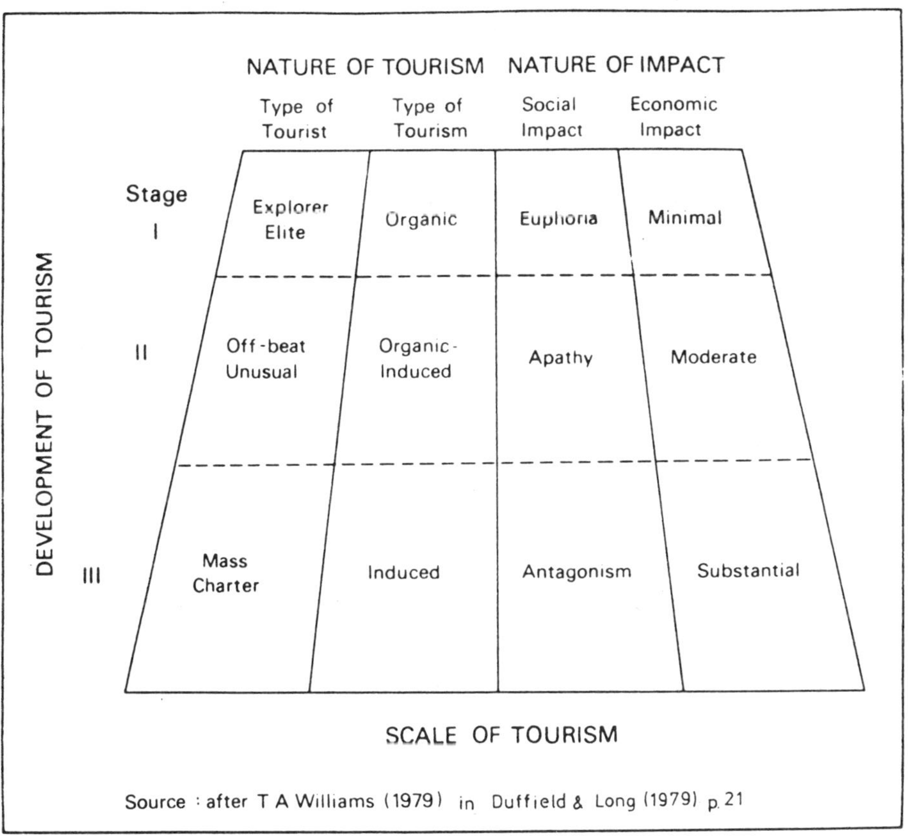

Figure 1: Scale and nature of tourism

As Kloke (1977, p. 3) has asserted, any assessment of the changes which may result from tourism needs to look beyond economic effects and be based upon 'the net result of the welfare of its people, comprising the sum of the desirable and undesirable effects over the full range of social conditions'. If the changes associated with tourism are to be properly understood, they have to be placed within a consistent framework or evaluation which recognises that such changes are but one facet of a general process of change within any area. Although each area is unique, it is possible to isolate certain key factors which are likely to condition the processes of change in any locality, thus permitting a consistent evaluation. Duffield and Long (1979) have devised a schema which attempts to provide a general framework for assessing tourism impact in the context of specific situations and the social and economic forces that characterise them (Figure 2).

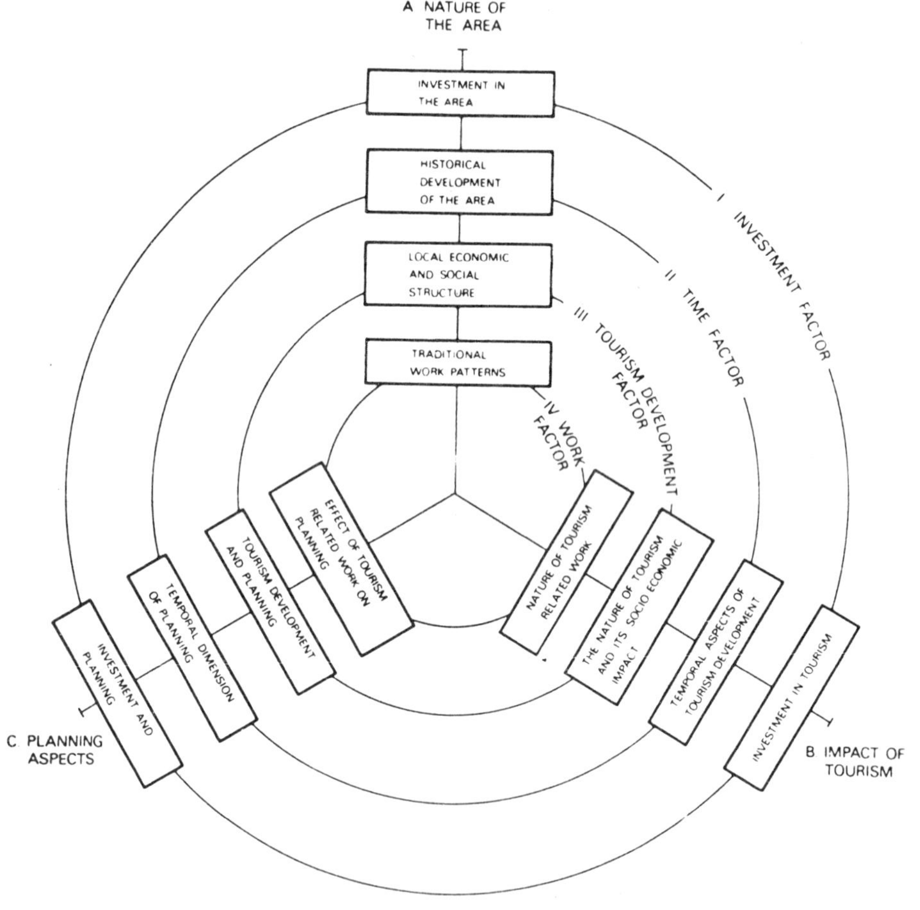

Source: Duffield & Long, 1979, p. 7.
Figure 2: A Schema for key areas and issues of tourism impact

They argue for an interactive model which relates the impact of tourism, not only to the nature of the area concerned, but also to the planning environment which conditions the character of the associated physical, social and economic response. The keynote of this conceptual framework is to emphasize the impossibility of abstracting questions relating to tourism from the physical, spatial, economic and cultural environment within which they are set.

3. A CASE STUDY

Studies of tourism impact, like other Environmental Impact Assessments, have invariably been made as part of planning schemes. One such study is that undertaken by the Tourism and Recreation Research Unit (TRRU), of tourism and recreation in the Chichester area of Southern England (TRRU, 1977). This study, while intended to assist the formulation of a local plan for tourism on behalf of the local authorities concerned, has as its primary objective, the development of a model approach to the collection, sifting and interpretation of data on tourism which would be capable of application in other areas. Assembling such data is a necessary preliminary in any assessment of tourism impacts. The Chichester study, therefore, provides an appropriate case study, and will be used, along with other examples, to illustrate both the types of impacts associated with tourism and the methods employed for determining these impacts.

The conceptual framework adopted for the Chichester study ensures that the identification and evaluation of tourism 'impacts' are made against the background of the supply of and demand for resources in the area and, as such, recognises the importance of considering not only the provision within an area, that is, the nature of tourism development, but also the levels of actual and anticipated use. An assessment of the demand generated by particular tourism developments is, therefore, a necessary concomitant to assessing tourism impacts. Figure 3 illustrates the framework which forms the basis of this supply and demand approach. The schema contains four main areas of investigation:

(i) current demand;
(ii) future demand;
(iii) supply;
(iv) impact.

Defining the objectives of an impact assessment within a disciplined framework of this kind also enables the associ-

Figure 3: Schema for Determining Tourism Impacts

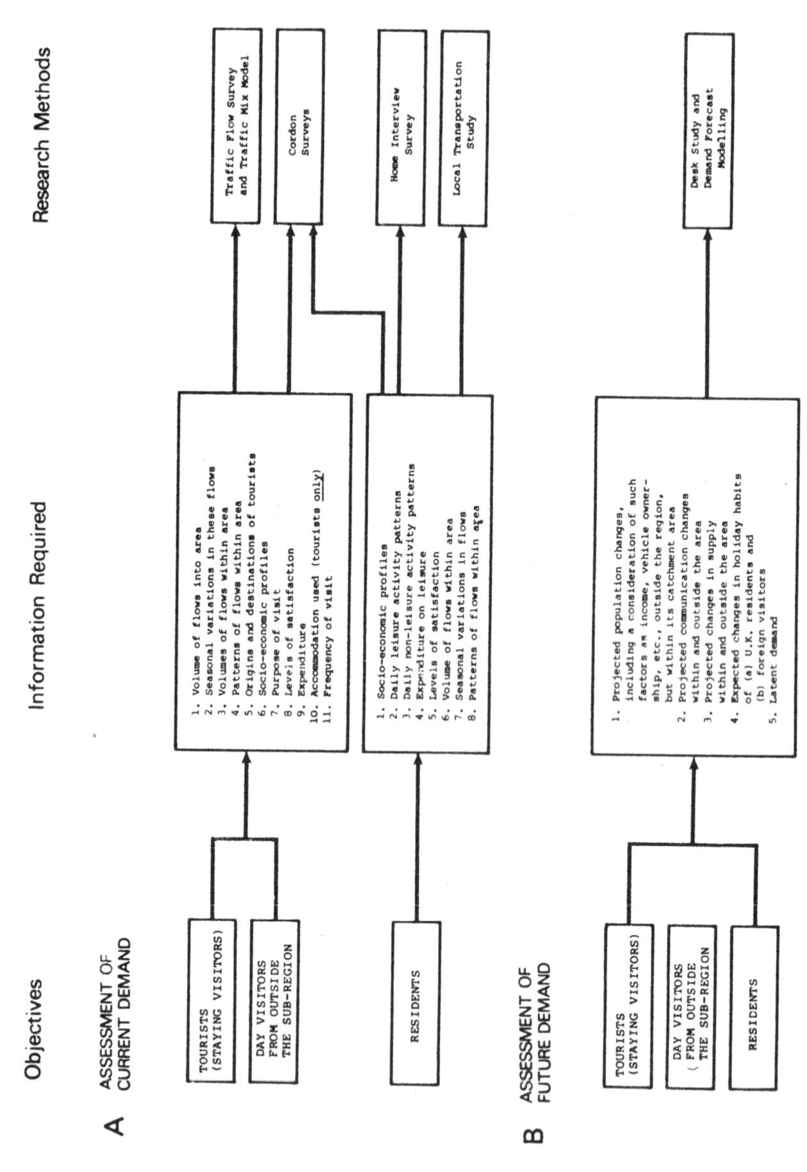

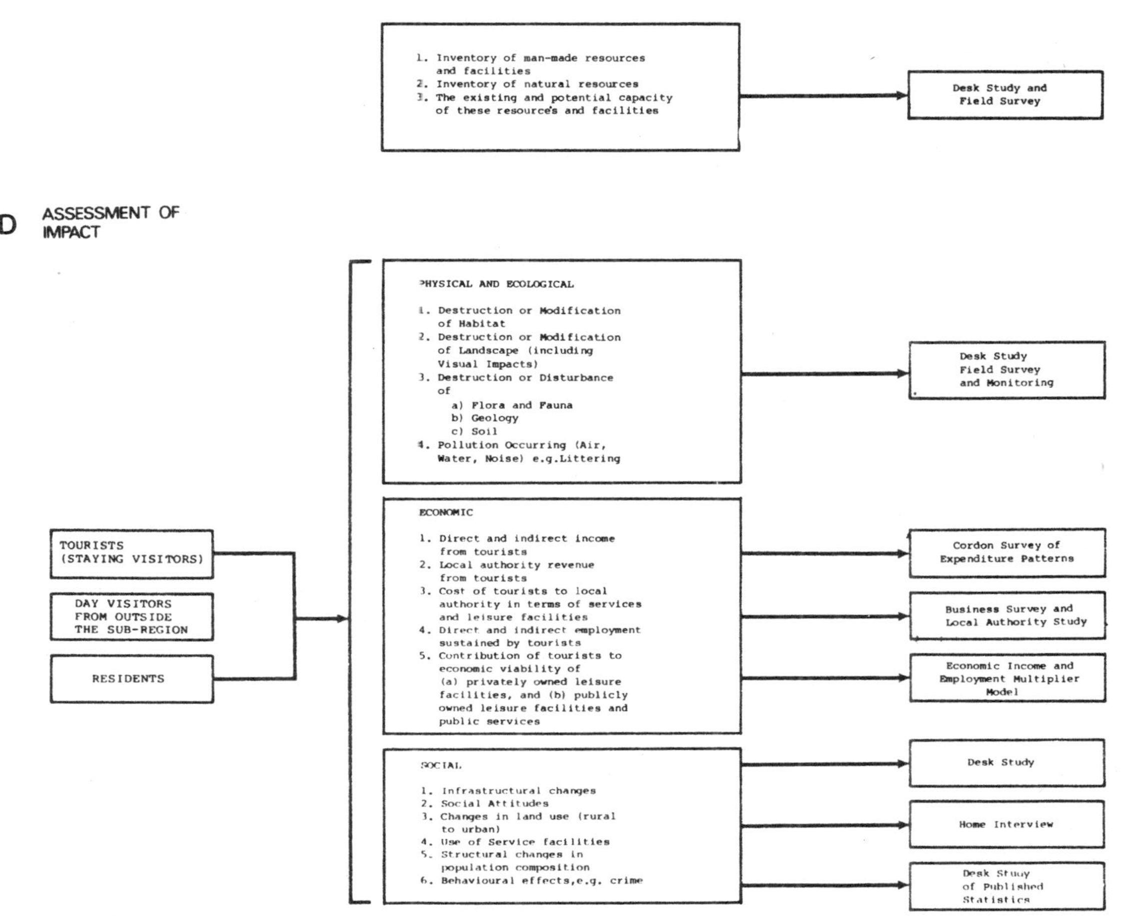
C ASSESSMENT OF THE SUPPLY OF LEISURE FACILITIES AND RESOURCES
1. Inventory of man-made resources and facilities
2. Inventory of natural resources
3. The existing and potential capacity of these resources and facilities
Desk Study and Field Survey
D ASSESSMENT OF IMPACT
TOURISTS (STAYING VISITORS)
DAY VISITORS FROM OUTSIDE THE SUB-REGION
RESIDENTS
PHYSICAL AND ECOLOGICAL
1. Destruction or Modification of Habitat
2. Destruction or Modification of Landscape (including Visual Impacts)
3. Destruction or Disturbance of
a) Flora and Fauna
b) Geology
c) Soil
4. Pollution Occurring (Air, Water, Noise) e.g.Littering
Desk Study Field Survey and Monitoring
ECONOMIC
1. Direct and indirect income from tourists
2. Local authority revenue from tourists
3. Cost of tourists to local authority in terms of services and leisure facilities
4. Direct and indirect employment sustained by tourists
5. Contribution of tourists to economic viability of (a) privately owned leisure facilities, and (b) publicly owned leisure facilities and public services
Cordon Survey of Expenditure Patterns
Business Survey and Local Authority Study
Economic Income and Employment Multiplier Model
SOCIAL
1. Infrastructural changes
2. Social Attitudes
3. Changes in land use (rural to urban)
4. Use of Service facilities
5. Structural changes in population composition
6. Behavioural effects,e.g. crime
Desk Study
Home Interview
Desk Study of Published Statistics

ated information requirements and the appropriate research methods to be identified.

Tourism developments, as already noted, impinge not only on the physical environment but also on the economic and social environment of an area. The remainder of this paper, therefore, following on from a discussion of the key characteristics of the demand generated by tourism, looks at the nature of the impacts in each of these three spheres and at the methods appropriate for identifying such impacts. It is important to acknowledge, however, that the techniques adopted within tourism impact studies are essentially similar to those utilised in other impact assessments. What this paper intends to demonstrate, therefore, is the application of such techniques within the context of the appraisal of tourism developments.

3.1 Assessment of Demand

In a British context at least, it is seldom possible to consider tourism in isolation from the broader aspects of the general recreational use of tourist facilities. It is, therefore, important to distinguish between demand generated by:

(i) the residents of a region;
(ii) day visitors to a region;
(iii) tourist visitors from other areas within a country (domestic tourists);
(iv) tourist visitors from outside a country (international tourists).

Each of the four categories differs with respect to:

(i) timing and frequency of visits;
(ii) the motivation for the visit;
(iii) awareness of opportunities;
(iv) the types of facilities and accommodation used;
(v) levels and patterns of expenditure;
(vi) their responsiveness to planning controls;
(vii) the nature and scale of associated impacts.

(Middleton, 1977).

The factor which distinguishes tourism from other recreational use, and shapes the distinctive nature of its impact, is the need for accommodation; the tourist being the visitor who is staying away from home. The need for accommodation provision not only amplifies the physical, economic and social impact of tourism, but distinguishes tourism from most other forms of economic activity in that the products of tourism are consumed at the point of production. Moreover, almost by definition, areas which attract tourists will have distinctive scenic and cultural landscapes which are particularly vulnerable to ecological, economic and social change (Forster, 1964).

In assessing the demand generated by tourism, it is important to seek answers to the following key questions:

(i) How many visitors were there?
(ii) Where did they come from?
(iii) What type of transport did they use?
(iv) Where did they stay?
(v) What type of accommodation did they use?
(vi) What was the length of stay?
(vii) What activities were undertaken?
(viii) What seasonal variations were there?

To enable the investigator to answer these questions, there exists a range of techniques largely based on home interview and site surveys (TRRU, 1977, p. 6).

3.1.1 Future demand

No less important than the present patterns of tourist demand is an understanding of future trends in tourism. Such exercises in prediction are inevitably speculative undertakings, relating as they do to highly variable factors such as changes in population, patterns of transport, as well as national and international economic conditions. To some extent the patterns of holiday-making are bound by established facts of human geography (such as settlement patterns and communications). Yet tourism and recreation are also susceptible, within limits, to planned development and to some extent, the 'future' is in the hands of those currently responsible for planning at all levels.

In Britain exercises in forecasting future demand have varied from simple extrapolation of past trends, to more sophisticated mathematical projections based on observed relationships between levels of tourism demand and factors such as:

(i) population;
(ii) disposable leisure time;
(iii) disposable income and Gross Domestic Product;
(iv) travel restrictions.

In Britain regular forecasts are now made available by the consultants, Business and Economic Planning (BEP), acting on behalf of the national tourist boards. The resultant

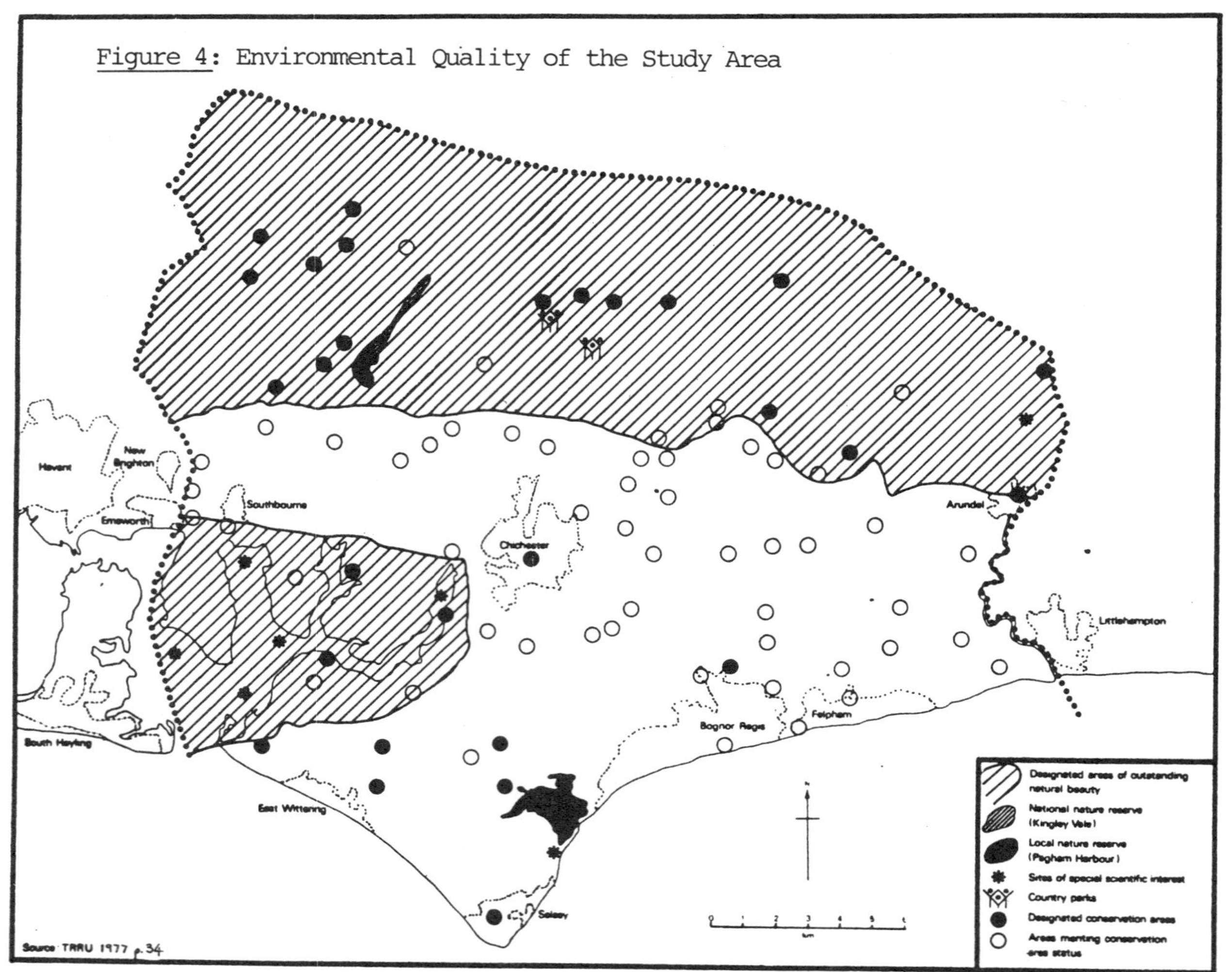

Figure 4: Environmental Quality of the Study Area

forecasts must be treated with extreme caution, for not only are they dependent on factors which are themselves difficult to predict, but also tourism itself is notoriously susceptible to fluctuations in circumstances (for example, in the international economic climate, in energy supplies) not readily given to prediction.

3.2 Physical and Ecological Impacts

A great deal of research has been carried out to investigate the impact of recreation and tourist activity on the physical environment for particular activities and for particular sites, but for the most part these studies have been partial, incorporating only a few of the components of the environment. Few studies have adopted a holistic approach and, in particular, as Wall (1979) points out, few have attempted to measure both cause and effect, that is to measure the degree and intensity of use as well as evaluating ecological change. Moreover, the results have lacked synthesis, although valuable reviews are available (Speight, 1973; Marren, 1974; Liddle, 1975; and, Wall and Wright, 1977). There has also been a major focus on *post hoc* studies, initiated in response to particular problems, which provide relatively few insights for prescriptive planning. Nevertheless, for those concerned to evaluate the possible physical impact of tourist developments and the choice between alternative sites for development, these studies provide a basis for identifying sensitive environments and for devising policies and policy instruments which will help to minimise any adverse effects of visitors. A complementary thrust has been to identify, in positive terms, areas suitable for recreational and tourist developments where conservation and ecological factors are taken into account during the evaluation process.

3.2.1 Resource evaluation

In every region, there are areas formally and informally designated, as being of important conservation value. Figure 4 maps the distribution of such sites in the Chichester study area, with Areas of Outstanding Natural Beauty, national and local nature reserves, Sites of Special Scientific Interest and conservation areas covering a large proportion of the area under study. However, such areas are not necessarily the most, or only, sensitive areas and many local authorities and other agencies have attempted to evaluate the ecological and scenic quality of the territory for which they are responsible.

The underlying concept of these approaches is that of environmental capacity: the ability of an area to accommodate pressures without undergoing unacceptable levels of detrimental change. The concept of environmental capacity implies a relationship between 'resources' and 'people' and in formulating policy for the use of resources for tourism it is necessary to evaluate management objectives against the constraints placed on this use by the resources themselves and then to adjudicate on the acceptable limits of change (Duffield, 1978). It is important, however, to distinguish between value choices and technical decisions. Wagar (1968) has pointed out that carrying capacity 'tends to obscure an essential distinction between technical issues (involving what <u>can</u> be) and value choices (involving which of various possibilities <u>might</u> be) ... defining what is acceptable is a value choice rather than a technical issue' or as Sidaway and O'Connor observed 'capacity is after all what we care to make it' (1978, p. 131).

The concept of capacity is, therefore, useful but a large gap exists between the ranking of areas according to their ecological or scenic quality and the determination of a safe recreational capacity which can, in turn, be influenced by the way in which the area is managed. As Burton notes:

> 'This (demand/supply) relationship is incomplete unless a third term, capacity, is added. The capacity of the countryside to absorb the rapidly increasing recreational pressures that fall upon it is little understood; and so the translation of factors of demand and supply into terms of a policy for the management and development of the countryside as a recreational resource is frustrated'. (Burton, 1974, p. 1).

Nevertheless, the grading of land according to its environmental value is a useful first step in devising strategies and attempting to assess the impact of tourism.

3.2.2. Ecological survey of West Sussex

In addition to charting the existence of sites of conservation value in the Chichester area, an ecological survey of West Sussex was undertaken by the County Council in 1971 which provided a more detailed assessment of the physical environment. In this survey, 1 kilometre squares were employed as the survey units and air photographs were used in recording the distribution of types of habitat on a map. A scoring range of 0-10 was considered adequate to compare habitats in terms of ecological value, and a system of weighting which was devised by staff of the Nature Conservancy Council was checked in the field and subsequently modified to give consistent results for each square throughout the county.

Factors, other than type of vegetation, which were considered in assessing the value for each square included:

(i) the number of ditches, hedgerows and copses;
(ii) the existence of any continuous area exceeding half a square kilometre of ecologically significant habitat;
(iii) sites designated as nature reserves;
(iv) Sites of Special Scientific Interest.

The results of this exercise are shown in Figure 5. Relating these areas to the distribution of recreational facilities showed that the areas of very high and medium ecological quality tend to be subject to pressure by visitors, while areas of rare ecological quality are less affected, except on the coast where pressure is often very heavy.

In contrast to ecological evaluations, attempts to delineate areas of high recreational potential are more wide-ranging. TRRU, for example, has used a synthetic approach to identify areas of 'environmental worth' within which ecological quality is considered along with scenic resources and other physical and cultural factors. This technique makes use of an integrated data base and exploits the data handling and mapping facilities of the modern computer (Duffield and Coppock, 1975).

The basic premise of the approach employed by TRRU was that individual resources could be assessed and then brought together at increasing levels of aggregation to define different resource surfaces, culminating in a map which would indicate, against selected criteria, 'areas of environmental worth' (Figure 6). Similarly, planning considerations that were likely to influence policies can also be incorporated in the analysis to produce planning constraint surfaces. The application of computer techniques which can output maps at great speed to show different combinations of surfaces and the effects of giving different weight to these surfaces to reflect known or desired assessments of

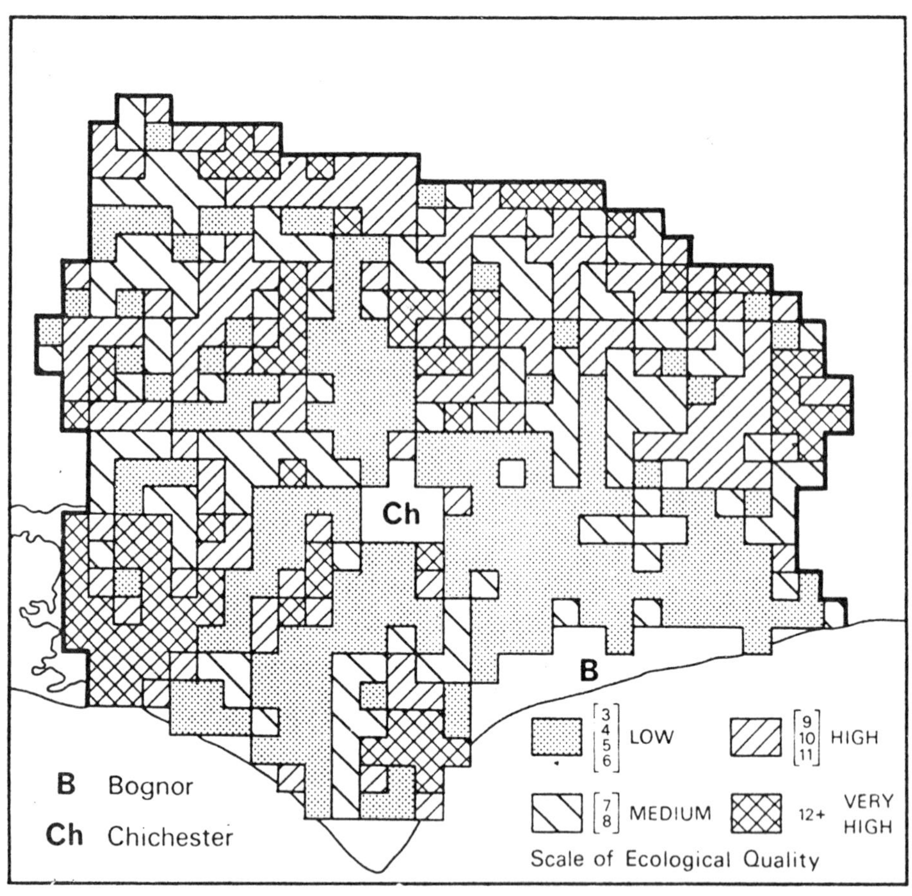

Source: TRRU 1977 p. 136

Figure 5: Ecological Quality of the Study Area

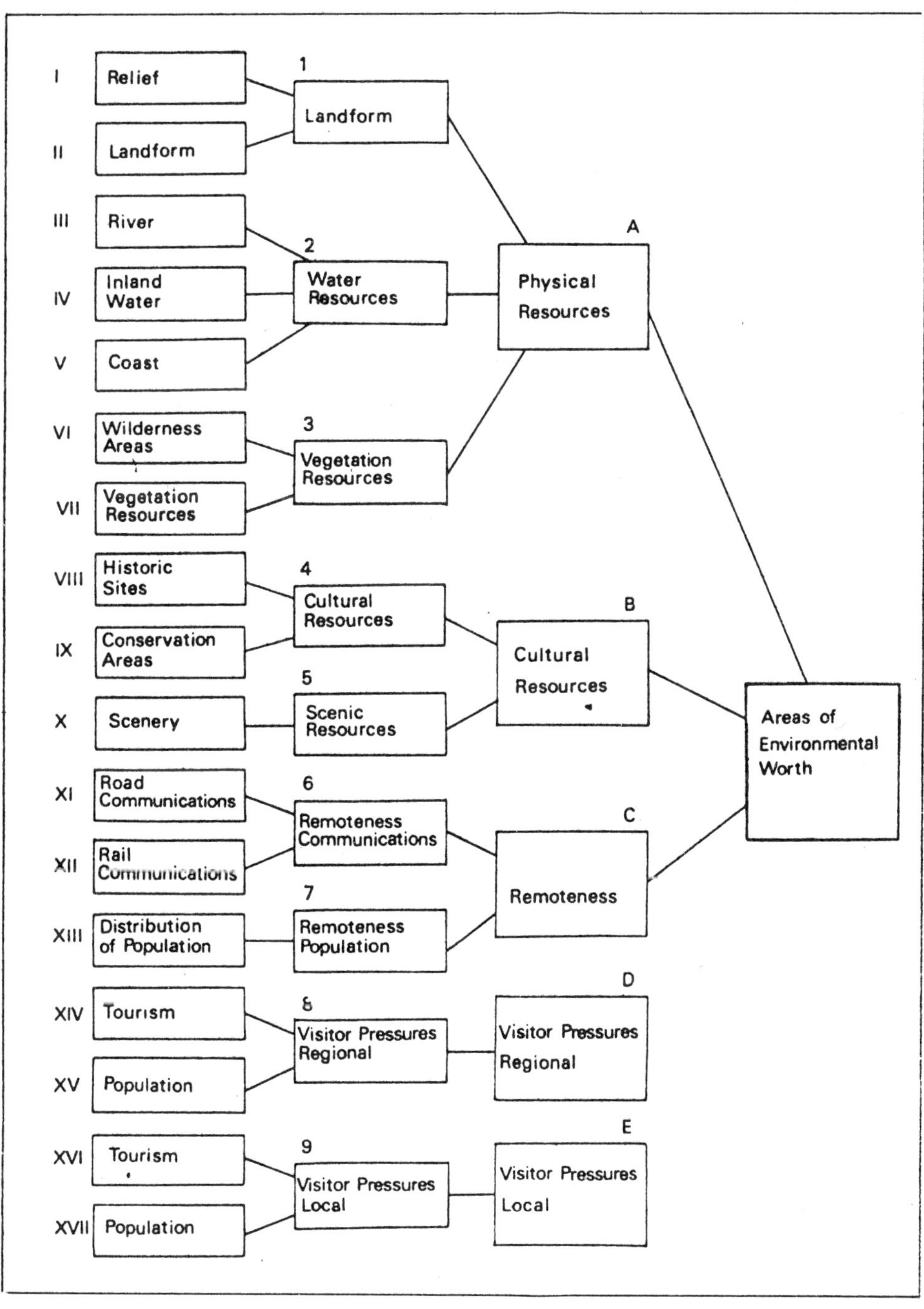

Source: TRRU 1977 p. 110

Figure 6: Definition of Areas of Environmental Worth

impacts, gives this approach great flexibility.

More recently, similar evaluation techniques have been utilised as part of a study to classify areas within the European Economic Community on the basis of their environmental characteristics (South Yorkshire County Council, 1979). Given the limited resources that would probably be available for assessments of the impact of tourism, the computer-based approach to ecological and environmental assessment is to be preferred in the first instance, to be supplemented at a later stage by detailed site surveys as required.

Deriving resource surfaces, however, is not without its difficulties and the assessment of scenic value is a particularly contentious area. The assessment of scenic value is essentially subjective and the problems of devising suitable methods have been subject to considerable critical evaluation in Britain (Robinson et al., 1976; Penning-Rowsell, 1973). Two main kinds of approach have been adopted towards the assessment of scenic value, those that depend on information from topographic maps and aerial photographs and those that depend on perspective and field assessment. (A third approach, which draws largely on assessments made by members of the public, has received much less attention (Penning-Rowsell et al., 1977)). In the former the assessment is synthetic, in that individual components are identified and scored. The second type is conceptually more appealing, but is very much more demanding in terms of human and other resources.

3.2.2 Site studies

It is at the individual site level that studies of tourist and recreational impact have been at their most prolific. From a review of these studies, and as a result of the accumulation of field information and a detailed investigation of five study areas, Coppock (1980) concluded that there were four main potential areas of tourism impact on the physical environment:

(i) destruction or modification of habitat;
(ii) destruction or modification of landscape (including visual impacts);
(iii) destruction or disturbance of flora and fauna; geology and soil;
(iv) pollution (air, water and noise).

The study further demonstrated that despite the voluminous

literature on this topic, concern has primarily been with vegetation, and within vegetation studies the focus is mainly on the pressure of human feet and vehicles.

The components of the natural environment are interrelated and interdependent. While often abstracted for purposes of research, it should be stressed that the impacts of recreational activity are rarely confined to just one aspect of the environment. Wall and Wright (1977) have usefully emphasised these impact interrelationships in schematic terms (Figure 7) and as they stress, the complex interactions between different aspects of the environment make an assessment of total impact almost impossible to measure. For example, changes in soil structure may lead to adjustments in vegetation which, in turn, may precipitate changes in water quality and wildlife. Further, it is extremely difficult to isolate the effects of tourist and recreational activity from naturally occurring fluctuations and changes.

In terms of techniques developed to assess physical and ecological impacts, three main types of approach have been adopted:

(i) 'after the fact', i.e. retrospective studies;
(ii) monitoring change through time;
(iii) simulation exercises.

3.2.4 Retrospective studies

Most of the studies of physical and ecological impact have been retrospective. The basic assumption inherent in 'after the fact' investigations is that the area under study was homogeneous before the introduction of tourist activities, and that there has not been an overall change in the environment since the area was made available to users. The differences between the impacts on sites in adjacent areas have been attributed solely to the effects of human recreational activity. There are several untenable assumptions associated with this approach, connected with the variations in environmental, ecological and human processes over time and space. Nevertheless, studies of this kind can be undertaken quickly and relatively cheaply and the results are certainly useful in influencing management decisions for the area in which the investigation has taken place.

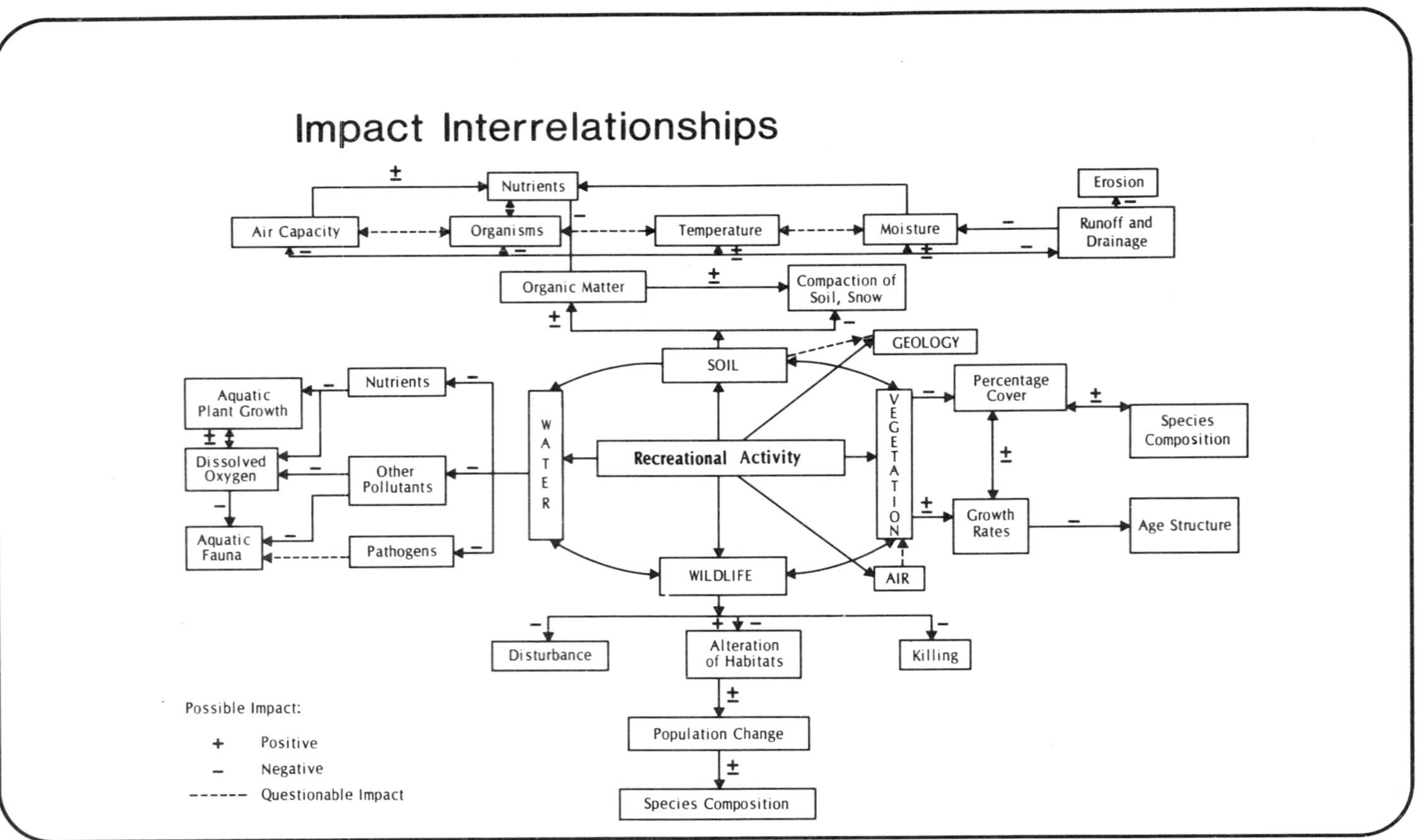

Figure 7: Impact Interrelationships

Source: Wall & Wright, 1977

3.2.5 Monitoring studies

Although adopting the same technical procedures as 'after the fact' studies, monitoring studies allow much more valuable insights into the physical and ecological effects of tourist activity. The undertaking of regular and repeated surveys, spanning various stages in the development of tourist activity, allows a simultaneous study of both cause (tourist activity) and effect (physical and ecological impact). Short-term studies are vulnerable to other processes at work in the environment. Thus, for example, weather conditions can prejudice the ability of an area to withstand use. Over the longer term, however, such studies can yield important information concerning rates of change and the results generated can be used to predict the likely impact of future levels of use or the effect of tourist activity on similar areas. However, this method is time-consuming and therefore costly, and prolonged studies of this kind, while academically satisfying, may be self-defeating in terms of influencing the actual pattern of tourist developments.

3.2.6 Simulation

When there exists a concern to evaluate the effects of a proposed tourist development, simulation techniques have occasionally been used to re-create the effects of human activity in experimental conditions. Obtaining adequate measures of use to relate to the observed impacts remains one of the major operational problems in studies of the environmental impact of tourist and recreation activities. Simulation experiments are therefore, to be commended in that they allow accurate and predetermined measures of use to be related to the observed ecological impacts.

In Scotland, for example, Bayfield (1971a) has used artificial tramplers to simulate the effect of pedestrian pressure on both vegetation and soil. Bayfield (1971b) has also attempted, although with less success, to simulate the slicing and sliding effect of skis and their impact on vegetation in the Cairngorm area of Scotland. Human ingenuity, however, seldom replicates accurately the recreational impacts. Also, agreement has yet to be reached as to the most appropriate indices for measuring change in these experiments. Harrison and Walker (1979), for example, suggest that the ability of a site to recover is a more useful indicator of its vulnerability than the more commonly used 'one-off' measures such as the reduction in live cover or the increase in bare ground.

Similarly, it is possible to assess in controlled conditions, varying management policies and their ability to limit the ecological damage resulting from tourist activity.

Each of the three types of approach has contributed to a greater understanding of the interactive effects of particular recreational activities on the environment (e.g. trampling) or, alternatively, has indicated the various impacts upon a particular component of the ecosystem.

Most studies of the environmental impact of tourism and recreation emphasise the detrimental effects of tourist and recreation activities on the environment, but beneficial effects can also occur. Human trampling of medium intensity, for example, can encourage diversity among vegetation. It is also argued by some authors that tourism serves to meet the cost of nature conservation (Akoglu, 1971), the tourist use of an area being the justification for expenditure on conserving the natural environment which would not otherwise take place. A further bias in the research undertaken so far has been a concern with the effects of tourist and recreation activities on the natural environment. More recently, however, there has been evidence to suggest that man-made environments in the form of historic buildings and sites are also under pressure from tourist use (Wingerson, 1979).

3.3 Economic Impacts

3.3.1 Economic benefits

In contrast to the studies of physical impacts which emphasise the detrimental effects of tourist use, studies of economic impacts have focused predominantly on the benefits derived. Knetsch and Var (1976) distinguish between two types of benefit. Firstly, the so-called primary benefits (the measure of welfare gain accruing to the visitors) and secondly, the benefits which are derived from the actual expenditure by tourists. Means of measuring the former has spawned a whole field of recreational economics and a number of reviews exist, see for example, Gibson (1974) and Baxter (1979). It is in respect of the latter, however, that the greatest assumptions have been made as to the benefits of tourism development. The assertions made, rest on the effect of the spending by holiday visitors upon local economies. In 1979 in Britain, over £4,000 million were spent by tourists, while at an international level, the gross receipts from international tourism have now become the largest single item in international trade (Bird, 1973, p. 83).

At a local level, however, the level of tourism expenditure does overestimate the financial benefits accruing to the local community. The movement of money beyond the region for the purchase of goods from elsewhere, the movement of profits, and government taxation, all serve to reduce the economic benefits accruing locally. Nevertheless, it is argued that the proportion of tourist expenditure retained locally has a disproportionate impact upon the local economy because of the so-called 'multiplier' effect. This multiplier effect is represented in Figure 8 and is based on the Keynesian concept of the circulation of money. The fundamental tenet of the concept is that the impact of any expenditure on the local economy extends far beyond the initial recipient. Thus, as the first round of spending works its way through the local economy, the general turnover within the area increases, jobs are created and personal incomes rise. Successive rounds of spending, both indirect and induced, spread transactions through the economy as businesses and households re-spend within the area some of the income they have earned, directly or indirectly from tourism.

The multiplier coefficient accumulates all the local income resulting from these series of transactions within the local economy which were triggered-off by the initial injection of tourist spending (Vaughan, 1979). Mathematically, the coefficients are expressed in one of two forms:

(i) the orthodox multiplier: multiplier values are expressed as the ratio of the direct, indirect and induced effects to the direct effect;

(ii) the unorthodox multiplier: the multiplier value is expressed as the ratio of the direct, indirect and induced effects to the initial injection of money.

The two formulations require similar data; they do, however, offer some different levels of advice to policy makers. The orthodox multiplier merely states that if one pound of direct income is created, 'n' other pounds will be created in other parts of the economy. Unorthodox multipliers, on the other hand, indicate how much income will be created by a given level of expenditure. In tourism studies, it is the unorthodox multipliers that have generally been adopted.

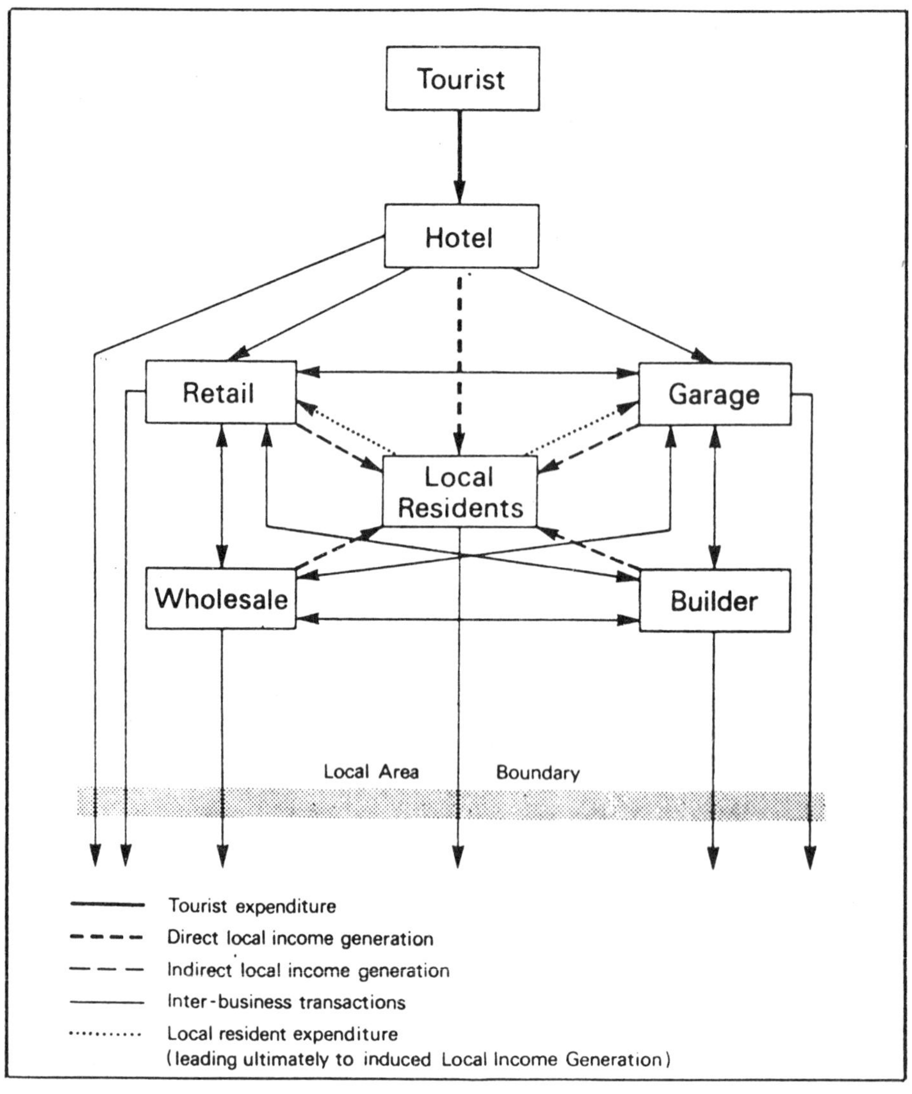

Source: TRRU 1977 p. 51
Figure 8: Local Income Generation from Tourist Expenditure

Multipliers usually take one of three forms:

(i) the income multiplier - indicates the rise in

local income (as represented by local wage, rent and profit payments) per unit of additional money coming into the economy;

(ii) the employment multiplier - indicates the employment supported within the local economy per unit of additional money coming into the economy;

(iii) the purchases multiplier - indicates the purchases generated per unit of additional money entering the economy.

Each of these multiplier formulations provides a different insight into the socio-economic benefits of tourism; until recently, research has concentrated on the first two types. But what does all this mean in practice?

In terms of the income multiplier, for example, it means the measurement of the local income derived directly from tourist expenditure, plus the measurement of the local income generated at each round of exchange as part or all of the original expenditure is re-spent locally. In Figure 8, for example, tourists spend money in hotels and the local income resulting from this will consist of:

(i) the direct income - wages, rent and profit paid locally by the hotels;

(ii) the indirect income - wages, rent and profit paid locally which can be attributed to the spending by tourists in hotels, but which arise out of the purchase of supplies by the hotels from, for example, local retailers; and,

(iii) the induced income - wages, rent and profit paid locally, which arise from the spending by local residents (e.g. hotel staff) of income earned from tourism.

In order to arrive at such figures for all tourists and all tourist expenditures, a large and varied amount of information is collected, consisting of the following elements:

(i) Interviews with tourists to obtain not only details of their expenditures over a retrospective period, but also information relating to the nature of the tourist visitors, e.g., origin, accommodation used, means of transport, length of stay and group composition.

(ii) Interviews with local businesses (including utilities and local authorities) to obtain details of their accounts, including information relating to turnover, wages and profits, and also relating

Type of Tourist Accommodation	Per £1 of Tourist Expenditure
Hotel	.300
Guest House	.342
Bed & Breakfast	.515
Touring caravan	.277
Static caravan	.284
Camping	.286
Rented accommodation	.355
Friends & relatives	.298
Other	.293
Touring coach	.333
Transit visitors	.276
Day trippers	.292
All Tourists - Average	.319

Source: TRRU, 1975 p. 44

Table 1: Tourist Income Multiplier Coefficients by Accommodation

Type of Tourist Accommodation	Per £1,000 of Tourist Expenditure Standardised	Unstandardised
Hotel	.220	.359
Guest House	.225	.516
Bed & breakfast	.256	.770
Touring caravan	.162	.367
Static caravan	.144	.310
Camping	.164	.361
Rented accommodation	.198	.499
Friends & relatives	.164	.359
Other	.153	.368
Touring coach	.217	.386
Transit visitors	.149	.326
Day trippers	.180	.412
All Tourists - Average	.192	.412

Source: TRRU, 1975 p. 48

Table 2: Tourist Employment Multiplier Coefficients by Accommodation

to their patterns of expenditure to various other businesses both within and outside the area under study.

(iii) Estimates of visitor numbers and lengths of stay, based either on specific surveys of the region concerned, or more usually obtained from national surveys capable of regional disaggregation (for example, the British Home Tourism Survey and the International Passenger Survey, both of which are carried out on an annual basis in Great Britain). Statistics from these surveys are used to translate sample information to represent the aggregate effect of tourism impacts.

Surveys of this kind are, of course, expensive and more recently researchers have been concerned to develop techniques which would allow the transference of results of multiplier studies to other areas. An examination of a range of studies undertaken in Britain demonstrates a high degree of consistency from one area to another, and the Tourism and Recreation Research Unit has been able to demonstrate that, as long as visitor information is available for the area under study, it is possible to mobilise multiplier analyses against tenable assumptions relating to the patterns of tourist spending and the structure of local businesses (TRRU, 1977, p. 113).

3.3.2 The nature of the results

The results of multiplier analysis can be presented in many different ways, with multiplier coefficients related to:

(i) the accommodation used by tourists;
(ii) mode of transport;
(iii) the length of time spent by the tourist in an area;
(iv) the origin of the tourist;
(v) the type of community in which the tourist stayed.

Tables 1 and 2, for example, show for the Tayside Region of Scotland both the income and employment multiplier coefficients associated with different forms of tourist accommodation.

The value of multiplier analysis to policy makers arises largely out of the incremental nature of the analysis. This is illustrated in Table 3, which shows that multiplier analysis not only measures the total volume of economic benefits accruing from different forms of tourist activity, but also indicates the nature of that impact.

Type of Tourist Accommo-dation	Average Daily Expend-iture	Local Multi-plier Coeffi-cient	Local Income Created per Visi-tor Day	Visitor Bed-nights (000s)	Total Expend-iture (£000s)	Total Local Income (£000s)
Hotel	£13.12	.29	3.84	590	7738	2267
Guest House	£ 7.42	.33	2.45	437	3245	1070
Caravan	£ 3.62	.22	0.81	171	619	138
Friends & Relatives	£ 3.42	.20	0.70	715	2452	499

Source: Vaughan, 1979 p. 5

Table 3: Tourist Income Generation by Accommodation - An Example*

* The statistics are drawn from a study of the economic impact of holiday visitors to the City of Edinburgh, 1976. It is estimated that holiday visits totalled 2.5 m nights and resulted in tourist spending of £17.4 m, of which £4.9 m was retained as income to residents of the city.

The results of multiplier analysis are valuable, not only in determining the income (or employment) generated for given patterns of tourist activity, but also in guiding the development of tourism in an area. Thus, for example, the information in Table 3 could be utilised as a basis of a promotion campaign geared to maximising income benefits from tourism whilst keeping visitor numbers at a minimum. Alternatively, the likely economic benefits of grant aid to particular sectors of the tourist economy can be assessed. If used in conjunction, income and employment multipliers can be used by those concerned in tourism planning to optimise their development objectives. As Table 4 indicates, for example, different development options would be chosen dependent upon whether income maximisation or employment maximisation was the key objective.

Also of major significance is the ability of multiplier analysis to indicate the economic benefits resulting from different geographic types of host community (Table 5).

Type of Tourist Accommodation	Local Income Multiplier Value (per £1)	Local Employment Multiplier Value (per £1,000)
Hotel	0.29	0.19
Guest House	0.33	0.21
Caravan	0.22	0.11
Friends & Relatives	0.20	0.12
Bed & breakfast	0.26	0.28

Source: Vaughan, 1979 p. 5 & 8

Table 4: Tourist Income and Employment Generation by Accommodation

Type of Community	Local Income Multiplier Value (per £1)	Local Employment Multiplier Value (per £1,000)
Major Nodal Towns	0.274	0.159
Highland Centres	0.245	0.157
Seaside Towns	0.226	0.139
Special Activity Centres	0.227	0.165
Rural Areas	0.185	0.121

Source: TRRU, 1975 pp. 94-96 & 99

Table 5: Tourist Income and Employment Generation by Community Type

Multiplier analysis supplies a range of information of value to policy makers on the positive economic impacts of tourist activity. Unfortunately, up to the present time, the methodologies devised for measuring the economic dis-benefits of tourism are considerably less sophisticated.

3.3.3 Economic costs of tourism

One of the positive spinoffs of tourist activity to host communities is the improvement to infrastructure which usually accompanies tourist activity. It must, however, be remembered that the costs of infrastructural improvements are usually borne by local and/or national government bodies, rather than by the tourist entrepreneurs themselves. Tourist-related expenditure on services and utilities is

County Council	% Total Per Service Expenditure Attributable to Tourism	Other Agencies	% Total Expenditure Attributable to Tourism	District Councils	% Locally-borne Service Expenditure Attributable to Tourism*
Beach safety	86	Water supply	8	Beaches & cafes	86
Advertising & publicity	100	Sewage conveyance	12	Beach safety (net)	86
Planning (net)	13	Sewage purification & disposal	17	Advertising & publicity	100
Parks	18	Ambulance service	13	Car parks	15
Coastal paths (net)	100	Police	0	Decorative lighting/ wardens	100
Oil pollution	50			Public conveniences	18
Refuse disposal	16			Planning	13
Trading standards	18			Parks	18
Fire service	10			Beach cleaning/ Oil Pollution (net)	50
				Swimming pools/leisure Centres/zoos	31
				Piers, boats etc.	31
				Museums/theatres	31
				Street cleansing	10
				Refuse collection	16

* For District Councils, varying proportions of total expenditure were met from the Resources Element of the Rate Support Grant, so figures for individual sources in Districts were reduced by the appropriate proportion in order to obtain the locally borne cost of each service, and then estimates were made of the tourism proportion of the total cost of each item. Therefore, only columns one and two are directly comparable.

Source: TRRU, 1977 p. 122

Table 6: Local Authority Service Expenditure Attributable to Tourism

incurred as a result of the seasonal fluctuations of tourist activity which place additional demands on services designed to meet more modest residential needs. In a study of tourism in Cornwall (Cornwall County Council, 1976) estimates were made of the proportion of expenditure incurred by local authorities and other agencies which was attributable to tourism. Particular attention was paid to services where it seemed likely that there would be some element attributable to tourism within the total cost. The services, and estimates of the tourism component, are listed in Table 6.

A pioneer attempt to quantify more exactly the extra public authority costs of tourism was developed in a study of the tourist industry in South West England (Edwards et al., 1976). Services considered included those with 'specific' tourism-related costs (such as tourist information services and publicity) and those with 'indirect' tourism components (such as public conveniences and sewage disposal, refuse collection and disposal, parks and open spaces, highways and street lighting, car parks, fire services, police, water supply and health care). In order to calculate these extra public authority costs, both specific and indirect costs were combined to give a total which was then translated into a per capita estimate for the settlement concerned. Using multiple regression, Edwards et al. established a relationship between the level of tourism in an area and the additional costs incurred. The technique is capable of adaptation to local circumstances, and in its application to the Chichester area of Southern England, TRRU calculated that in 1976, specific and indirect costs associated with tourism amounted to approximately £500,000. Part of these costs will, of course, be offset indirectly by the rate revenue raised from tourist-related establishments, such as hotels, restaurants, holiday camps, etc., and from revenue income from tourist facilities, for example, car parks.

Useful though this technique is, doubts as to its accuracy prevail and the technique, as yet, does not allow the disaggregation of costs between different types of tourist which multiplier analysis most usefully provides on the other side of the 'balance sheet'.

Specific information made available by public agencies can offer accurate insights into the additional costs resulting from tourist activity. Thus, for example, the demand on health services creates a high additional workload during the summer months in British tourist areas. Table 7 records for the Chichester Study Area the changing demands of temporary residents on family practitioner services; more than half of such demands are made in the summer quarter.

	30/6/1976	30/9/1976	31/12/1976	31/3/1977
Arundel and Barnham	167	332	91	208
Bognor	3,438	5,049	682	733
Chichester	550	528	424	430
Selsey and Witterings	2,085	5,030	786	246
Total	6,130	10,939	1,983	1,617

Source: TRRU, 1977 p. 57

Table 7: Temporary Residents - Claims Received by Family Practitioner Services (Year quarter ended)

Without doubt other public authorities, for example, hospitals and water supply agencies, are in a position to make available similar indicators of the pressure that an influx of tourists can make on public services.

Economic costs borne more directly by the residents of the areas concerned are the induced inflation that tourists can bring. This is particularly evident, for example, in the costs of private property, where the competition from second home owners and those retiring to tourist areas can not only result in disproportionately high purchase costs, but also displaces local residents from the housing market. Utilising information from newspaper sources, it has been possible to demonstrate for example, that house prices within a national park in England were not only higher than prices for similar houses in the surrounding region and in the United Kingdom, but, in a period of three years, the differential increased by between 60 and 100 per cent (Table 8).

3.4 Social Costs and Benefits

Analysis of the economic costs and benefits of tourism include only those tangible elements that are susceptible to measurement. Another, and important, aspect of impact is the social costs and benefits which are both difficult to measure and less susceptible to control by intervention on the part of local authorities. The sheer numbers of tourists during the summer months can create particular problems in areas which have evolved to serve a much smaller resident population.

The social impact of tourism is largely under-researched, compared with studies of its physical and economic impact. Like the rest of environmental impact studies, social impact assessment has come late to the agenda (Carley and Derow, 1980). Social impacts arise because of 'the differences in attitudes, perceptions, values, and expectations between visitors and the visited population' (Butler, 1979, p. 372). Charting the nature of social impacts is not easy; in particular changes may occur through time and Thomason et al. (1979) stress the need to monitor impacts rather to rely on one-off studies. Nevertheless, despite the operational problems recent surveys of residents have demonstrated one useful method by which social impacts can be studied.

Pizam (1978), for example, interviewed both residents and entrepreneurs to measure their perception of tourism impacts on Cape Cod, Massachesetts. Pizam sought opinions on a range of variables relating to tourism: for example, the environment, social aspects, economic factors, availability factors, and quality factors. The response to each question was rated on an eleven-point scale ranging from -5 to +5. An Attitude Index was then created by averaging the scores of each respondent. In Table 9, some of the more significant responses of both residents and entrepreneurs are presented. It is interesting to note the generally more favourable assessment provided by entrepreneurs connected with tourist activity as opposed to resident reactions. Interviewee responses to specific factors confirms more recent research in Scotland (TRRU, 1975 and Brougham and Butler, 1976), that it is issues relating to the intangible social, cultural and environmental disbenefits that arouse the strongest complaints, whilst perceived advantages relate more to the concrete benefits of increased trade, income and employment that tourism brings.

Doxey (1975), arising out of comparative studies of Barbados, West Indies and Niagra-on-the-Lake, Ontario, has devised (using similar techniques) an 'irridex': index of tourist irritation. This index seeks to measure the levels of reaction to tourist activity on a scale progressing through thresholds of euphoria, apathy and annoyance to antagonism, consequent upon increasing levels of tourist activity.

As with the physical environment where those areas most

	1976 Prices	Index of Prices (UK=100)	1979 Prices	Index of Prices (UK=100)	Percentage Change 1976-1979
	£		£		%
Lake District National Park:					
South East*	19,500	160	40,400	216	+107
North East*	15,700	129	40,900	219	+160
Windermere**	13,400	110	30,300	162	+126
Selected Control Areas:					
Northern Region+	10,500	80	15,000	80	+ 43
United Kingdom+	12,200	100	18,700	100	+ 53

* Westmoreland Gazette/Lake District Herald (average house prices)
** Lake District Special PLanning Board (Average selling price of a three-bedroomed semi-detached)
\+ Building societies

Source: TRRU, 1981 p. 213
Table 8: House Prices in the Lake District and Selected Areas (1976-1979)

Factor	Residents	Entrepreneurs
	Average Reaction	
Availability of Land/Housing	-0.5	+0.1
Cost of Land/Housing	-2.3	-1.7
Traffic conditions	-4.0	-4.0
Availability of Recreational Facilities	+1.0	+1.3
Job Opportunities	+0.9	+2.6
Employment fluctuations	-0.8	-0.9
Increased Incomes	+2.1	+3.3
Shopping opportunities	+1.4	+2.4
Increased prices	-2.7	-2.1
Noise	-2.7	-2.6
Litter	-3.5	-3.5
Standard of living	+1.5	+2.3
Vandalism	-2.7	-2.5
Understanding different people	+0.8	+1.5
Changes in values, norms, etc	-0.3	+0.2

\+ indicates a positive effect) on a range +5 to -5
\- indicates a negative effect)

Source: Pizam, 1978 p. 10
Table 9: Resident and Entrepreneur Responses to Tourism Effects on Cape Cod, U.S.A.

vulnerable to change are often the most attractive tourist areas, so too in the social sphere it is societies in the less developed sectors of the world which attract large numbers of tourists which have cultures and social settings susceptible to change. An increasing number of studies have been undertaken in tourist paradises in the Pacific and the West Indies. The results of these studies are far from conclusive, for example, while Wall and Ali (1977) suggest that the impact of tourism on Trinidad and Tobago is negative, Boissevain (1977) argues that in Malta tourism has encouraged cultural identity and independence.

Further useful measures of the social effects of tourist activity can be provided by published statistics. Although providing inconclusive results as far as the Chichester area is concerned, the statistics in Table 10 demonstre an attempt to monitor the seasonal incidence of crime in a holiday area.

	Chichester		Bognor		Total	
Month	1975	1976	1975	1976	1975	1976
	Percentage of total					
January	9	9	5	8	7	9
February	9	6	7	8	8	7
March	7	9	9	8	9	8
April	8	9	8	9	9	9
May	8	11	10	8	9	9
June	8	8	8	7	8	8
July	7	11	11	8	9	9
August	9	9	12	12	11	11
September	9	8	9	8	9	8
October	7	9	8	7	7	8
November	10	8	8	7	9	8
December	8	4	5	10	7	7
Total Number	2530	2408	2815	2682	5345	5089

Source: TRRU, 1977 p. 58

Table 10: Recorded Crimes in Chichester and Bognor Region 1975/76

Similar techniques can be used to plot the seasonal variations in, for example, traffic accidents, emergency treatments in hospitals and other aspects of the pathology of the social impact of tourism.

In monitoring longer term trends, particularly structural

changes in the composition of the population in an area dominated by tourism, census data are invaluable as are other regular government-sponsored social surveys (for example, the Annual Census of Employment and the General Household Survey).

4. CONCLUSION

This paper has reviewed, in brief, the impacts of tourism on the environment and some of the methodologies adopted to investigate the nature of these impacts. To obtain a fuller understanding of the environmental impact of tourism, research will be complex and need to involve an interdisciplinary approach. Further, if more precise information on net benefits and costs of tourism are to be obtained then integrated studies of the physical, economic and social impacts are required. It is unlikely that such research will explicitly determine tourism-use limits, but it should be possible to indicate to planners and managers the range of capacities of different environments for tourist activity and the accompanying physical, economic and social impacts of any particular development option.

REFERENCES

Akoglu, T.: 1971, 'Tourism and the problem of environment: Relations between environment, nature and tourism', The Tourist Review, 26(1) pp. 18-20.

Baxter, M.J.: 1979, Measuring the benefits of recreational site provision: A review of techniques related to the Clawson method. London: Sports Council.

Bayfield, N.G.: 1971a, Ecological effects of recreation on Cairngorm. Progess Report No. 1 Monks Wood: Nature Conservancy Range Ecology Group, p. 47-55.

Bayfield, N.G.: 1971b, 'Some effects of walking and skiing on vegetation at Cairngorm', in E. Duffey and A.S. Watts (eds.), The Scientific Management of Plant Communities for Conservation. Oxford: Blackwell.

Bird, R.A.: 1973, A new approach to tourism planning. Economic Salon.

Boissevain, J.: 1977, 'Tourism and development in Malta', Development and Change, 8, pp. 523-538.

Bosselman, F.P.: 1978, In the wake of the tourist, Washington, D.C.: Conservation Foundation.

Butler, R.W.: 1979, 'The social impact of tourism and recreation', in E.M. Avedon, M. LeLievre and T.O. Stewart (eds.) Proceedings of the Second Canadian Conference on Leisure Research, Ontario: Ontario Research Council on Leisure.

Butler, R.W.: 1980, 'The concept of a tourist area cycle of evolution: implications for management of resources', Canadian Geographer, 24 (1), pp. 5-12.

Brougham, J.E. and R.W. Butler: 1976, The social and cultural impact of tourism: A case study of Sleat, Isle of Skye , Edinburgh: Scottish Tourist Board.

Burton, R.C.J.: 1974, The recreational carrying capacity of the countryside. Keele University Library Occasional Paper No. 11, Keele, University of Keele.

Carley, M.J. and E.O. Derow: 1980, Social impact assessment: a cross-disciplinary guide to the literature, Policy Studies Research Institute research paper, London: Policy Studies Institute.

Cohen, E.: 1972, 'Towards a sociology of international tourism', Social Research, 39 (1), pp. 164-182.

Coppock, J.T.: 1980, 'Nature conservation and tourism in Great Britain', Paper given at a conference 'Nature and Tourism', April 1980, Perth, Scotland.

Cornwall County Council: 1976, County Structure Plan: The holiday industry, Topic Report, Truro: The Council.

Doxey, G.V.: 1975, 'A causation theory of visitor resident irritants: methodology and research inferences', in The Impact of Tourism, The Travel Research Association Proceedings, Sixth Annual Conference, Septamber 1975, San Diego, California.

Duffield, B.S. and J.T. Coppock: 1975, 'The delineation of recreational landscapes: the role of a computer-based information system', Transactions of the Institute of British Geographers, 66, pp. 141-148.

Duffield, B.S.: 1978, The capacity of our resources: when do we stop promoting growth? Tourism and Recreation Research Unit, Working Paper No. 3, Edinburgh: TRRU.

Duffield, B.S. and J.A. Long: 1979, Reward and conflict associated with tourism in the Highlands and Islands of Scotland, Paper given at the 10th Conference of the European Society for Rurual Sociology, April 1979, Cordoba, Department of Agriculture, Spain.

Edwards, S.L., B.G. Jackson, M.G. Ankers and S.J. Dennis: 1976, Tourism in the South West Region: Methodological report, Bristol, Department of the Environment.

Forster, J.: 1964, 'The sociological consequences of tourism', International Journal of Comparative Sociology, 5(12), pp. 217-227.

Gibson, J.G.: 1974, 'Recreational Cost Benefit Analysis: A Review of English Case Studies', Planning Outlook Special Issue, Planning for Recreation, pp. 28-46.

Harrison, C.M. and S.E. Walker: 1979, Preliminary results of field wear trials of amenity grasslands and heathlands in South-East England, London: Department of Geography, University College.

Kloke, C.: 1977, 'South Pacific economics and tourism', in B.R. Finney and K.A. Watson (eds.), A new kind of sugar, tourism in the Pacific, Honolulu: The East-West Culture Learning Institute.

Knetsch, J.L. and T. Var: 1976, 'The impacts of tourism and recreational facility development', Tourist Review, 31 (4) pp. 5-9.

Liddle, M.J.: 1975, 'A selective review of the ecological effects of human trampling on natural ecosystems', Biological Conservation, 7, pp. 17-34.

Marren, P.R.: 1974, Ecology and recreation: A review of the European literature, London: University College.

Middleton, V.T.C.: 1977, 'Some implications of overseas tourism for regional development', in B.S. Duffield (ed.) Tourism: A tool for regional development, Leisure Studies Association Conference, Edinburgh, 1977, Edinburgh: Tourism and Recreation Research Unit.

Pizam, A.: 1978, 'Tourism's impacts: the social costs to the destination community as perceived by its residents', Journal of Travel Research, Spring, pp. 8-12.

Penning-Rowsell, E.C.: 1973, Alternative approaches to landscape appraisal and evaluation, Middlesex Polytechnic Planning Research Group Report 11, Enfield: Middlesex Polytechnic.

Penning-Rowsell, E.C., G.H. Gullet, G.H. Searle and S.A. Witham: 1977, Public evaluation of landscape quality, Middlesex Polytechnic Planning Research Group Report 13, Enfield: Middlesex Polytechnic.

Rajotte, F.: 1978, A method for the evaluation of tourism impact on the Pacific, Santa Cruz: Center for South Pacific Studies, University of California.

Robinson, D.G. et al., (eds.): 1976, Landscape evaluation Report of the Landscape Evaluation Research Project to the Countryside Commissions, Manchester, Manchester University.

Sidaway, R.M. and F.B. O'Connor: 1978, 'Recreation pressures in the countryside', in Countryside Recreation Advisory Group Conference, York University, September 1978, pp. 141-155, Cheltenham: Countryside Commission.

South Yorkshire County Council, Environment Department: 1979, Environmental mapping of the European Community: South Yorkshire Case Study, Vol. 1 & 2, Barnsley: South Yorkshire County Council.

Speight, M.C.D.: 1973, Outdoor recreation and its ecological effects: A bibliography and review, Discussion Papers in Conservation No. 4 , London: University College.

Thomason, P., J.L. Crompton and D. Kamp: 1979, 'A study of the attitude of impacted groups within a host community towards prolonged stay tourist visitors', Journal of Travel Research, 17(3), pp. 2-6.

Tourism and Recreation Research Unit: 1975, The economic impact of tourism: A case study in Greater Tayside, TRRU research report no. 13, Edinburgh: TRRU.

Tourism and Recreation Research Unit: 1977, A research study into tourism and recreation in the Chichester area: A basis for planning, TRRU research report no. 75, Edinburgh: TRRU.

Tourism and Recreation Research Unit: 1981, The Economy of Rural Communities in the National Parks of England and Wales, TRRU Research Report No 47, Edinburgh, TRRU.

Vaughan, D.R.: 1979, What have we learnt from tourism multipliers? Some lessons from Scotland, Paper given at the proceedings of the 1st National Conference on Local Government and Tourism, 1978.

Wagar, J.A.: 1968, Quoted in Bury, R. (1976) Recreation carrying capacity - hypothesis or reality, Parks and Recreation, 115, pp. 23-25, 56-58.

Wall, G.: 1979, 'Ecological impacts of outdoor recreation', in E.M. Avedon, M. Le Lievre and T.O. Stewart (eds.), Proceedings of the Second Canadian Conference on Leisure Research, Ontario: Ontario Research Council on Leisure.

Wall, G. and I.M. Ali: 1977, 'The impact of tourism in Trinidad and Tobago', Annals of Tourism Research, 5, pp. 43-49.

Wall, G. and C. Wright: 1977, The environmental impact of outdoor recreation, Department of Geography Publication Series No. 11, Waterloo, Ontario: University of Waterloo.

Williams, T.A.: 1979, 'Impact of tourism on host populations: the evolution of a model', Tourism Recreation Research, Dec., pp. 15-21.

Wingerson, L.: 1979, 'Heritage under seige', New Scientist, 83, No. 1174, pp. 962-965.

AFFILIATION

Mr B. Duffield is the Director and Dr S. Walker a Research Fellow, of the Tourism and Recreation Research Unit at The University of Edinburgh, Scotland.

INDEX